ADDISON-WESLEY MATHEMATICS

Robert E. Eicholz

Phares G. O'Daffer

Charles R. Fleenor

Randall I. Charles

Sharon Young

Carne S. Barnett

Addison-Wesley Publishing Company
Menlo Park, California   Reading, Massachusetts   London   Amsterdam   Don Mills, Ontario   Sydney

Illustration Acknowledgments

by courtesy of **Bay Area Rapid Transit District:** 374 bottom
Lisa French: 402
Linda Harris-Sweezy: 12, 38, 44
Tom Hickey: 18, 48
Masami Miyamoto: 135, 250, 262, 264, 321, 368, 380, 400, 403, 416
Sandra Popovich: 11, 17, 32, 58, 59, 72, 96, 98, 127, 278, 329, 334, 400
Valerie Randall: 59, 78, 102, 142, 150, 210, 212, 227, 277, 299
Cynthia Swann-Brodie: 29, 47, 271, 286, 384, 386
Margaret Tisdale: 27
Susan Wollum: 2, 3
from *The World Book Encyclopedia.* © 1982 **World Book-Childcraft International, Inc.:** 107

Photograph Acknowledgments

Jeff Albertson/Stock, Boston: 313
Erik Anderson/Stock, Boston: 60 right
© 1979 **Jim Anderson/Woodfin Camp & Associates:** 101
© **Mark Antman/Stock, Boston:** 379 bottom
© 1980 **Craig Aurness/Woodfin Camp & Associates:** 393 top right
© **Craig Aurness/West Light:** 396
© 1976 **Craig Aurness/West Light:** 234
David Austen/Stock, Boston: 223 top left
© 1980 **Frank Balthis:** 362
John Barnett: 115 top left
Julian Baum/Bruce Coleman Inc.: 326–327, 372
Tom Bean/Tom Stack & Associates: 367 bottom right
© 1980 **C. Walsh Bellville/Photo Library:** 206
© 1981 **Nathan Benn/Woodfin Camp & Associates:** 117
© **The Bettmann Archive, Inc.:** 31 top right, 85, 271, 293 top right
© **John Blaustein/Woodfin Camp & Associates:** 23
Elihu Blotnick*: 322
© 1980 **Dennis Brack/Black Star:** 293 top left
Tom Brakefield/Bruce Coleman Inc.: 276 left
© 1982 **Sisse Brimberg/Woodfin Camp, Inc.:** 354
Alexander Cameron/Tandem Computers, Inc.: 113
© **Alan Clifton/After-Image:** 40
© **Stuart Cohen/Stock, Boston:** 255
Lois and George Cox/Bruce Coleman Inc.: 323
Culver Pictures: 31 top left, 74, 133, 275 bottom right
© 1982 **Doris DeWitt/Atoz Images:** 130
© **Donald Dietz/Stock, Boston:** 264
David Em, artist; computer software written by Dr. James F. Blinn: 221
Milton Feinberg/Stock, Boston: 241, 256
© **Patricia Lanza Field/Bruce Coleman Inc.:** 410
© 1981 **Frank Fisher/After-Image:** 245
© **Jeff Foott/Bruce Coleman Inc.:** 144, 332–333, 343 top left
Lee Foster/Bruce Coleman Inc.: 78–79, 122
George B. Fry III*: 31 bottom left, 60 left, 62, 115 top right, center right, bottom left and bottom right, 129, 168
© **Paul Fusco/Magnum Photos:** 247 center right, 393 bottom left
© **Peter Garfield/After-Image:** 367 bottom center
© **Kenneth Garrett/Woodfin Camp, Inc.:** 275 center right
© **Burt Glinn/Magnum Photos:** 280
© 1979 **Luise Graff/Atoz Images:** 263 top
M. W. Grosnick/Bruce Coleman Inc.: 184–185
© 1980 **Gerhard Gscheidle/Peter Arnold, Inc.:** 190
© 1980 **George Hall/Woodfin Camp & Associates:** 68–69
© **Charles Harbutt/Archive Pictures:** 135 (both)
© **Erich Hartmann/Magnum Photos:** 148
© **Michal Heron/Woodfin Camp & Associates:** 232–233
© **D. P. Hershkowitz/Bruce Coleman Inc.:** 268
J. R. Holland/Stock, Boston: 120
© **Thomas Hopker/Woodfin Camp & Associates:** 121
© 1980 **Richard Howard/Black Star:** 367 top left and center left
© **Manfred Kage/Peter Arnold, Inc.:** 140
Stanley King Collection: 24 (all)
© **Pierre Kopp/West Light:** 393 center right
Larry Lee/West Light: 284 (both)
Wayland Lee*/Addison-Wesley Publishing Company: 1 bottom right, 2 (all), 4, 10 left and center, 14, 16, 17, 20, 31 bottom right, 44 left, 45, 50, 58 (both), 59 (both), 64, 66, 72, 73, 82, 92, 94, 95, 96, 98, 100 left, 102 bottom, 103, 118, 124, 127, 167, 172, 178 (both), 179, 180 (both), 194 (both), 196, 204 (all), 205, 213, 224 (all), 225 (all), 227, 228 left, 229 (both), 258 bottom, 262, 263 bottom, 265, 278, 293 bottom left, 294, 296, 297, 299, 329, 344 (all), 345 (all), 346 bottom left and bottom right, 347 (both), 348 (both), 349 (all), 351 left and right, 361, 364 (all), 365 (all), 375, 385, 388, 391, 398 (both), 400
Library of Congress: 1 top left

continued on page 456

Copyright © 1985 by Addison-Wesley Publishing Company, Inc. All rights reserved. No part of this publication may be reproduced, stored in a retrieval system, or transmitted, in any form or by any means, electronic, mechanical, photocopying, recording, or otherwise, without the prior written permission of the publisher. Printed in the United States of America. Published simultaneously in Canada.

ISBN 0-201-24700-3

CDEFGHIJKL-VH-87654

Contents

CHAPTER 1

Addition and Subtraction of Whole Numbers, 1

- 2   Ancient Numeration Systems
- 4   The Decimal Numeration System
- 6   Comparing and Ordering Whole Numbers
- 8   Rounding Whole Numbers
- 10 Estimating Sums and Differences
- 12 Adding Whole Numbers
- 14 Subtracting Whole Numbers
- 16 Problem Solving: The 5-Point Checklist
- 18 Adding and Subtracting Units of Time
- 20 Problem Solving: Understanding the Question
- 21 Problem Solving: Answering the Question
- 22 Skills Practice
- 23 Problem Solving: Using a Data Bank
- 24 Problem Solving: Guess and Check
- 25 **CHAPTER REVIEW/TEST**
- 26 **ANOTHER LOOK**
- 27 **ENRICHMENT: Discovering A Pattern**
- 28 **TECHNOLOGY: Using a Calculator**
- 30 **CUMULATIVE REVIEW**

CHAPTER 2

Addition and Subtraction of Decimals, 31

- 32 Tenths and Hundredths
- 34 Decimal Place Value
- 36 Comparing and Ordering Decimals
- 38 Rounding Decimals
- 40 Adding Decimals
- 42 Subtracting Decimals
- 44 Estimating Sums and Differences with Decimals
- 46 Skills Practice
- 47 Problem Solving: Using Data from an Advertisement
- 48 Problem Solving: Using Data from a Map
- 50 Problem Solving: Make an Organized List
- 51 **CHAPTER REVIEW/TEST**
- 52 **ANOTHER LOOK**
- 53 **ENRICHMENT: Place Value**
- 54 **CUMULATIVE REVIEW**

CHAPTER 3

Multiplication and Division of Whole Numbers, 55

- 56 Basic Properties of Multiplication
- 57 Special Products: Mental Math
- 58 Estimating Products
- 60 Multiplying by a 1-digit Factor
- 62 Multiplying by a 2-digit Factor
- 64 Multiplying by Larger Factors
- 66 Exponential Notation
- 68 Problem Solving: Practice
- 70 Special Quotients: Mental Math
- 71 Estimating Quotients
- 72 Dividing by a 1-digit Divisor
- 73 Short Division
- 74 Dividing by a 2-digit Divisor
- 76 Using Larger Divisors
- 78 Problem Solving: Choosing the Operation
- 80 Skills Practice
- 81 Problem Solving: Using a Formula
- 82 Problem Solving: Find a Pattern
- 83 **CHAPTER REVIEW/TEST**
- 84 **ANOTHER LOOK**
- 85 **ENRICHMENT: History of Mathematics**
- 86 **CUMULATIVE REVIEW**

CHAPTER 4

Multiplication of Decimals, 87

- 88 Decimal Multiplication
- 90 Finding Products Using Estimation
- 92 Multiplying Decimals
- 94 Zeros in Products
- 96 Problem Solving: Using Data from a Table
- 98 Multiplying Decimals: Mental Math
- 100 Estimating Products Using Decimals
- 101 Problem Solving: Using Estimation
- 102 Scientific Notation
- 104 Skills Practice
- 105 Problem Solving: Using Data from a Table
- 106 Problem Solving: Practice
- 108 Problem Solving: Draw a Picture
- 109 **CHAPTER REVIEW/TEST**
- 110 **ANOTHER LOOK**
- 111 **ENRICHMENT: Number Theory**
- 112 **TECHNOLOGY: Computer Programs and Flowcharts**
- 114 **CUMULATIVE REVIEW**

CHAPTER 5

Division of Decimals, 115

- 116 Dividing a Decimal by a Whole Number
- 118 Rounding Decimal Quotients
- 120 Dividing Decimals: Mental Math
- 121 Problem Solving: Practice
- 122 Problem Solving: Using Data from a Table
- 124 Dividing by a Decimal
- 126 Problem Solving: Finding Averages
- 128 Skills Practice
- 129 Estimating Quotients with Decimals
- 130 Problem Solving: Work Backward
- 131 **CHAPTER REVIEW/TEST**
- 132 **ANOTHER LOOK**
- 133 **ENRICHMENT: History of Mathematics**
- 134 **CUMULATIVE REVIEW**

CHAPTER 6

Geometry, 135

- 136 Basic Geometric Figures
- 138 Angles
- 140 Triangles
- 142 Quadrilaterals
- 144 Polygons
- 146 Circles
- 148 Congruent Figures
- 150 Lines of Symmetry
- 152 Constructing Parallel Lines and Perpendicular Lines
- 154 Bisecting Segments and Angles
- 156 Space Figures
- 158 Problem Solving: Solve a Simpler Problem
- 159 **CHAPTER REVIEW/TEST**
- 160 **ANOTHER LOOK**
- 161 **ENRICHMENT: Space Perception**
- 162 **CUMULATIVE REVIEW**

CHAPTER 7

Number Theory and Equations, 163

- 164 Factors
- 166 Divisibility Rules
- 168 Prime and Composite Numbers
- 170 Prime Factorization
- 172 Greatest Common Factor (GCF)
- 174 Least Common Multiple (LCM)
- 176 Variables and Expressions
- 178 Addition and Subtraction Equations
- 180 Multiplication and Division Equations
- 182 Problem Solving: Choosing the Right Equation
- 184 Problem Solving: Writing and Solving an Equation
- 186 Problem Solving: Make a Table
- 187 **CHAPTER REVIEW/TEST**
- 188 **ANOTHER LOOK**
- 189 **ENRICHMENT: Number Relationships**
- 190 **TECHNOLOGY: Using INPUT in a Computer Program**
- 192 **CUMULATIVE REVIEW**

CHAPTER 8

Addition and Subtraction of Fractions, 193

- 194 Fractions
- 196 Equivalent Fractions
- 198 Lowest-Terms Fractions
- 200 Improper Fractions and Mixed Numbers
- 202 Comparing Fractions
- 204 Using Fractions in Estimation
- 205 Problem Solving: Using Data from a Table
- 206 Adding Fractions
- 208 Subtracting Fractions
- 210 Adding Mixed Numbers
- 212 Subtracting Mixed Numbers
- 214 Subtracting Mixed Numbers with Renaming
- 216 Skills Practice
- 217 Problem Solving: Practice
- 218 Problem Solving: Choose the Operations
- 219 **CHAPTER REVIEW/TEST**
- 220 **ANOTHER LOOK**
- 221 **ENRICHMENT: Fractions**
- 222 **CUMULATIVE REVIEW**

CHAPTER 9

Multiplication and Division of Fractions, 223

- 224 Finding a Fraction of a Whole Number
- 226 Multiplying Fractions
- 227 Using a Multiplication Shortcut
- 228 Multiplying with Mixed Numbers
- 230 Skills Practice
- 231 Reciprocals
- 232 Dividing Fractions
- 234 Dividing with Mixed Numbers
- 236 Problem Solving: Using Data from a Table
- 238 Fractions and Decimals
- 240 Skills Practice
- 241 Terminating and Repeating Decimals
- 242 Problem Solving: Use Logical Reasoning
- 243 **CHAPTER REVIEW/TEST**
- 244 **ANOTHER LOOK**
- 245 **ENRICHMENT: Converting Fractions to Decimals**
- 246 **CUMULATIVE REVIEW**

CHAPTER 10

Measurement: Metric Units, 247

- 248 Units of Length
- 250 Changing Units of Length
- 252 Perimeter
- 254 Area of Rectangles and Parallelograms
- 255 Problem Solving: Practice
- 256 Area of Triangles and Trapezoids
- 258 Surface Area
- 259 Problem Solving: Practice
- 260 Volume
- 262 Capacity
- 263 Problem Solving: Practice
- 264 Units of Weight
- 265 Problem Solving: Using Data from a Table
- 266 Celsius Temperature
- 267 Problem Solving: Practice
- 268 Problem Solving: Using the Strategies
- 269 **CHAPTER REVIEW/TEST**
- 270 **ANOTHER LOOK**
- 271 **ENRICHMENT: Measurement**
- 272 **TECHNOLOGY: Computer Decisions**
- 274 **CUMULATIVE REVIEW**

CHAPTER 11

Ratio and Proportion, 275

- 276 Ratio
- 278 Equal Ratios
- 280 Solving Proportions
- 282 Problem Solving: Using the 5-Point Checklist
- 284 Similar Figures
- 286 Problem Solving: Using Data from a Map
- 287 Problem Solving: Using Estimation
- 288 Problem Solving: Using the Strategies
- 289 **CHAPTER REVIEW/TEST**
- 290 **ANOTHER LOOK**
- 291 **ENRICHMENT: Finding the Golden Ratio**
- 292 **CUMULATIVE REVIEW**

CHAPTER 12

Percent, 293

- 294 Percent
- 296 Percents and Fractions
- 298 Percents and Decimals
- 299 Estimating Percents
- 300 Finding Percents for Other Fractions
- 302 Finding a Percent of a Number
- 304 Finding Simple Interest
- 305 Problem Solving: Using Simple Interest Formulas
- 306 Finding the Percent One Number Is of Another
- 308 Problem Solving: Using Percents for Test Scores
- 310 Finding a Number when a Percent of It Is Known
- 312 Skills Practice
- 313 Discounts and Sale Prices
- 314 Proportions and Percents
- 316 Applied Problem Solving
- 317 **CHAPTER REVIEW/TEST**
- 318 **ANOTHER LOOK**
- 319 **ENRICHMENT: Calculating Percents**
- 320 **CUMULATIVE REVIEW**

CHAPTER 13

Circles and Cylinders, 321

- **322** Circumference
- **324** Area of Circles
- **326** Problem Solving: Practice
- **328** Area of Cylinders
- **330** Volume of Cylinders
- **332** Problem Solving: Practice
- **334** Cross Sections
- **336** Applied Problem Solving
- **337** CHAPTER REVIEW/TEST
- **338** ANOTHER LOOK
- **339** ENRICHMENT: Hex Numeration System
- **340** TECHNOLOGY: Loops in Computer Programs
- **342** CUMULATIVE REVIEW

CHAPTER 14

Probability, Statistics, and Graphs, 343

- **344** Equally Likely Outcomes
- **346** Chance and Probability
- **348** Ordered Pairs in Probability
- **350** Problem Solving: Practice
- **352** Frequency, Range, and Mode
- **354** Arithmetic Mean and Median
- **355** Problem Solving: Using Data from a Table
- **356** Bar Graphs
- **357** Line Segment Graphs
- **358** Pictographs
- **359** Circle Graphs
- **360** Problem Solving: Using Data from a Graph
- **362** Applied Problem Solving
- **363** CHAPTER REVIEW/TEST
- **364** ANOTHER LOOK
- **365** ENRICHMENT: Probability
- **366** CUMULATIVE REVIEW

CHAPTER 15

Integers, 367

- **368** Positive and Negative Integers
- **370** Basic Properties for Integers
- **372** Adding Integers
- **374** Adding Positive and Negative Integers
- **376** Subtracting Integers
- **378** Problem Solving: Using Data from a Graph
- **379** Problem Solving: Practice
- **380** Multiplying Integers
- **382** Dividing Integers
- **384** Integer Coordinates
- **386** Graphing Equations
- **388** Applied Problem Solving
- **389** CHAPTER REVIEW/TEST
- **390** ANOTHER LOOK
- **391** ENRICHMENT: Function Rules
- **392** CUMULATIVE REVIEW

CHAPTER 16

Measurement: Customary Units, 393

- **394** Units of Length
- **396** Area
- **398** Volume
- **400** Liquid Measure
- **401** Problem Solving: Practice
- **402** Units of Weight
- **403** Fahrenheit Temperature
- **404** Problem Solving: Using a Calculator
- **406** Applied Problem Solving
- **407** **CHAPTER REVIEW/TEST**
- **408** **ANOTHER LOOK**
- **409** **ENRICHMENT: Other Number Systems**
- **410** **TECHNOLOGY: Computer Graphics**
- **412** **CUMULATIVE REVIEW**

APPENDIX

- **413** Data Bank
- **417** More Practice
- **442** Table of Measures
- **443** Mathematical Symbols
- **444** Glossary
- **451** Index
- **457** Selected Answers

1
Addition and Subtraction of Whole Numbers

Stretch, pre-washed, baggy, designer—everywhere you look, people of all ages are wearing jeans! The history of jeans goes back to 1853 and the Gold Rush. A man named Levi Strauss went West to sell tent canvas to the miners. In California, he found that people needed pants strong enough for work in the mines. So, the canvas was used to make the pants that became America's first jeans.

After World War II, the demand for denim pants that shrunk to a perfect fit swept across the country. Faded blue jeans became the fashion in the 1950s, and they were followed by wheat-colored jeans in the 1960s. By this time, jeans were everyday wear for American youth. Jeans became a status symbol in the 1970s when highly-styled designer jeans that cost as much as sixty dollars were introduced.

The sale of denim clothing is an important business. Over 1 billion pairs of jeans have been sold since the Gold Rush days.

Ancient Numeration Systems

Mathematics began in early times as counting. Before recorded history, people probably counted on their fingers. The Kamba people of southern Kenya use finger counting today.

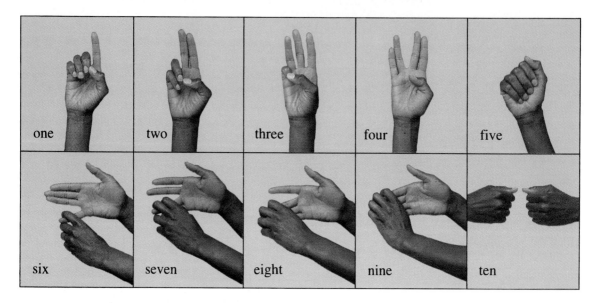

Symbols for numbers, called **numerals,** were invented to keep a record of numbers. Some ancient numerals are shown below.

Egyptian 1700 B.C.	1	∩	೨		(					
	1	10	100	1,000	10,000	100,000	1,000,000			
Chinese 500 B.C.	一	二	三	四	五	六	七	八	九	十
	1	2	3	4	5	6	7	8	9	10
Roman 50 B.C.	I	V	X	L	C	D	M			
	1	5	10	50	100	500	1,000			
Mayan A.D. 1600	⊙	·	—	·̄	=	=̇	≡	≡̈		
	0	1	5	6	10	11	15	17		

In the Egyptian numeration system, the number for a symbol was found by adding.

$$9∩∩|||| = 100 + 20 + 4 = 124$$

$$↟99∩||||| = 1{,}000 + 200 + 10 + 5 = 1{,}215$$

Give the number for each Egyptian numeral.

1. ∩∩|||||
2. 99∩|||||||
3. ↟↟999∩∩||
4. ↟↟↟∩∩∩
5. ||↟999||||
6. ⌇⌇||↟↟↟

Write the Egyptian numeral for each number.

7. 28
8. 231
9. 1,416
10. 12,247
11. 110,001
12. 2,204,034

The Roman numeration system used both addition and subtraction to find the number for a symbol.

IV = 4   (5 − 1)
VI = 6   (5 + 1)
IX = 9   (10 − 1)
XI = 11 (10 + 1)
XL = 40 (50 − 10)
LX = 60 (50 + 10)

XC = 90   (100 − 10)
CX = 110 (100 + 10)
CM = 900 (1,000 − 100)
MC = 1,100 (1,000 + 100)

Give the number for each Roman numeral.

13. V
14. XII
15. VIII
16. LXI
17. CD
18. MCLIX
19. CLXVII
20. MCCLIX
21. MMDCXCIV

Write the Roman numeral for each number.

22. 7
23. 12
24. 20
25. 53
26. 200
27. 1,300
28. 1,500
29. 1,984
30. 3,099

31. Write the present year using Roman numerals.
32. Write the year of your birth using Roman numerals.

3

The Decimal Numeration System

Our system of writing and reading numerals is called a **decimal numeration** system because we group by tens. The Latin word *decem* means ten. Some computers may use systems that group by twos, eights, or sixteens.

The decimal numeration system has place value because the value of each digit in a numeral depends upon its place in the numeral.

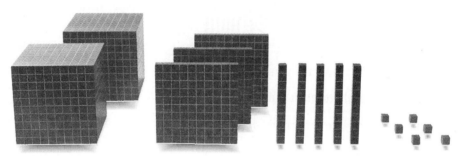

Expanded numeral: $2,000 + 300 + 50 + 6$

Standard numeral: $2,356$

Read: "two thousand, three hundred fifty-six"

The digits for large numbers are usually separated into groups of three called **periods.** We use commas to mark off the periods. This makes the numeral easier to read. The names of the first five periods are shown in the chart below.

Place-value names	hundreds	tens	ones	hundreds	tens	ones	hundreds	tens	ones	hundreds	tens	ones	hundreds	tens	ones
Numeral						2,	7	4	6,	3	8	9,	5	0	1
Period names	Trillions			Billions			Millions			Thousands			Units		

We read 2,746,389,501 as "two billion, seven hundred forty-six million, three hundred eighty-nine thousand, five hundred one."

Warm Up

Read each numeral aloud.

1. 38,625
2. 1,076
3. 125,744
4. 3,000,000
5. 618,746
6. 84,500,000
7. 185,344,107
8. 100,100
9. 7,000,000,000
10. 55,725,000
11. 40,500,030,100
12. 504,844,623

Give the place value of each underlined digit.

Example: 6,739   Answer: 700

1. 5,349
2. 25,339
3. 167,923
4. 16,894
5. 376,518
6. 750,100,570
7. 6,235,000
8. 87,367,950
9. 39,799,947,499

Write the numeral.

10. three thousand, two hundred forty-seven
11. eighteen million, five hundred twenty-three thousand, one hundred twenty
12. seven hundred seventy-five thousand, sixty-six
13. eleven thousand, eleven
14. five billion, nine hundred million
15. two hundred seventeen million, eighty-four thousand, two hundred thirty-nine

Write each standard numeral in words.

16. 47,200
17. 42,591
18. 2,763
19. 500,065
20. 3,575,000
21. 6,280,000,000

Write an expanded numeral for each standard numeral.

22. 6,723
23. 50,468
24. 487,083
25. 17,609,203
26. 34,589,403
27. 200,042,000,892

Write a standard numeral for each expanded numeral.

28. $4,000 + 300 + 50 + 7$
29. $5,000 + 400 + 60 + 1$
30. $20,000 + 7,000 + 900 + 60 + 2$
31. $300,000 + 5,000 + 2$
32. $7,000,000 + 60,000 + 2,000 + 80$
33. $4,000,000 + 700 + 9$
34. $8,000,000 + 300,000 + 500 + 4$
35. $8,000,000,000 + 500,000 + 1,000$

Comparing and Ordering Whole Numbers

The planet Saturn has nine moons. One of its moons, Dione, is 378,000 kilometers (km) from Saturn. The average distance between Earth and its moon is 384,500 km. Which distance is greater?

To compare two numbers, we compare digits in the same places of the two numbers.

Start at the left and compare the digits in the same place.

384,500
378,000

Compare the digits in the first place where they are different.

384,500
378,000

8 is greater than 7.

The numbers compare the same way the digits compare.

384,500 > 378,000

is greater than

378,000 < 384,500

is less than

The distance between Earth and its moon is greater.

Other Examples

73,629 < 73,820

6 < 8

103,483 > 98,734

6 digits   5 digits

2,754,366 > 1,754,636

2 > 1

Warm Up

Which number is greater?

1. 54,291,733
   59,066,927

2. 13,992,248
   13,920,929

Which number is smaller?

3. 1,071,429
   1,073,934

4. 966,773
   966,769

List the numbers in order from least to greatest.

5. 2,763
   2,918
   2,664
   2,876

6. 368,904
   363,719
   365,035
   367,738

7. 7,482,649
   7,500,000
   7,482,630
   7,484,119

8. 38,746,239
   3,742,811,266
   438,927,988
   93,647,149

Write > or < for each ●.

1. 1,875 ● 1,862
2. 2,193 ● 2,175
3. 6,470 ● 6,490
4. 92,364 ● 92,634
5. 26,083 ● 29,095
6. 186,025 ● 188,012
7. 3,004,700 ● 3,004,693
8. 7,000,000 ● 700,000
9. 867,010 ● 867,001
10. 666,675 ● 666,667
11. 90,004,063 ● 90,004,036
12. 999,000,000 ● 1,000,000,000

Give the number that is 1 more.

13. 99
14. 329
15. 1,999
16. 400,342
17. 9,999
18. 399,999,999

Give the number that is 1 less.

19. 400
20. 1,000
21. 3,800
22. 2,780,780
23. 51,000,000
24. 1,000,000,000

Give the number.

25. 100 more than 1,000
26. 1,000 less than 100,000
27. 10 more than 99
28. 1 more than 999,999

Write the numbers in order from least to greatest.

29. 25   13   245
30. 1,400   1,329   1,321
31. 18,908   17,809   17,909
32. 14,874,000   14,874   148,740,000

33. Neptune's moon, Triton, is 354,000 km from its planet. Our moon is 384,500 km from Earth. Which distance is less? Use > or < to compare the numbers.

34. **DATA HUNT** What are the four largest moons of Jupiter? Give their names and list their distances from Jupiter in order from nearest to farthest.

THINK MATH

Digit Patterns

How many digits would you use if you wrote all of the whole numbers from 1 through 1,000?

More Practice, page 417, Set A

Rounding Whole Numbers

The Seattle Kingdome has 64,752 seats. A ticket agency wants to know the approximate number of seats.

To find the approximate number, we round. What is 64,752 rounded to the nearest thousand?

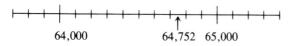

64,752 is closer to 65,000 than to 64,000.

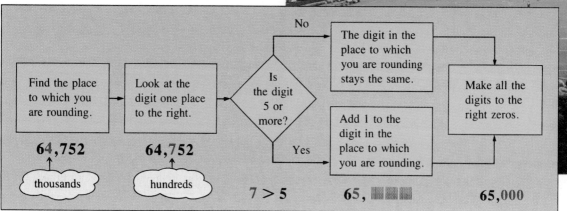

64,752 **rounded to the nearest thousand** is 65,000.

Other Examples

396 rounded to the nearest **ten** is **400**.

839 rounded to the nearest **hundred** is **800**.

64,752 rounded to the nearest **ten thousand** is **60,000**.

39,684,922 rounded to the nearest **million** is **40,000,000**.

Warm Up

Round to the nearest ten; to the nearest hundred.

1. 264
2. 2,055
3. 6,247
4. 781

Round to the nearest ten thousand; to the nearest million.

5. 752,293
6. 1,234,809
7. 29,792,874
8. 34,419,369

Round to the nearest hundred.

1. 742
2. 2,375
3. 18,492
4. 350
5. 2,550
6. 5,641
7. 37,719
8. 614,189,960
9. 663,097,985
10. 126,855

Round to the nearest thousand.

11. 6,239
12. 8,574
13. 10,807
14. 74,682
15. 19,854
16. 8,500
17. 7,449
18. 998,998,992
19. 1,817,296
20. 225,500

Round to the nearest ten thousand.

21. 28,847
22. 59,163
23. 417,900
24. 96,432
25. 94,432
26. 109,955
27. 81,957
28. 509,867
29. 862,385
30. 547,658

Round to the nearest million.

31. 1,654,821
32. 9,946,275
33. 1,284,715
34. 5,555,187
35. 3,982,133
36. 508,837
37. 15,897,446
38. 967,681,000
39. 34,767,100
40. 5,555,555

41. The Atlanta-Fulton County Stadium has a seating capacity of 60,489. What is the seating capacity rounded to the nearest hundred?

42. The Milwaukee County Stadium has a seating capacity of 55,958. Is the seating capacity 55,900, 56,000, or 55,960 when rounded to the nearest hundred?

SKILLKEEPER

1. 7 + 5
2. 13 − 6
3. 9 + 7
4. 17 − 8
5. 5 + 9
6. 9 + 4
7. 11 − 2
8. 8 + 1
9. 9 + 9
10. 17 − 9
11. 8 + 7
12. 10 − 7
13. 6 + 7
14. 9 + 8
15. 6 + 6
16. 11 − 6
17. 7 + 3
18. 4 + 8
19. 12 − 8
20. 5 + 6
21. 12 − 9
22. 5 + 8
23. 18 − 9
24. 9 + 7

More Practice, page 417, Set B

Estimating Sums and Differences

Tim Griffin manages a large record store. On Friday the store sold 376 records. On Saturday, 519 records were sold. Tim uses a calculator to add the two numbers.

First try

Second try

Which sum seems more reasonable? We can make an **estimate** of the sum by rounding each addend to the nearest hundred.

376 + 519   (400 + 500 = 900
                  estimate)

The estimate 900 is close to 895, so 895 seems more reasonable.

Estimates may vary because of the way the numbers are rounded.

Estimate the difference. 2,261 − 914

Round to the nearest thousand.

2,261 − 914   (2,000 − 1,000 = 1,000
                   estimate)

Round to the nearest hundred.

2,261 − 914   (2,300 − 900 = 1,400
                   estimate)

Warm Up

Estimate each sum or difference by rounding to the nearest ten.

1. 79 + 17
2. 64 − 29
3. 119 + 72
4. 209 − 87

Estimate each sum or difference by rounding to the nearest hundred.

5. 278 + 415
6. 930 − 666
7. 2,311 + 892
8. 1,747 − 1,288

Estimate each sum or difference by rounding to the nearest thousand.

9. 2,774 + 1,828
10. 6,105 − 1,978
11. 4,915 − 3,282
12. 12,835 + 5,166

Estimate each sum by rounding.

1. 89 + 33
2. 76 + 82 + 39
3. 214 + 388 + 115
4. 772 + 226
5. 1,296 + 1,825
6. 2,775 + 3,148
7. 3,280 + 1,790
8. 813 + 629
9. 2,372 + 5,127
10. 6,904,381 + 7,870,468
11. 336,639 + 290,714
12. 334,441 + 206,027

Estimate each difference by rounding.

13. 81 − 37
14. 72 − 38
15. 627 − 196
16. 915 − 388
17. 707 − 185
18. 914 − 299
19. 7,124 − 1,958
20. 9,213 − 5,838
21. 5,856 − 2,935
22. 6,274,399 − 3,812,939
23. 883,270 − 264,449
24. 530,627 − 218,963

25. A record store sold 83 rock records, 71 country-western records, 38 classical records, and 66 other kinds of records in one day. Estimate the total number of records sold.

26. Estimate how much less the January sales were than the December sales.

Spinner Record Store

December sales	$19,714
January sales	$12,048
February sales	$13,541

THINK MATH

Number Patterns

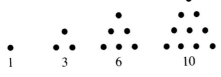

1   3   6   10

The ancient Greeks called numbers like 1, 3, 6, and 10 **triangular numbers.**

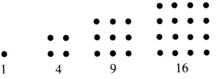

1   4   9   16

Numbers like 1, 4, 9, and 16 are called **square numbers.**

List all the triangular numbers up to 100.

List all the square numbers up to 100.

Choose a triangular number. Add it to the next larger triangular number. What do you notice about the sum?

Try this again with any triangular number.

Adding Whole Numbers

The Mississippi-Missouri-Red Rock river system is the longest in the United States and the third longest in the world. What is the total length of the system?

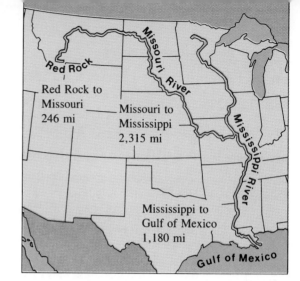

To find the total of the numbers, we add.

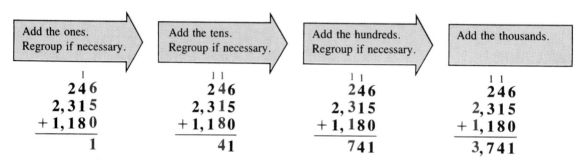

Add the ones. Regroup if necessary.	Add the tens. Regroup if necessary.	Add the hundreds. Regroup if necessary.	Add the thousands.
1 246 $2,315$ $+1,180$ 1	11 246 $2,315$ $+1,180$ 41	11 246 $2,315$ $+1,180$ 741	11 246 $2,315$ $+1,180$ $3,741$

The total length of the system is 3,741 miles (mi).

Other Examples

```
              2  2            1 1 1 1    1
  235       2,839           20,348,209
 +963         748            1,461,601
 ----      +1,507         +    500,268
 1,198      -----            ----------
            5,094           22,310,078
```

Warm Up

Find the sums.

1. 282
    +355

2. 817
    +629

3. 8,809
    +6,477

4. 22,746
    +19,815

5. 6,384
    +5,616

6. 14,392,406
    + 2,690,000

7. 5,714
    6,073
    +9,215

8. 862
    749
    688
    +977

9. 4,378
    925
    2,833
    +   95

10. 31,428
    66,773
    90,306
    +42,177

Add.

1. 83 + 74
2. 619 + 352
3. 884 + 776
4. 375 + 275
5. 8,141 + 3,776

6. 23,548 + 66,395
7. 9,337 + 6,129
8. 7,075 + 6,848
9. 30,542 + 75,598
10. 8,627,150 + 7,490,980

11. 11,324 + 1,776 + 22,974
12. 10,854 + 12,321 + 13,664
13. 97,291 + 829 + 1,089
14. 3,581,380 + 6,287,402 + 5,749,015 + 1,255,102
15. 62,847,421 + 59,417,342 + 88,093,106 + 52,898,830

Find the sums.

16. 247 + 129 + 547
17. 7,140 + 8,337 + 2,950
18. 853 + 94 + 119 + 87
19. 7,075 + 3,834 + 2,966
20. 14,562,000 + 13,013,691
21. 2,716,593 + 36,829,720
22. 1,822,745 + 19,091 + 33,706
23. 57,626 + 84,188 + 93,727,013

24. The Arkansas River and the Colorado River are each 1,450 mi in length. The Columbia River is 1,214 mi long. What is the total length of these rivers?

25. The Allegheny River is 325 mi long. The Ohio River is 656 mi longer than the Allegheny River. What is the combined length of the two rivers?

THINK MATH

Mental Math

You can find the sum of this column of numbers mentally. Follow the arrow to add the tens, then the ones, of each number. The sum is 119.

```
 2  6
 3  8
+5  5
```

$20 + 6 = 26$
$26 + 30 = 56$
$56 + 8 = 64$
$64 + 50 = 114$
$114 + 5 = 119$

Find the sums using mental addition.

1. 32 + 15 + 43
2. 47 + 26 + 35
3. 19 + 66 + 49
4. 62 + 84 + 27 + 23
5. 33 + 46 + 81 + 25

More Practice, page 418, Set A

Subtracting Whole Numbers

Throughout the world, people use the telephone many times during a year. How many more telephone calls per person per year are made in Norway than in the United States?

Telephone Calls Per Person Per Year	
Norway	2,641
Sweden	2,302
Finland	1,493
United States	1,092
Switzerland	886
Canada	825

To find out how much more one number is than another, we subtract.

Subtract the ones. Regroup if necessary.

$$\begin{array}{r} \overset{3\ 11}{2{,}6\cancel{4}\cancel{1}} \\ -1{,}092 \\ \hline 9 \end{array}$$

Not enough ones. Regroup 1 ten for 10 ones.

Subtract the tens. Regroup if necessary.

$$\begin{array}{r} \overset{5\ 13\ 11}{2{,}\cancel{6}\cancel{4}\cancel{1}} \\ -1{,}092 \\ \hline 49 \end{array}$$

Not enough tens. Regroup 1 hundred for 10 tens.

Subtract the hundreds. Regroup if necessary.

$$\begin{array}{r} \overset{5\ 13\ 11}{2{,}\cancel{6}\cancel{4}\cancel{1}} \\ -1{,}092 \\ \hline 549 \end{array}$$

Subtract the thousands.

$$\begin{array}{r} \overset{5\ 13\ 11}{2{,}\cancel{6}\cancel{4}\cancel{1}} \\ -1{,}092 \\ \hline 1{,}549 \end{array}$$

There are 1,549 more telephone calls made per person per year in Norway.

Check:
$$\begin{array}{r} 1{,}092 \\ +1{,}549 \\ \hline 2{,}641 \end{array}$$ It checks.

Other Examples

$$\begin{array}{r} \overset{6\ 14\ 14}{\cancel{7}\cancel{5}\cancel{4}} \\ -695 \\ \hline 59 \end{array} \quad \begin{array}{r} \overset{10}{\overset{7\ \cancel{0}\ 13}{\cancel{8}{,}\cancel{1}\cancel{3}4}} \\ -\ \ 672 \\ \hline 7{,}462 \end{array} \quad \begin{array}{r} \overset{8\ 9\ 10}{\cancel{9}\cancel{0}\cancel{0}} \\ -127 \\ \hline 773 \end{array} \quad \begin{array}{r} \overset{6\ 9\ 9\ 14}{\cancel{7}{,}\cancel{0}\cancel{0}\cancel{4}{,}764} \\ -3{,}268{,}352 \\ \hline 3{,}736{,}412 \end{array}$$

Warm Up

Subtract. Check your answers.

1. $\begin{array}{r} 636 \\ -318 \\ \hline \end{array}$
2. $\begin{array}{r} 7{,}129 \\ -4{,}572 \\ \hline \end{array}$
3. $\begin{array}{r} 1{,}278 \\ -\ \ 844 \\ \hline \end{array}$
4. $\begin{array}{r} 2{,}713 \\ -1{,}957 \\ \hline \end{array}$

5. $\begin{array}{r} 600 \\ -375 \\ \hline \end{array}$
6. $\begin{array}{r} 1{,}014 \\ -\ \ 769 \\ \hline \end{array}$
7. $\begin{array}{r} 1{,}200{,}618 \\ -\ \ \ 351{,}509 \\ \hline \end{array}$
8. $\begin{array}{r} 10{,}000{,}352 \\ -\ 8{,}472{,}163 \\ \hline \end{array}$

Subtract.

1. 73 − 28
2. 305 − 82
3. 482 − 96
4. 6,208 − 4,796
5. 4,816 − 2,379
6. 3,500 − 1,765
7. 5,600 − 2,349
8. 2,714 − 1,876
9. 18,401,783 − 301,641
10. 17,105 − 3,440
11. 27,000 − 13,427
12. 16,243 − 5,685
13. 75,846 − 31,976
14. 9,000,188 − 6,385,029
15. 86,683 − 4,974

16. 315 − 96
17. 800 − 355
18. 2,117 − 854
19. 704 − 227
20. 1,259 − 877
21. 6,006 − 3,727
22. 8,396 − 5,227
23. 26,105 − 19,704
24. 30,000,000 − 712,982

Add or subtract. Do the operations inside the parentheses first.

25. (827 − 454) + 197
26. 683 + (1,027 − 366)
27. 700 − (847 − 689)
28. 2,346 − (548 + 727)
29. 9,005 − (7,213 − 69)
30. 88,176 − (954 − 778)
31. 2,818 + (6,274 − 2,818)
32. (72,421 − 628) − 33,249
33. 72,421 − (29,628 + 3,249)

34. How many more telephone calls per person per year are made in the United States than in Canada? Use the table on page 14.

35. There are about 185,200,000 telephones in use in North America. South America has about 16,900,000 telephones in use. How many fewer telephones are in use in South America?

SKILLKEEPER

1. 3×7
2. $40 \div 5$
3. 8×4
4. $48 \div 6$
5. 9×5
6. 7×7
7. $18 \div 2$
8. $36 \div 9$
9. 6×9
10. $27 \div 3$
11. 9×3
12. 6×6
13. $42 \div 6$
14. 8×7
15. $0 \div 7$
16. $32 \div 8$
17. $81 \div 9$
18. 1×9
19. 3×8
20. $20 \div 4$
21. $56 \div 7$
22. $4 \div 1$
23. 6×4
24. 9×7
25. 7×4
26. $72 \div 9$
27. 0×8
28. $18 \div 6$
29. 9×6
30. $63 \div 7$

More Practice, page 418, Set B

PROBLEM SOLVING: The 5-Point Checklist

| QUESTION |
| DATA |
| PLAN |
| ANSWER |
| CHECK |

To Solve a Problem

1. Understand the QUESTION
2. Find the needed DATA
3. PLAN what to do
4. Find the ANSWER
5. CHECK back

These 5 points give an organized way of solving problems. Follow them to solve this problem.

Ralph Martin decided to count the calories in his meals. What is the total calorie count in his breakfast?

1. **Understand the QUESTION**
   You are asked to find the number of calories in the breakfast.

2. **Find the needed DATA**
   This is the number of calories in each item of food: 90; 115; 160; and 60.

3. **PLAN what to do**
   Since you are asked to find the total, you must add.

4. **Find the ANSWER**
   Do the computation.
   The breakfast has 425 calories.

   ```
     90
    115
    160
   + 60
   ----
    425
   ```

5. **CHECK back**
   See if your answer is reasonable.
   Estimate or check your computation.

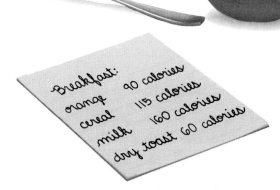

Solve. Use the 5-Point Checklist.

1. Ralph wants to compare the number of calories in the milk and the cereal. How many more calories are in the milk than in the cereal?

2. One morning Ralph ate a scrambled egg (95 calories) instead of cereal and milk. How many calories were in this breakfast?

Solve.

1. What is the total number of calories in the lunch menu?

2. What is the total number of calories in the dinner menu?

3. What is the total number of calories for all three meals?

4. For lunch Van Thu ate everything but carrot cake. How many calories were in her lunch?

5. How many more calories are in a serving of cocoa than in a serving of tomato juice?

6. Janet Goodwing is trying to gain weight. For lunch she had 2 sandwiches, soup, and carrot cake. How many calories were in her lunch?

7. Frank Federico wanted to eat meals that did not total more than 1,850 calories per day. He counted the calories one day and found that they totaled 2,210. How many more calories than 1,850 is this?

8. How many calories are in a dinner of chicken slices, green beans, and salad?

9. What would be the total calories in the lunch if milk were substituted for carrot cake?

★ 10. Jason Johnson ate all but one of the lunch items. The total number of calories for his lunch was 545. What item on the lunch menu did he not eat?

Adding and Subtracting Units of Time

Jenny works part time after school. She works from 4:45 p.m. to 8:15 p.m. How many hours (h) and minutes (min) does she work?

To find the time she works, we can subtract the starting time from the stopping time.

> Write the problem in hours and minutes. ⟹ Regroup 1 hour to 60 minutes if necessary. ⟹ Subtract the minutes. Subtract the hours.

```
   8 h  15 min              7  75                    7  75
 − 4 h  45 min            8̸ h 1̸5̸ min              8̸ h 1̸5̸ min
                        − 4 h  45 min            − 4 h  45 min
                                                   3 h  30 min
```

(Need more minutes.)   (1 h = 60 min; 15 + 60 = 75)

Jenny works 3 hours and 30 minutes.

Other Examples

Add 3 h 30 min and 2 h 55 min.

```
   3 h 30 min
 + 2 h 55 min
   5 h 85 min = 6 h 25 min
```

(85 min = 1 h 25 min)

How many hours and minutes is it from 8:15 a.m. to 2:50 p.m.?

From 8:15 a.m. to noon:
```
    12 h          =   11 h 60 min
 −  8 h 15 min    = −  8 h 15 min
                       3 h 45 min
```

From noon to 2:50 p.m.:
```
                     + 2 h 50 min
          Total =      5 h 95 min
                    or 6 h 35 min
```

Warm Up

Add or subtract the hours and minutes.

1. 3 h 20 min
  + 4 h 35 min

2. 11 h 25 min
  − 7 h 19 min

3. 7 h 37 min
  + 8 h 46 min

4. 8 h 45 min
  − 6 h 59 min

5. How many hours and minutes is it from 9:10 a.m. to 2:25 p.m.?

6. How many hours and minutes is it from 10:35 p.m. to 7:45 a.m.?

Find the total hours and minutes.

1. 1 h 25 min
   + 3 h 35 min

2. 4 h 17 min
   + 6 h 54 min

3. 8 h 42 min
    3 h 9 min
   + 2 h 26 min

4. 2 h 45 min
    5 h 17 min
   + 4 h 28 min

Find the differences in hours and minutes.

5. 6 h 47 min
   − 3 h 39 min

6. 8 h 10 min
   − 4 h 20 min

7. 5 h 15 min
   − 1 h 28 min

8. 10 h
   − 5 h 39 min

Find the number of hours and minutes between the two times.

9. From 6:50 a.m. to 11:30 a.m.

10. From 2:30 p.m. to 10:55 p.m.

11. From 9:15 a.m. to 3:15 p.m.

12. From 8:05 a.m. to 3:15 p.m.

13. From midnight to 6:27 a.m.

14. From 1:25 p.m. to 12:10 a.m.

15. From 6:35 a.m. to 2:10 p.m.

16. From 5:09 a.m. to 1:37 p.m.

17. One day Jenny went to work at 3:25 p.m. She worked until 6:10 p.m. How many hours and minutes did she work?

18. A store is open for business from 8:00 a.m. to 5:30 p.m., Monday through Friday. On Saturday it is open from 8:00 a.m. to noon. How many hours and minutes is the store open during one week?

19. **DATA HUNT** What time does your school start? What time do you get out of school? How many hours and minutes are you in school each week?

THINK MATH

Mental Math

You can do subtraction exercises like 1,000 − 299 in your head.

Think: **1,000 − 300 is 700.**
I subtracted 1 too many, so I must add 1 back.
700 + 1 = 701   1,000 − 299 = 701

Solve mentally.

1. 600 − 399
2. 700 − 198
3. 5,000 − 1,999
4. 2,000 − 299
5. 362 − 99
6. 1,500 − 499
7. 875 − 199
8. 1,750 − 49
9. 162 − 63
10. 375 − 76

More Practice, page 418, Set C

PROBLEM SOLVING: Understanding the Question

QUESTION
DATA
PLAN
ANSWER
CHECK

To solve any problem, you must know what it is you are trying to find. The question always asks you to find something.

Make up a question about the information given in each group of sentences below. Find the answer to each question.

1. Brian is in the 7th grade and is 157 centimeters (cm) tall. His sister Maxine is a 9th grader and is 171 cm tall.

2. Roberta has saved $57 to buy a new bicycle. The model she would like to buy costs $139.

3. Lamont got 27 people to sign for his walk in a charity "Walk-a-thon." Louise got 32 people to sign for her, and Heather got 39 people to sign.

4. Libby has 87¢ but spends 69¢ for a ballpoint pen for school. Later on, Libby earns 50¢ more.

5. Arianna put 23 new stamps in her collection. She counted all her stamps and found that she had collected 207 stamps.

6. Carl wants to put up 3 shelves, each 118 cm long. He has one piece of wood that is 360 cm long.

7. Ed Johnson was on a business trip 425 km away from his home city of Daviston. He started driving home at 2:00 p.m. At 5:15 p.m. he passed a highway sign that read, "Daviston 139 km."

8. Elvia starts school at 8:05 a.m. School ends at 3:50 p.m.

PROBLEM SOLVING: Answering the Question

QUESTION
DATA
PLAN
ANSWER
CHECK

You can write a short sentence that answers the question in a problem.

Airport Code Letters

CLE	Hopkins Field, Cleveland
DFW	Dallas-Ft. Worth
ATL	Atlanta
JFK	Kennedy, New York
SFO	San Francisco
MIA	Miami
DCA	National, Washington, D.C.
IAD	Dulles, Washington, D.C.
LAX	Los Angeles
LGA	LaGuardia, New York
ORD	O'Hare Field, Chicago

JFK had 22,545,000 passengers in one year. LGA had 15,087,000 passengers. ATL had 29,977,000 passengers that year.

How many more passengers did the two New York airports have together than the Atlanta airport?

```
JFK    22,545,000      Total    37,632,000
LGA  + 15,087,000      ATL   - 29,977,000
Total  37,632,000                7,655,000
```

The New York airports had 7,655,000 more passengers.

Do the computation to solve the problems.
Write a short sentence that answers each question.

1. SFO had 20,249,000 passengers in one year. DFW had 2,931,000 fewer passengers than SFO that year. How many passengers did DFW have?

2. IAD had 2,300,000 passengers in a year, and DCA had 14,200,000 passengers the same year. The total for the two Washington, D.C. airports is about the same as the number of passengers for MIA. About how many passengers did MIA have?

3. JFK had 22,545,000 passengers. ORD had 852,000 less than double this number. How many passengers did ORD have?

4. DCA had double the number of passengers of CLE. CLE had 7,100,000 passengers. What was the total number of passengers for the two airports?

5. MIA had 16,500,000 passengers. LGA had 15,087,000 passengers. How many fewer passengers did LGA have?

6. The number of passengers for three busy airports are given below.
   ORD   44,238,000
   ATL   29,977,000
   LAX   28,361,000
   What is the total number of passengers for the three airports?

Skills Practice

Add.

1. 374 + 259
2. 658 + 467
3. 889 + 602
4. 406 + 394
5. 5,253 + 8,957

6. 4,675 + 9,326
7. 27,982 + 90,635
8. 3,044,689 + 9,276,042
9. 7,813,621 + 6,782,394
10. 14,962,153 + 15,347,964

11. 2,603 + 84 + 575
12. 13,909 + 6,724 + 10,688
13. 683,011 + 1,429 + 3,462,914
14. 74,991,620 + 31,374,826 + 80,367,555
15. 93,286,523 + 81,103,675 + 44,632,193

Subtract.

16. 39,451 − 2,301
17. 622,210 − 3,101
18. 70,398 − 62,179
19. 96,457 − 18,368
20. 427,561 − 238,462

21. 358,200 − 269,147
22. 721,165 − 630,476
23. 365,926 − 70,977
24. 62,100 − 3,172
25. 4,983,576 − 1,897,649

26. 23,078,210 − 4,556,701
27. 80,031,001 − 37,741,269
28. 92,613,172 − 37,985,275
29. 9,300,010 − 4,768,201
30. 101,000,101 − 33,918,652

Add or subtract.

31. 42,760,130 + 979,893 + 4,250,618
32. 22,009,876 + 1,998,032 + 460,171
33. 65,532,988 + 10,738,642 + 9,473,590
34. 8,643,201 − 7,896,469
35. 5,551,328 − 3,875,219

36. 33,761,002 − 32,694,554
37. 56,439,802 − 50,429,809
38. 85,770,341 − 29,872,403
39. 1,632,104 + 7,522,462 + 8,196,344 + 3,010,257
40. 23,501,269 + 7,116,330 + 11,408,791 + 986,032

41. 36,029,410 + 59,983,570
42. 6,103,451 − 6,004,988
43. 25,003,681 − 24,855,392
44. 947,305 + 956,895
45. 7,000,931 − 6,989,744
46. 7,116,330 + 177,033

PROBLEM SOLVING: Using a Data Bank

QUESTION
DATA
PLAN
ANSWER
CHECK

A **data bank** is any source of information or **data.**

Modern microcomputers and computers can be used for storing and retrieving data. Large amounts of data can be stored on magnetic discs. The computer can quickly retrieve data. It can also direct a printer to print any stored data.

Information can be passed from one computer to another over telephone lines. Thus, great amounts of data stored in a central library, or **data bank,** can be retrieved and used by people all over the world.

On pages 413–416, a data bank has been printed for your use. You will find Data Bank problems throughout this book that ask you to go to the data bank to find the information you need.

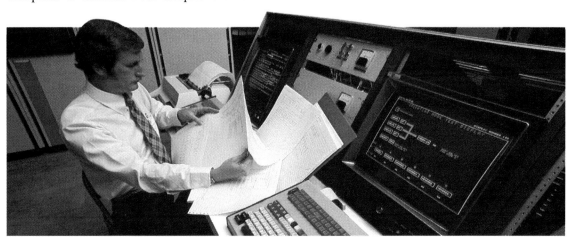

Use the Data Bank on page 413 to find the answers to these questions.

1. An advertising manager of a company wants to contact each AM radio station in the state of Colorado. How many stations must be contacted?

2. A company wishes to advertise a new product on TV stations in Ohio, Indiana, and Michigan. What is the total number of TV stations in these states?

3. Which state is listed as not having a TV station?

4. Which five states have the most TV stations? List the states and the number of TV stations from largest to smallest.

5. Which state has more FM radio stations, New York or Pennsylvania? How many more?

6. What is the number of AM, FM, and TV stations in your home state?

PROBLEM SOLVING: Guess and Check

QUESTION
DATA
PLAN
ANSWER
CHECK

Try This

Kris and Gary collect election campaign buttons. Kris has collected 13 more than Gary. Together they have collected 125 buttons. How many buttons has Gary collected?

Can you solve this problem? If you do not see how to do it, you can use a **strategy** called **Guess and Check**. Here is the way it works.

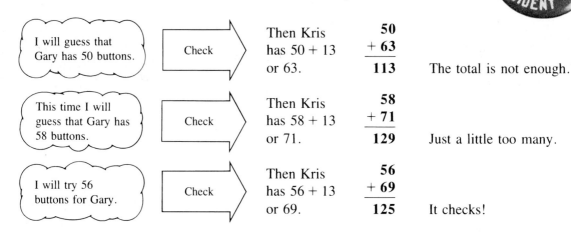

I will guess that Gary has 50 buttons.	Check →	Then Kris has 50 + 13 or 63.	50 + 63 / 113	The total is not enough.
This time I will guess that Gary has 58 buttons.	Check →	Then Kris has 58 + 13 or 71.	58 + 71 / 129	Just a little too many.
I will try 56 buttons for Gary.	Check →	Then Kris has 56 + 13 or 69.	56 + 69 / 125	It checks!

Gary has collected 56 buttons.

Solve.

1. In an election campaign button collection, there are 15 more modern buttons than old-time buttons. There are 121 buttons all together. How many old-time buttons are there?

2. Lakisha and Jay have a total of 210 stamps in their stamp collections. Lakisha has 24 fewer stamps than Jay. How many stamps are there in Jay's collection?

CHAPTER REVIEW/TEST

Write the numeral.

1. four thousand, six hundred seventeen
2. seventeen thousand, nine hundred forty-three
3. three hundred five million, sixty-three thousand
4. forty-eight billion

Write > or < for each ●.

5. 9,673 ● 9,681
6. 16,279 ● 16,729
7. 315,073 ● 314,073

Estimate each sum or difference by rounding to the nearest hundred.

8. 673 − 420
9. 2,628 + 1,233
10. 49,986 − 1,200

Estimate each sum or difference by rounding to the nearest thousand.

11. 8,709 − 1,397
12. 4,378 + 2,691
13. 69,284 − 28,748

Find the sums or differences.

14. 278 + 753
15. 3,851 − 1,627
16. 1,391 + 429
17. 12,800 − 490
18. 8,748 + 7,284

19. 36,869 + 5,295
20. 43,076 − 41,582
21. 65,135 + 67,040
22. 54,172,000 − 27,358,000
23. 9,000,846 − 2,637,797

Find the sums or differences in hours and minutes.

24. 3 h 15 min + 4 h 24 min
25. 8 h 53 min − 3 h 35 min
26. 9 h 27 min + 8 h 45 min
27. 2 h 40 min + 3 h 25 min
28. 6 h 10 min − 3 h 30 min

29. What is the total number of calories in the lunch?

Lunch	
salad	115 calories
sandwich	235 calories
juice	85 calories

30. There are about 3,712,407 telephones in use in Mexico. There are about 5,583,330 telephones in use in Australia. How many more telephones are in use in Australia than in Mexico?

ANOTHER LOOK

3,652 > 3,462
6 > 4

Write > or < for each ●.

1. 2,942 ● 3,942
2. 64,827 ● 68,427
3. 903,414 ● 930,414
4. 862,374 ● 862,574

Round to the nearest hundred.
7,931 → 7,900
less than 5 hundreds

Round to the nearest hundred.

5. 2,381
6. 64,729
7. 128,552

Round to the nearest thousand.

8. 7,726
9. 59,607
10. 554,398

Round to the nearest ten.

64 → 60
29 → 30
+18 → +20
      110 estimate

Estimate each sum or difference by rounding.

11. 29
     39
    +42

12. 389
    −115

13. 6,388
    −4,426

14. 227
     519
    +194

15. 3,096
    +4,779

16. 51,628
    −19,742

```
  1 1
  6 3 8
+ 3 7 6
-------
1,0 1 4
```

Find the sums.

17. 394
    +762

18. 1,809
    +2,776

19. 231,792
    +469,347

20. 35
     76
    +42

21. 129
     317
    +662

22. 1,247
     3,374
    +2,908

```
        Check.
  5 9 13
  6̶0̶3̶      417
−1 8 6    +186
-------   -----
  4 1 7    603
```

Find the differences.

23. 836
    −782

24. 6,021
    −4,227

25. 8,004
    −5,176

26. 23,719
    −16,446

27. 936,042
    −666,781

28. 809,107
    −365,386

ENRICHMENT

Discovering a Pattern

The numbers in the sequence below are called **Fibonacci numbers.** Fibonacci was the nickname of the medieval mathematician Leonardo of Pisa (c. 1170–1250).

1, 1, 2, 3, 5, 8, 13...

Can you guess what the next few numbers would be? Write down your guesses before going on. Study the pattern below.

1 + 1 = 2	1 + 2 = 3	2 + 3 = 5
3 + 5 = 8	5 + 8 = 13	8 + 13 = ?
13 + ? = ?		

Continue this pattern to find more numbers in the Fibonacci sequence.

Try these problems.

1. List the first 30 numbers in the Fibonacci sequence.

2. Study the pattern below.

    $$1 + 1 = \quad 3 - 1 = 2$$
    $$1 + 1 + 2 = \quad 5 - 1 = 4$$
    $$1 + 1 + 2 + 3 = \quad 8 - 1 = 7$$
    $$1 + 1 + 2 + 3 + 5 = 13 - 1 = 12$$

    Write the next three rows in this pattern.

3. Use the pattern shown in problem 2 to give the sum of the first ten numbers in the Fibonacci sequence without actually adding the numbers.

4. Find the sum of the first 30 numbers in the Fibonacci sequence.

Biologists find patterns which show Fibonacci numbers in nature.

TECHNOLOGY

Using a Calculator

Some calculators have memories. A number can be stored in the memory of a calculator and used at a later time.

Store, recall, and clear a number. Use 39,483.

Enter	Press	Display	Comments
39483	M+	M 39483.	The number is stored.
	C	M 0.	The display is cleared.
	MR	M 39483.	The number is recalled.
	MC	39483.	The calculator memory is cleared.

Store the number 4,186 and use the memory of the calculator to solve this exercise.

$$\frac{3 \times 4{,}186 \times 4{,}186 + 4{,}186}{4{,}186}$$

Enter	Press		Display	Comments
4186	M+		M 4186.	4,186 is stored.
3	×	MR	M 4186.	$3 \times 4{,}186$
	×	MR	M 4186.	$12{,}558 \times 4{,}186$
	+	MR	M 4186.	$52{,}567{,}788 + 4{,}186$
	÷	MR	M 4186.	$52{,}571{,}974 \div 4{,}186$
	=		M 12559.	The answer is 12,559.

Store 45,788 in the memory of the calculator. Add 45,788 to each number.

1. 56,845 **2.** 123,377 **3.** 60,583 **4.** 987,445

Store 3,094 in the memory of the calculator. Subtract 3,094 from each number.

5. 4,096 **6.** 27,999 **7.** 5,000 **8.** 400,004

Solve.

9. $(482 \times 492) + 482$

10. $(11,284 \div 1,612) + 1,612 + 1,612$

11. $\dfrac{(1,788 \times 1,788) + 1,788}{1,788}$

12. $\dfrac{(3 \times 2,491) + 2,491 + 2,491}{5}$

Use the map to solve the problems below.

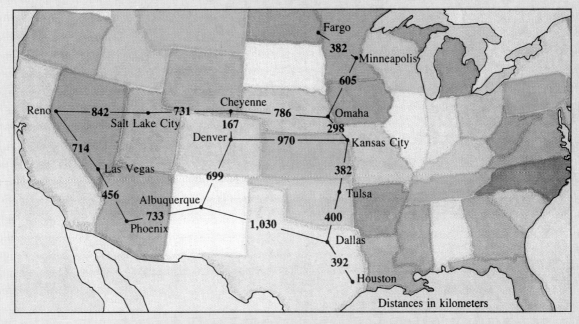

Automobile Touring Distances Between Some Western Cities

Distances in kilometers

13. How much greater is the distance from Minneapolis to Reno than the distance from Fargo to Houston? Hint: Find the distance from Fargo to Houston first and store it in the memory of the calculator.

14. How much greater is the distance of a journey from Reno to Cheyenne and Phoenix and on to Reno than the round trip from Omaha by way of Cheyenne, Albuquerque, Dallas, and Kansas City?

CUMULATIVE REVIEW

1. What is each group of three digits in a numeral called?

   A a period   B a comma
   C a unit   D not given

2. What is the numeral for twenty thousand, eighty-six?

   A 2,086   B 20,086
   C 20,860   D not given

3. Which is correct?

   A 2,748 < 2,497   B 27,500 > 25,709
   C 83,409 < 81,999   D not given

4. Round 386,729 to the nearest thousand.

   A 387,000   B 390,000
   C 400,000   D not given

5. Estimate the sum. 38 + 42

   A 70   B 80
   C 90   D not given

6. Estimate the difference. 640 − 281

   A 200   B 900
   C 400   D not given

Find the sums.

7. 838
   576
   + 647

   A 2,051
   B 1,961
   C 2,061
   D not given

8. 9,275
   1,866
   4,773
   + 8,495

   A 2,249
   B 24,409
   C 21,049
   D not given

Find the differences.

9. 1,608
   − 923

   A 685
   B 785
   C 325
   D not given

10. 6,100
    − 2,738

    A 4,638
    B 3,462
    C 4,662
    D not given

11. Add.

    6,245,621
    + 4,952,912

    A 10,198,533
    B 11,197,533
    C 11,198,533
    D not given

12. Subtract.

    10,655,961
    − 9,296,523

    A 1,369,448
    B 1,359,438
    C 1,459,448
    D not given

13. During one year, a state issued 719,000 fishing licenses and 423,000 hunting licenses. What is the total number of licenses?

    A 1,142,000   B 1,132,000
    C 11,420,000   D not given

14. A state issued 483,000 driver's licenses and 315,000 learner's permits. How many more driver's licenses were issued than learner's permits?

    A 68,000   B 172,000
    C 168,000   D not given

2
Addition and Subtraction of Decimals

Johannes Gutenberg invented movable type in the 1400s, and the printed word became the world's most important means of mass communication. Today it is still at the base of advertising, business, and education. Printing is the eighth largest industry in the United States and the ninth largest in Canada.

Alphabet letters for typesetting were first made of raised metal or wood and the typesetter arranged them by hand. Machines that set type with molten metal were invented in the late nineteenth century. Now computers are used to set type. Alphabet letters as small as 0.5 mm each can be formed by the digital strokes of a computer.

Type comes in a wide variety of sizes, shapes, and styles. The words in this book are set in Times Roman type. The typesetter uses a unit called a pica to measure the height and width of a page. There are 2.3 letters per pica on this page. The word **pica** is about 2 picas long.

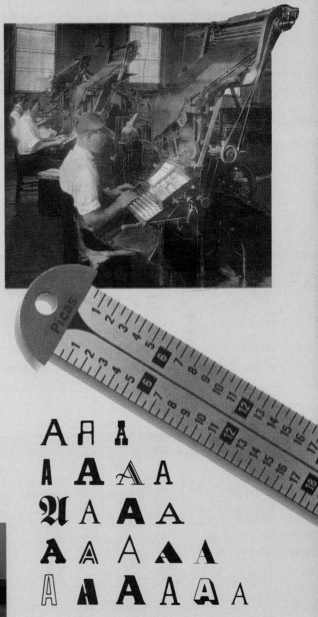

Tenths and Hundredths

There is often a need for numbers other than whole numbers. The newspaper headlines show some numbers that are called **decimals**.

10 equal parts
3 parts shaded
3 **tenths** shaded
0.3 shaded

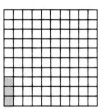

100 equal parts
3 parts shaded
3 **hundredths** shaded
0.03 shaded

1 and 2 tenths shaded
1.2 shaded

2 and 54 hundredths shaded
2.54 shaded

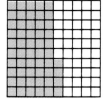

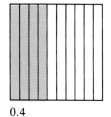 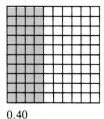

0.4   0.40

The same part is shaded.
0.4 = 0.40

100 hundredths shaded
10 tenths shaded
1 whole unit shaded
1.00 = 1.0 = 1

Write a decimal for the part shaded.

1.
2.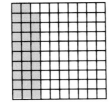
3.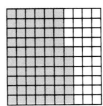
4.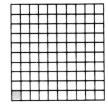

5.
6.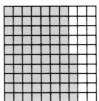

Write the decimal.

7. 3 and 5 tenths
8. 16 and 52 hundredths
9. 2 tenths
10. 3 and 16 hundredths
11. 90 hundredths
12. 47 hundredths
13. 10 and 7 hundredths
14. 15 and 5 tenths
15. 2 and 21 hundredths
16. 2 hundredths
17. 32 and 2 hundredths
18. 32 hundredths

19. How many tenths are shaded?
    How many hundredths are shaded?

20. What part is shaded?
    What part is not shaded?

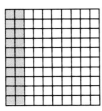

★ 21. What part is shaded? Write the decimal.

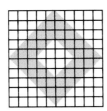

22. **DATA HUNT** What part of the letters used in printing newspaper or magazine articles are vowels? To find out, count off 100 letters in a newspaper or magazine. Count the number of vowels in the 100 letters. Write your answer as a decimal.

Decimal Place Value

Martin Thomas is a genetic scientist. With the aid of an electron microscope, Martin can see parts of cells magnified thousands of times. The ribosome particle found in a cell is 0.000002 cm in diameter.

To read this number, we can extend the idea of place value to decimals. Each place in the decimal is 10 times the place to its right.

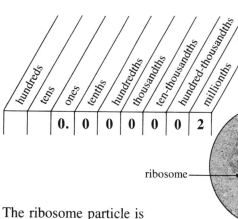

ribosome

The ribosome particle is
2 millionths cm in diameter.

Decimal numeral: 36.284

Read: "thirty-six **and** two hundred eighty-four **thousandths**"

We do not always have to use place-value names. We can read each digit and read the decimal as **point**.

6.9547 "six **point** nine five four seven"

0.0339 "zero **point** zero three three nine"

Other Examples

5.2840   five **and** two thousand, eight hundred forty **ten-thousandths**

0.0100   one hundred **ten-thousandths**

0.03912   three thousand, nine hundred twelve **hundred-thousandths**

100.010   one hundred **and** ten **thousandths**

Warm Up

Read both ways.

1. 2.73
2. 0.6128
3. 19.73764
4. 0.08395
5. 0.0650
6. 37.037
7. 243.010
8. 0.000093

Write the decimal.

1. sixteen and four hundred twenty-five thousandths
2. three thousand, sixty-six ten-thousandths
3. fifty-two thousand, eight hundred nine hundred-thousandths
4. fourteen ten-thousandths
5. sixty-three thousand, four hundred twelve hundred-thousandths
6. one hundred ninety-five millionths
7. eighteen point four six two seven
8. one thousand, twenty-seven and eight hundredths
9. eight hundred one and one hundred eight thousandths
10. sixty-five hundred-thousandths
11. nine hundred forty-one ten-thousandths
12. seventy millionths

Give the place value of each underlined digit.

Example: 2.13<u>5</u>6   Answer: 0.005

13. 6.2<u>7</u>
14. 0.<u>7</u>14
15. 0.187<u>5</u>
16. 5.6<u>9</u>2
17. 23.09<u>1</u>
18. <u>2</u>84.7
19. 0.00001<u>9</u>6
20. 3.3458<u>7</u>

21. The human eye can see particles that are about one ten-thousandth of a meter in diameter. Write this number as a decimal.

★ 22. A field ion microscope can magnify a million times so that an atom that is twenty-seven billionths of a centimeter in diameter can be seen. Write this number as a decimal.

SKILLKEEPER

| 1. | 629
+ 597 | 2. | 804
− 369 | 3. | 567
827
+ 904 | 4. | 2,663
6,801
+ 5,353 | 5. | 54,683
95,347
+ 38,845 |
| 6. | 742
− 356 | 7. | 9,304
− 4,665 | 8. | 2,384
+ 6,388 | 9. | 23,000
− 18,374 | 10. | 512,508
− 129,742 |

Comparing and Ordering Decimals

The Cougars made 0.322 of their shots in a basketball game. The Lions made 0.325 of their shots. Which team made the greater part of their shots?

To compare the decimals, follow the steps below.

Start at the left and compare the digits in the same place.

0.322
0.325

Compare the digits in the first place where they are different.

0.322
0.325

5 > 2

The numbers compare the same way the digits compare.

0.325 > 0.322

The Lions made the greater part of their shots.

Other Examples

1.07 < 1.6   0.273 > 0.27   26.3175 < 26.3775
   0 < 6   3 > 0   1 < 7

Warm Up

Write > or < for each ●.

1. 2.736 ● 2.716   **2.** 0.660 ● 0.693   **3.** 0.007 ● 0.070   **4.** 24.95 ● 23.95

Give the decimals in order from least to greatest.

5. 0.068   **6.** 3.155   **7.** 0.0084   **8.** 5.7762
    0.065   3.182   0.0089   5.7759
    0.605   3.15   0.0189   5.7852

Write >, =, or < for each ▦.

1. 0.83 ▦ 0.87
2. 0.4 ▦ 0.2
3. 0.715 ▦ 0.725
4. 2.73 ▦ 2.92
5. 0.667 ▦ 0.677
6. 0.35 ▦ 0.350
7. 0.04 ▦ 0.40
8. 12.46 ▦ 12.046
9. 6.46 ▦ 6.5
10. 2.1 ▦ 2.100
11. 0.0628 ▦ 0.06281
12. 0.014 ▦ 0.140
13. 0.2678 ▦ 0.2695
14. 0.5 ▦ 0.55
15. 3 ▦ 3.00
16. 0.99 ▦ 1.00
17. 0.009 ▦ 0.01
18. 634.275 ▦ 634.268

Which decimal is greatest?

19. 0.086
    0.083
    0.806
20. 5.143
    5.185
    5.147
21. 9.35
    9.3
    0.93
22. 0.0084
    0.0089
    0.0189

Write each set of decimals in order from least to greatest.

23. 0.72   0.67   0.79
24. 0.628   0.678   0.648
25. 0.043   0.34   0.24
26. 0.09   0.009   0.9
27. 2.73   2.39   2.180
28. 0.4348   0.3448   0.4483

29. Jones of the Cougars made 0.723 of his free throws. McAllen of the Lions made 0.698 of his free throws. Which player made the greater part of his free throws?

30. Desterman of the Lions made 0.4 of his field goal attempts. Garcia made 0.524, and Shane made 0.50. Which player made the greatest part of his field goal attempts?

31. Make your calculator display the smallest possible decimal, greater than 0, in the ordinary decimal notation. Write the word name for this decimal.

32. **DATA BANK** (Use the Data Bank on page 415.) Which professional basketball player made the greater part of his free throws, Rick Barry or Bill Sharman? Compare the decimals. Use > or <.

THINK MATH

Calculator Puzzle

How does William earn his living? Find the answers on a calculator. Turn the calculator upside down to read the words.

$$\begin{array}{r} 2{,}139 \\ 1{,}982 \\ +\,3{,}597 \\ \hline \end{array}$$

$$\begin{array}{r} 100{,}000 \\ -\,42{,}265 \\ \hline \end{array}$$

$$\begin{array}{r} 255 \\ 827 \\ 974 \\ +\,989 \\ \hline \end{array}$$

$$\begin{array}{r} 19{,}742 \\ 28{,}887 \\ +\,5{,}076 \\ \hline \end{array}$$

Rounding Decimals

Astronomers call the distance from Earth to the sun 1 **astronomical unit.** They use this unit to measure the distance other planets are from the sun. The planet Mercury is 0.387 astronomical units from the sun. What is this number rounded to the nearest tenth?

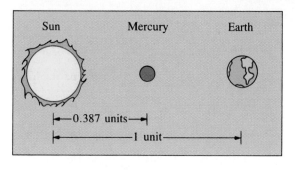

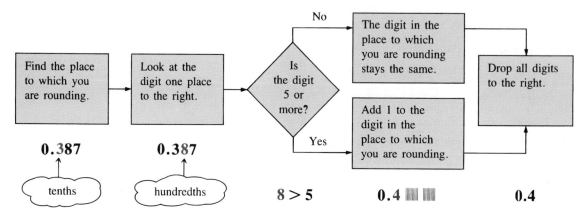

0.387 rounded to the nearest tenth is 0.4.

Other Examples

Decimal	Nearest tenth	Nearest hundredth	Nearest thousandth	Nearest whole number
3.4196	3.4	3.42	3.420	3
16.9635	17.0	16.96	16.964	17
0.70692	0.7	0.71	0.707	1

Sometimes we round numbers to **1-digit accuracy.** Working from left to right, round to the first place that is not 0.

Number	1-digit Accuracy
8.632	9
27.48	30
0.07712	0.08

Warm Up

Round to the nearest tenth.

1. 7.864   **2.** 12.072

Round to the nearest hundredth.

3. 0.695   **4.** 0.0372

Round to the nearest thousandth.

5. 0.4563   **6.** 2.9996

Round to 1-digit accuracy.

7. 0.7719   **8.** 35.8

Round to the nearest tenth.

1. 3.728
2. 6.922
3. 0.5514
4. 16.362
5. 0.4935
6. 8.4809
7. 12.082
8. 0.1482

Round to the nearest hundredth.

9. 23.028
10. 3.4857
11. 0.5538
12. 0.29721
13. 0.4350
14. 1.9967
15. 0.07296
16. 148.6464

Round to the nearest thousandth.

17. 0.38824
18. 10.7066
19. 5.031532
20. 0.00848
21. 12.43309
22. 0.129852
23. 0.70950
24. 0.010762

Round to the nearest whole number.

25. 42.66
26. 6.3994
27. 18.55
28. 382.089
29. 1,278.5
30. 469.883
31. 8.4589
32. 23,507.664

Round to 1-digit accuracy.

33. 72.48
34. 8.68
35. 0.742
36. 0.0695
37. 69.814
38. 0.00522
39. 0.1875
40. 0.9628

41. Mars is 1.524 astronomical units from the sun. Round the number to the nearest tenth.

42. Saturn is 9.539 astronomical units from the sun. Round the number to the nearest hundredth.

43. Astronomers have measured the length of one year on Earth as 365.25636 days. Round the number of days to the nearest hundredth and to the nearest whole number.

44. **DATA BANK** Is Uranus or Neptune farther from the sun? Round the two distances in astronomical units to the nearest tenth. (See page 415.)

SKILLKEEPER

1. 7×8
2. $18 \div 3$
3. $27 \div 9$
4. 6×5
5. 9×8
6. $45 \div 5$
7. 5×9
8. 6×6
9. $48 \div 6$
10. 8×8
11. $64 \div 8$
12. $42 \div 7$
13. 3×8
14. $32 \div 4$
15. $54 \div 6$
16. 7×6
17. 7×7
18. $49 \div 7$
19. 4×8
20. $40 \div 8$

More Practice, page 419, Set B

Adding Decimals

Jason visited some European countries. He has 1 British pound note, 1 German mark, and 1 Greek drachma. What is their total value in U.S. dollars?

Since we want to find the total amount, we can add.

Decimals can be added just like whole numbers if the decimal points are lined up.

Country	Unit of money	Value in U.S. dollars*
Germany	deutsche mark (DM)	0.4075
Great Britain	pound (£)	1.7412
Greece	drachma (Dr)	0.0145
*Values in a recent year		

Write the problem with the decimal points in line.

```
  0.4075
  1.7412
+ 0.0145
```

Add as with whole numbers.

```
   1  1 1
  0.4075
  1.7412
+ 0.0145
  ——————
  2 1632
```

Place the decimal point in line with the others.

```
   1  1 1
  0.4075
  1.7412
+ 0.0145
  ——————
  2.1632
```

The total value is $2.1632 or $2.16, rounded to the nearest hundredth (cent).

Other Examples

$3.6 + 1.28 + 9$

```
   3.60
   1.28
 + 9.00
  ——————
  13.88
```

Annex zeros to line up the decimal places.

$12.795 + 0.693 + 5.759$

```
  12.795
   0.693
 + 5.759
  ——————
  19.247
```

Neuschwanstein, West Germany

Warm Up

Add.

1. 32.8
   + 27.4

2. 0.26
   + 0.58

3. 1.385
   + 2.776

4. 0.96
   + 1.488

5. 0.72595
   + 0.68867

6. $12.4 + 6 + 8.49$

7. $\$25 + \$3.98 + \$0.65$

8. $9.5 + 0.772 + 0.9$

Add.

1. 5.94
   + 6.87

2. 0.467
   + 5.39

3. 6.875
   + 27.436

4. 0.087
   + 9.689

5. 2.043
   + 0.857

6. 4.6817
   + 9.3548

7. 59.634
   + 24.86

8. 7.492
   + 0.685

9. 59.379
   + 86.458

10. 33.286
    + 77.975

11. 4.96
    8.75
    + 7.68

12. 6.597
    8.68
    + 3.046

13. 0.798
    0.697
    + 0.483

14. 54.967
    38.482
    + 9.768

15. 0.63497
    0.57218
    + 0.84956

16. 25.63
    13.8
    + 9.09

17. $8.47
    1.96
    + 0.78

18. $15.84
    19.36
    + 23.99

19. 0.681
    0.883
    + 0.393

20. $349.50
    277.54
    + 194.85

21. 0.3
    7.92
    + 26.0

22. $ 9.57
    10.35
    + 16.80

23. $0.92
    0.34
    + 1.87

24. 0.3806
    0.4009
    + 0.5760

25. $688.23
    492.32
    + 711.96

26. 4.56 + 9.4 + 3.768
27. 0.975 + 4.38 + 6
28. 497.6 + 0.786 + 4.975
29. 74.96 + 3.875 + 0.684
30. 27.0 + 19.0 + 0.28
31. 342.5 + 57.5 + 0.24
32. 0.101 + 10.01 + 0.001
33. 37.6 + 67.3 + 84.9
34. $27.56 + $9.34 + $16
35. $97.99 + $7.06 + $12.79

Use the table on page 40 to do problems 36 and 37.

36. How much are 2 German marks, 1 British pound, and 2 Greek drachmas worth in U.S. dollars?

37. Erik has 3 German marks and 2 British pounds. Does he have money that is worth as much as $5.00 in U.S. money? How much is Erik's money worth?

38. Lucy exchanged $3.00 in U.S. money for Greek drachmas. She got 68.7156 drachmas for each dollar. To the nearest hundredth, how many drachmas did she get?

39. Use the information in the table on page 40 to write your own problem. Solve the problem.

More Practice, page 419, Set C

Subtracting Decimals

Using an electronic timer, a tennis serve was clocked at 62.483 meters per second (m/s). A baseball pitch had a speed of 41.556 m/s. How much greater was the speed of the tennis ball?

Since we want to find how much more one number is than another, we subtract.

Decimals can be subtracted just like whole numbers if the decimal points are lined up.

Write the problem with the decimal points in line.	Subtract as with whole numbers.	Place the decimal point in line with the others.
62.483 − 41.556	1 14 7 13 6̸2.4̸8̸3̸ − 41.556 20 927	1 14 7 13 6̸2.4̸8̸3̸ − 41.556 20.927

The speed of the tennis ball was 20.927 m/s greater than the speed of the baseball.

Other Examples

52 − 37.83

  52.00
− 37.83
  14.17

(Annex zeros to line up the decimal places.)

29.196 − 17.4

  29.196
− 17.400
  11.796

$25 − $18.75

  $25.00
− 18.75
  $ 6.25

Warm Up

Subtract.

1. 58.42 − 26.65
2. 9.27 − 5.969
3. 0.8062 − 0.5571
4. $601.09 − 233.46
5. 1.213 − 0.944
6. $16 − $2.75
7. 1.8 − 0.635
8. 23.55 − 1.978
9. 32.417 − 29
10. 17 − 0.34

Subtract.

1. 34.8 − 12.6
2. 84.6 − 29.8
3. 0.672 − 0.186
4. 9.68 − 4.39
5. 329.7 − 167.9
6. 17.45 − 8.21
7. 5.8623 − 3.4871
8. 56.725 − 38.463
9. 19.471 − 7.836
10. 40.70 − 26.36
11. 19.01 − 3.98
12. 5.63 − 2.48
13. 583.6 − 29.9
14. 0.697 − 0.248
15. 8.721 − 4.83
16. 10.7 − 8.69
17. 12.1398 − 8.6476
18. 25.865 − 16.9532
19. 10.07661 − 0.947
20. 2.00075 − 1.29564
21. $78.02 − 35.43
22. $50.06 − 37.69
23. $53.40 − 27.86
24. $1.00 − 0.69
25. $200.00 − 176.29

26. 9.14 − 4.715
27. 29.857 − 16.948
28. 5.76 − 3.042
29. 2 − 1.675
30. 182.943 − 49.685
31. 9.075 − 4.68
32. 6.175 − 3.9
33. 0.936 − 0.75
34. 100 − 84.97

35. A hockey puck had a speed of 44.556 m/s. The pelota, or ball, used in jai alai had a speed of 63.111 m/s. How much slower was the speed of the hockey puck?

36. When hit, a golf ball had a speed of 28.361 m/s. After three seconds in the air, the speed dropped to 19.5 m/s. How much slower was the speed after three seconds in the air?

THINK MATH

Logical Reasoning

Magic Squares have the same sum in each row, in each column, and in the two diagonals.

Find the missing numbers to make Magic Squares.

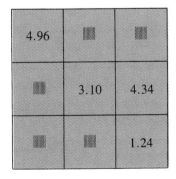

More Practice, page 420, Set A

Estimating Sums and Differences with Decimals

Hal needs to buy four items at the grocery store. He estimates the total cost by rounding each amount to the nearest dollar. What is his estimate?

$$
\begin{array}{rcr}
\$1.89 & \to & \$2 \\
5.95 & \to & 6 \\
2.68 & \to & 3 \\
+4.06 & \to & +4 \\
\hline
& & \$15
\end{array}
$$

His estimate is 15 dollars.

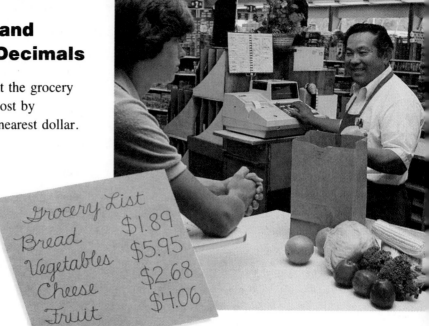

Other Examples

Estimate $\$62.95 - \$37.36$.
Round to the nearest ten dollars.

$$
\begin{array}{rcr}
\$62.95 & \to & \$60 \\
-37.36 & \to & -40 \\
\hline
& & \$20
\end{array}
$$

Estimate $\$219.75 + \$376.47$.
Round to the nearest hundred dollars.

$$
\begin{array}{rcr}
\$219.75 & \to & \$200 \\
+376.47 & \to & +400 \\
\hline
& & \$600
\end{array}
$$

When estimating sums or differences with decimals, round each decimal to the same place.

Estimate $0.614 + 0.78$

$$
\begin{array}{rcr}
0.614 & \to & 0.6 \\
+0.78 & \to & +0.8 \\
\hline
& & 1.4
\end{array}
$$

Estimate $0.07339 - 0.0492$

$$
\begin{array}{rcr}
0.07339 & \to & 0.07 \\
-0.0492 & \to & -0.05 \\
\hline
& & 0.02
\end{array}
$$

Warm Up

Estimate each sum or difference by rounding.

1. $\$1.87 + 2.98$
2. $\$88.19 - 39.75$
3. $\$709.48 - 285.50$
4. $0.0808 - 0.047$
5. $\$0.89 + 0.62 + 0.49$
6. $0.822 + 0.466 + 0.317$
7. $0.6261 - 0.1984$
8. $0.0122 + 0.0638 + 0.0771$

Estimate each sum or difference by rounding to the nearest dollar.

1. $1.88
   3.14
   + 2.75

2. $8.25
   − 6.99

3. $5.75
   6.22
   + 4.05

4. $9.09
   − 2.89

5. $8.50
   2.45
   + 7.27

Estimate each sum or difference by rounding to the nearest ten dollars.

6. $23.14
   + 58.85

7. $91.45
   − 56.50

8. $49.99
   27.75
   + 33.69

9. $77.25
   − 19.37

10. $83.12
    33.98
    + 47.45

Estimate each sum or difference by rounding to the nearest hundred dollars.

11. $719.09
    − 195.24

12. $209.17
    131.45
    + 377.66

13. $518.35
    − 195.00

14. $903.29
    − 419.83

15. $606.99
    478.10
    + 825.50

Estimate each sum or difference by rounding.

16. 0.313
    0.251
    + 0.482

17. 0.74
    − 0.198

18. 0.0612
    0.017
    + 0.05422

19. 0.4841
    − 0.29

20. 5.059
    3.8
    + 2.1774

21. 0.2884 + 0.442
22. 0.6923 − 0.3117
23. 0.89 + 0.78 + 0.43
24. 3.04 + 2.95 + 6.16
25. 0.071 − 0.0477
26. 29.124 − 18.88
27. 0.175 + 0.388 + 0.89
28. 609.4 − 189.29
29. 0.0072 − 0.0039

30. Jessica needs to buy grocery items that cost $2.77, $2.19, $1.68, $3.55, and $1.95. She estimates the total by rounding to the nearest dollar. What is her estimate?

31. Anthony bought grocery items that cost $1.89, $2.09, $1.19, and $3.84. He paid with a $20 bill. Estimate how much change he should get back.

32. **DATA HUNT** What is the estimate of the cost of eight grocery items? Find and list the prices of the items. Estimate the total cost. Find the exact total.

More Practice, page 420, Set B

Skills Practice

Add.

1. 27.30
   + 84.97

2. 8.092
   + 13.645

3. 0.039
   + 8.966

4. 95.842
   + 68.731

5. 49.051
   + 0.998

6. 9.87
   3.05
   + 6.21

7. 24.09
   26.1
   + 84.3

8. $36.95
   54.60
   + 0.38

9. $13.56
   88.17
   + 26.63

10. 57.167
    0.4
    + 3.6

11. 0.82971
    0.67553
    + 0.48401

12. $833.96
    917.42
    + 1,000.63

13. $268.71
    350.28
    + 140.93

14. 1.01365
    208.45
    + 2.369

15. 0.2873
    0.396
    + 0.5439

Subtract.

16. 79.1
    − 30.2

17. 63.4
    − 54.7

18. 0.4812
    − 0.3945

19. 83.52
    − 29.17

20. $50.37
    − 48.99

21. 0.6213
    − 0.6139

22. $94.60
    − 83.75

23. 3.0
    − 1.54

24. 56.48
    − 49.0

25. $418.63
    − 379.56

26. $7,000.03
    − 6,521.04

27. 84.1
    − 6.25

28. 0.53
    − 0.52999

29. 4.0007
    − 3.4695

30. $2,987.30
    − 2,986.49

Add or subtract.

31. 32.0
    + 51.167

32. 0.032917
    + 0.094655

33. 55.0
    − 44.41

34. $99.99
    + 90.05

35. 0.9826
    − 0.9817

36. 861.95 + 23.1 + 367.02

37. 56.0128 − 55.6437

38. 577,102 − 465,391.5

39. 457.92 − 412.99

40. 0.000051 − 0.0000483

41. 9,623.47 − 8,847.89

42. 33,015.6 + 24,782.5 + 101,621.9

43. 0.0257 + 0.026931 + 0.03804

44. 13,279.84 + 62,943.10 + 70,062.48

45. 592.43 + 1,167.05 + 43.29

PROBLEM SOLVING: Using Data from an Advertisement

QUESTION
DATA
PLAN
ANSWER
CHECK

Hair dryer $9.99
Bracelet $3.88
Electronic game $8.75
T-shirt $6.95
Duffel bag $13.95
Belt $3.75
Socks $1.29
Poster $4.99

SALE!

Solve.

1. Terry bought a hair dryer and 2 bracelets. What was the total cost?

2. Sharma bought a duffel bag. She paid for it with a $20 bill. How much money did she get back?

3. Larry wants to buy a pair of socks and a belt. He only has $5.00. Does he have enough money? Find the difference.

4. Carlos had $25.00. He bought 2 electronic games. How much money did he have left after paying for the games?

5. Luann bought a hair dryer, a poster, and a belt. She paid for her purchases with 2 ten-dollar bills. How much change should she get back?

6. The duffel bag usually sells for $18.50. How much less is the price in the advertisement?

7. Dustin bought a duffel bag and a belt. Jason bought a hair dryer and an electronic game. How much more did Jason spend than Dustin?

8. Estimate the total cost of all the articles in the advertisement. Find the exact total. Compare your estimate with the exact total.

9. **DATA HUNT** How much would it cost to buy four articles of clothing? List the articles with the prices and find the total.

10. **Try This** Julie and Margaret went shopping. Margaret spent $2.50 more than Julie. Together they spent a total of $28.60. How much did Julie spend? Hint: Guess and check.

PROBLEM SOLVING: Using Data from a Map

There are 24 time zones around the world. Standard time begins in Greenwich, England, and time zones are one hour apart. The map below shows four standard time zones, Eastern Standard Time (EST), Central Standard Time (CST), Mountain Standard Time (MST), and Pacific Standard Time (PST).

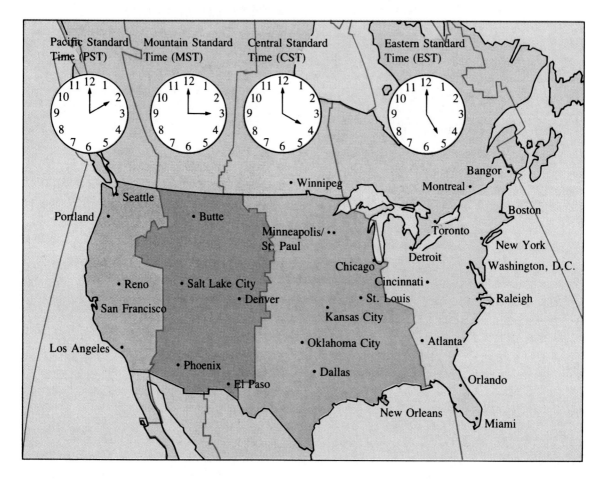

Carol Franklin lives in Seattle. At 6:00 p.m. (PST) she telephones her mother in Chicago. What is the time (CST) in Chicago?

The map shows that it is 2 hours later in Chicago than in Seattle. Therefore, the time is 8:00 p.m. (CST).

Solve.

1. It is 2:00 p.m. (CST) in Oklahoma City. What time (MST) is it in Phoenix?

2. What is the time (EST) in Cincinnati when it is midnight (PST) in Portland?

3. It is 7:00 a.m. (CST) in Minneapolis. What time (PST) is it in San Francisco?

4. What time is it in Miami when it is 11:00 a.m. (MST) in Salt Lake City?

5. It is 8:45 a.m. in Detroit. What time is it in Montreal?

6. It is 1:00 p.m. in Washington, D.C. What is the time in Los Angeles?

7. Rosana lives in Atlanta. At 10:15 a.m. (EST), she called a friend in Los Angeles. What was the time (PST) in Los Angeles?

8. Roger lives in Orlando. He called a friend in Chicago at 8:30 p.m. (EST). What was the time in Chicago?

9. Mrs. Wood has to make a business call to San Francisco from her office in New York City. She wants to make her call at 10:00 a.m. (PST) in San Francisco. What time (EST) should she make her call from New York?

10. Estelle lives in Reno. Her mother lives in Boston. Estelle calls her mother at 9:25 p.m. (PST). What time (EST) is it in Boston?

11. Larry lives in Denver. His brother Barry lives in Raleigh. Larry called Barry at 6:52 p.m. (MST). They talked on the telephone for 15 minutes. What time (EST) was it when Barry hung up the telephone in Raleigh?

12. A nonstop flight from Denver to San Francisco left Denver at 10:00 a.m. (MST). It arrived in San Francisco at 11:30 a.m. (PST). How long was the flying time?

13. The flying time for a nonstop trip from Atlanta to Seattle is 5 hours and 3 minutes. If a flight from Atlanta arrived in Seattle at 6:15 p.m. (PST), what time did the plane leave Atlanta?

14. A nonstop flight from Los Angeles to Chicago takes 3 hours and 50 minutes. What is the latest time (PST) you could leave Los Angeles to arrive in Chicago no later than noon (CST)?

15. **Try This** Mary talked for several minutes on the telephone. Her sister Michelle made another call that was 9 minutes longer than Mary's call. Their father told them, "You girls have been using the telephone for 55 minutes." How long was Mary's call?

PROBLEM SOLVING: Make an Organized List

QUESTION
DATA
PLAN
ANSWER
CHECK

Try This

Julia, Frank, Beth, Eduardo, and Linda were students suggested to the principal to serve on the honors committee. The principal put the five names into a hat and drew out three of them. Using the five names, how many different groups of three can be drawn from the hat?

In some problems you must find the number of different ways something can be done. To help you do this, try the strategy **Make an Organized List.** Here is a way the problem might be solved using this strategy.

To make the list simpler, use only the first letter of each person's name.

All groups of three with Julia as one of the members	All remaining groups with Frank as a member	Remaining group with Beth as a member
JFB JBE JEL JFE JBL JFL	FBE FEL FBL	BEL

There are ten different groups of three names that can be drawn from the hat.

Solve.

1. At the cafeteria there are three choices of sandwiches: hamburger, tuna salad, or peanut butter and jelly. There are three choices of drinks: milk, cocoa, or fruit juice. How many combinations of one sandwich and one drink are possible?

2. Ettalene saw the word STOP on a sign. She wondered how many different combinations could be made using the four letters in the word. How many combinations are possible?

CHAPTER REVIEW/TEST

Write a decimal for the part that is shaded.

1.
2.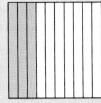
3.

Write the decimal.

4. two hundred fifty-four thousandths
5. twenty-four and sixty-eight thousandths

Write > or < for each ●.

6. 20.347 ● 23.047
7. 7.42 ● 7.22
8. 0.8 ● 0.09

9. Round 23.456 to the nearest tenth.
10. Round 0.7435 to the nearest hundredth.
11. Round 0.334725 to the nearest thousandth.

Round to 1-digit accuracy.

12. 3,816
13. 0.7628
14. 0.0462

Add or subtract.

15. 23.7 + 14.9
16. 9.6 − 1.88
17. 69.1 − 33.8
18. $24.95 + 72.56
19. $100.00 − 83.47

20. 2.15 + 4.9
21. 0.937 − 0.85
22. 12.59 + 71.556

Estimate each sum or difference by rounding to the nearest tenth.

23. $0.79 + 0.87 + 0.33 + 0.48
24. $79.89 − 49.99
25. 0.56 + 0.34 + 0.88
26. 0.625 + 0.292 + 3.567
27. 8.952 − 0.628

28. The height of the Columbia is 17.89 m. The height of a pilot is 1.83 m. What is the difference in the heights?

29. Leon bought running shoes for $23.95 and socks for $2.95. He paid for them with a $20 bill and a $10 bill. How much money should he get back?

ANOTHER LOOK

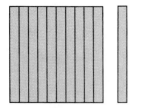

1.1 = one and one tenth

0.01 = one hundredth

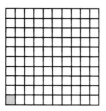

1.1 > 0.01

Write a decimal for the part that is shaded.

1. 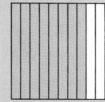   2.

Write the decimal.

3. twenty-five and four tenths
4. eighteen thousandths
5. nine thousand, six hundred fourteen ten-thousandths

Write > or < for each ●.

6. 0.3 ● 0.4
7. 2.8 ● 0.03

Round to the nearest hundredth.

8. 23.4872
9. 0.7539

Round to the nearest hundredth.

1.818 ← 5 or more
↑
hundredths
Round up to 1.82.

Round to the nearest thousandth.

10. 5.3557
11. 0.6295

Line up decimal points.

```
   1 1
  1 6.4    ← 0 hundredths
 +  8.97
 ───────
  2 5.3 7
```

Add.

12. 7.68
    + 3.89

13. 0.8276
    + 0.5775

14. $25.75
      32.95
    + 18.47

15. 3.89 + 0.618
16. 26.4 + 9.88
17. 0.0526 + 0.09

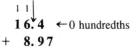

6.8 − 1.33

```
    7 10
  6.8̸0̸    ← 0 in the
 − 1.33     hundredths
 ──────     place
  5.47
```

Subtract.

18. 8.34
    − 3.18

19. 16.05
    − 9.77

20. 0.703
    − 0.096

21. 12.4 − 9.86
22. 3.17 − 0.5
23. 0.9572 − 0.83

52

ENRICHMENT

Place Value

Computers and calculators often use a numeration system based on two instead of ten. A base-two system uses only two digits: 0 and 1. Base-two numerals are called **binary** numerals.

Base-Ten Numerals

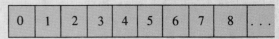

Binary Numerals

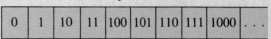

To find a base-ten numeral for a binary numeral such as 111011, we can use a place-value chart.

Add the values of the digits in each place.

The base-ten numeral for the binary numeral 111011 would be 59.

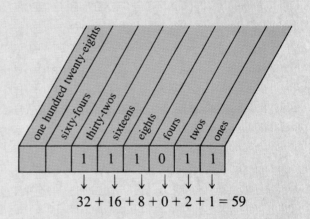

Copy and complete the table.

Base-Ten Numeral	Binary Numeral							
	128s	64s	32s	16s	8s	4s	2s	1s
10				1	0	1	0	
▓			1	1	0	0	0	1
17				▓	▓	▓	▓	▓
▓		1	1	0	1	0	0	0
▓	1	0	1	1	1	0	0	1

Give the base-ten numeral for each binary numeral.

1. 110
2. 1111
3. 11
4. 11001
5. 11111
6. 100000
7. 1000000
8. 1000001

Give the binary numeral for each base-ten numeral.

9. 12
10. 16
11. 20
12. 30
13. 193
14. 70
15. 128
16. 250

CUMULATIVE REVIEW

1. Which digit in 246,789 is in the ten thousands place?

   A 2
   C 6
   B 4
   D not given

2. What is the decimal for ninety and forty-eight thousandths?

   A 90,048
   C 90.048
   B 90.0048
   D not given

3. Which is correct?

   A 0.07 > 0.7
   C 0.07 = 0.7
   B 0.07 < 0.7
   D not given

4. Round 62,299 to the nearest hundred.

   A 62,300
   C 60,000
   B 62,000
   D not given

5. Round 195.975 to the nearest hundredth.

   A 200
   C 195.97
   B 195.98
   D not given

Find the sums.

6. 838 + 576 + 647

   A 2,051
   C 1,961
   B 2,061
   D not given

7. $16.98 + $23.44 + $17.66

   A $57.98
   C $58.08
   B $56.08
   D not given

8. 2.37 + 6.48 + 9.29 + 0.65

   A 18.80
   C 19.79
   B 18.79
   D not given

Find the differences.

9. 2,304 − 876

   A 1,428
   C 1,528
   B 1,438
   D not given

10. $500 − $39.47

    A $34.47
    C $570.53
    B $539.47
    D not given

11. 56.39 − 2.968

    A 53.438
    C 26.71
    B 53.422
    D not given

12. Estimate by rounding.

    $12.95 − $7.87

    A $5
    C $6
    B $4
    D not given

13. Jacklyn bought a dress that cost $49.75. The sales tax was $2.99. What was the total cost?

    A $46.76
    C $51.64
    B $52.74
    D not given

14. A baseball was pitched at a speed of 38.555 m/s. A tennis ball was hit at a speed of 56.5 m/s. How much faster was the speed of the tennis ball?

    A 22.055 m/s
    C 17.945 m/s
    B 27.945 m/s
    D not given

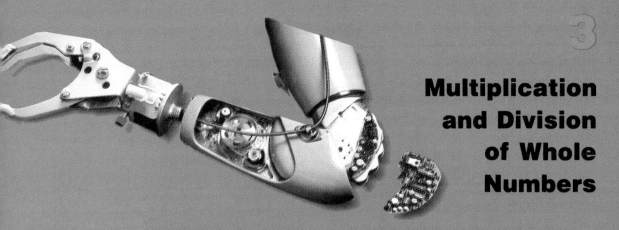

Multiplication and Division of Whole Numbers

Computers and robots are being used more and more for dangerous or routine jobs. They also have many uses outside the world of work and industry. Imagine speaking to a computer that understands you and speaks back. A robotic arm will one day be able to do many tasks by receiving simple word commands. Machines that scan a printed page and read words out loud are already being used. People who cannot speak can type words into a computer which will say the words for them. Scientists are making computers and robots like these to help people who have physical disabilities.

Scientists are always trying to make the devices work better. The computer that helps people speak may only be able to process 40 words a minute. The normal rate of speaking is about 5 times as many words a minute. Someday, these computers may be able to talk as fast as people.

Basic Properties of Multiplication

The Commutative Property

For every pair of numbers a and b,
$a \times b = b \times a$
Changing the order of factors does not change the product.

Example: $23 \times 35 = 35 \times 23$

The Associative Property

For any three numbers, a, b, and c,
$(a \times b) \times c = a \times (b \times c)$
Changing the grouping of factors does not change the product.

Example: $4 \times (2 \times 10) = (4 \times 2) \times 10$

The Property of One

For every number n,
$n \times 1 = n$
Any number times 1 is equal to the number.

Example: $32 \times 1 = 32$

The Distributive Property

For any numbers a, b, and c,
$a \times (b + c) = (a \times b) + (a \times c)$
Multiplying a sum by a number is the same as multiplying each addend by the number and then adding the products.

Example: $3 \times (40 + 2) = (3 \times 40) + (3 \times 2)$

The basic properties can be used to explain the rules we use to multiply numbers.

Since both the order and grouping of factors can be changed without changing the product, we can arrange three or more factors any way we choose.

$2 \times (3 \times 4) = 3 \times (2 \times 4)$
$= (4 \times 2) \times 3$
$= (2 \times 3) \times 4$

Name the property used.

1. $35 \times 1 = 35$
2. $276 \times 809 = 809 \times 276$
3. $(5 \times 2) + (5 \times 8) = 5 \times (2 + 8)$
4. $(6 \times 5) \times 2 = 6 \times (5 \times 2)$
5. $3,357 \times 1 = 3,357$
6. $(8 \times 4) + (8 \times 6) = 8 \times 10$

Find the missing numbers. The basic properties will help you.

7. $27 \times 26 = n \times 27$
8. $9 \times (8 \times 2) = 9 \times (n \times 8)$
9. $18 \times n = 18$
10. $n \times 6 = 6 \times 23$
11. $5 \times (9 + 3) = (n \times 9) + (5 \times 3)$
12. $6 \times (7 + 4) = (6 \times 7) + (6 \times n)$
13. $20 \times (n + 5) = (20 \times 30) + (20 \times 5)$
14. $n \times 1 = 47$

Special Products: Mental Math

The patterns below show multiplication of numbers by 10, 100, and 1,000.

8 × 10 = 80	35 × 10 = 350	167 × 10 = 1,670
8 × 100 = 800	35 × 100 = 3,500	167 × 100 = 16,700
8 × 1,000 = 8,000	35 × 1,000 = 35,000	167 × 1,000 = 167,000

When factors are multiples of 10, 100, or 1,000, the products can be found without using paper and a pencil. Study the patterns below.

30 × 2 = 60	20 × 4 = 80	700 × 8 = 5,600
300 × 2 = 600	20 × 40 = 800	700 × 80 = 56,000
3,000 × 2 = 6,000	20 × 400 = 8,000	700 × 800 = 560,000

Write the products only.

1. 7 × 10
2. 9 × 10
3. 4 × 100
4. 100 × 10
5. 36 × 10
6. 74 × 10
7. 86 × 100
8. 254 × 100
9. 40 × 10
10. 87 × 1,000
11. 7 × 50
12. 20 × 9
13. 40 × 80
14. 8 × 6,000
15. 80 × 90
16. 9 × 3,000
17. 20 × 300
18. 50 × 60
19. 4,000 × 3
20. 700 × 10
21. 70 × 70
22. 600 × 40
23. 9,000 × 6
24. 4,000 × 40
25. 80 × 300
26. 6 × 7,000
27. 60 × 90
28. 400 × 100
29. 70 × 500
30. 6,000 × 70

Find the products mentally.

31. 8 × 10 × 4
32. 50 × 100 × 6
33. 8 × 7 × 100
34. 40 × 20 × 10
35. 60 × 30 × 100
36. 2,000 × 3 × 10
37. 9 × 1,000 × 3
38. 20 × 20 × 20
39. 10 × 100 × 1,000

40. There are 60 seconds in 1 minute and 60 minutes in 1 hour. How many seconds are there in 1 hour?

41. There are 1,000 m in 1 km. How many centimeters are there in 10 km?

42. Have you lived 1 billion seconds? Estimate how old a person is who has lived 1 billion seconds.

43. **DATA HUNT** What is your weight in kilograms? What is your weight in grams? (There are 1,000 grams in 1 kilogram.)

Estimating Products

Gail Cooper finds cat food on sale. She wants to estimate the cost of a case of 48 cans of cat food.

Estimate 48 × 33¢ by rounding each factor to 1-digit accuracy.

48 × 33¢
↓   ↓
50 × 30¢ = 1,500¢ = $15.00

The case of 48 cans will cost about $15.00.

Other Examples

```
  418  →    400         691  →    700        174  →    200
×   8  →  ×   8       ×  49  →  ×  50      × 327  →  × 300
          ─────                 ─────                ─────
          3,200                35,000               60,000
```

Warm Up

Choose the best estimate of each product.

1. 93   A 1,000
    × 7   B 700
            C 630
            D 6,300

2. 69¢   A 480¢
    × 83   B 560¢
            C $48.00
            D $56.00

3. 519   A 20,000
    × 37   B 2,000
            C 24,000
            D 15,000

4. 284   A 6,000
    × 196   B 60,000
            C 40,000
            D 4,000

Estimate each product by rounding to 1-digit accuracy.

5.   87   6. 237   7.   53   8. 779
   × 6   ×   9   × 79   × 62

9. 208   10. 819   11. 984   12. 119
  × 57   × 206   × 327   × 784

Estimate each product by rounding to 1-digit accuracy.

1. 69 × 8	2. 23 × 37	3. 209 × 33	4. 73 × 18	5. 425 × 67
6. 63 × 7	7. 39 × 9	8. 83 × 6	9. 389 × 3	10. 5,089 × 6
11. 84 × 27	12. 63 × 41	13. 47 × 18	14. 38 × 82	15. 43 × 17
16. 51 × 69	17. 34 × 68	18. 67 × 23	19. 78 × 19	20. 38 × 43
21. 521 × 11	22. 588 × 61	23. 709 × 67	24. 892 × 77	25. 475 × 83

26. 95 × 25
27. 36 × 69
28. 72 × 66
29. 22 × 87
30. 47 × 95
31. 722 × 128
32. 643 × 315
33. 296 × 504
34. 726 × 194
35. 814 × 578

36. Estimate the cost of 6 boxes.

37. Estimate the change from $5 if Gina buys 12 cans of dog food.

★ 38. Estimate the total.

39. **DATA HUNT** Find the cost of a grocery item in an advertisement. How much would 36 of that item cost? Make an estimate.

Multiplying by a 1-digit Factor

Jamie likes to play the guitar. When she plucks the middle C string on her guitar, the string vibrates 256 times in 1 second. How many times will the string vibrate in 4 seconds?

Since the same number is repeated, we can use multiplication.

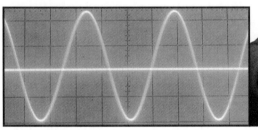

Multiply the ones. Regroup if necessary.	Multiply the tens. Add and regroup if necessary.	Multiply the hundreds. Add if necessary.

$$\begin{array}{r} \overset{2}{2}56 \\ \times 4 \\ \hline 4 \end{array} \qquad \begin{array}{r} \overset{2\,2}{2}56 \\ \times 4 \\ \hline 24 \end{array} \qquad \begin{array}{r} \overset{2\,2}{2}56 \\ \times 4 \\ \hline 1{,}024 \end{array}$$

← factor
← factor
← product

The string will vibrate 1,024 times in 4 seconds.

Other Examples

$$\begin{array}{r}\overset{7}{5}9\\ \times8\\ \hline 472\end{array} \qquad \begin{array}{r}\overset{4}{3}80\\ \times6\\ \hline 2{,}280\end{array} \qquad \begin{array}{r}\overset{2\,6\,4}{3{,}2}75\\ \times9\\ \hline 29{,}475\end{array} \qquad \begin{array}{r}\overset{4}{3}07\\ \times6\\ \hline 1{,}842\end{array}$$

Warm Up
Multiply.

1. 23
    × 6

2. 79
    × 8

3. 80
    × 5

4. 239
    × 9

5. 826
    × 4

6. 367
    × 9

7. 1,293
    × 7

8. 5,276
    × 3

9. 3,737
    × 7

10. 8,309
    × 8

Multiply.

1.	62 × 5	2.	37 × 7	3.	95 × 8	4.	53 × 9	5.	85 × 6
6.	224 × 6	7.	380 × 9	8.	604 × 8	9.	375 × 5	10.	642 × 9
11.	3,247 × 2	12.	26,093 × 4	13.	15,776 × 7	14.	8,200 × 9	15.	3,006 × 8
16.	5,881 × 3	17.	3,948 × 5	18.	6,275 × 6	19.	36,344 × 4	20.	8,826 × 7

21. 854 × 9   22. 64 × 7   23. 1,024 × 8   24. 7,142 × 4   25. 8,166 × 9

26. 952 × 7   27. 7,965 × 8   28. 654 × 3   29. 52,560 × 6   30. 2,927 × 4

31. The lowest note on a piano is made by a string vibrating 27 times a second. How many times will the string vibrate in 3 seconds?

32. The highest note on a piano is made by a string vibrating 3,473 more times per second than the string for the lowest note. How many vibrations will it make in 6 seconds?

SKILLKEEPER

1.	5.36 + 2.77	2.	8.1 − 0.20	3.	29.564 + 1.6	4.	0.567 + 6.439
5.	23.544 − 17.256	6.	7.629 − 3.542	7.	678.35 + 9.706	8.	0.766 − 0.379
9.	$51.97 − 0.79	10.	$0.98 + 0.29	11.	$5,979.60 − 2,596.70	12.	$288.00 + 199.00
13.	$905.20 + 617.90	14.	$5.67 − 2.39	15.	$0.75 + 0.38	16.	$6,095.77 − 5,966.98

More Practice, page 421, Set A

Multiplying by a 2-digit Factor

To cover a floor, it will take 24 rows of tiles. Each row will have 28 tiles. How many tiles will it take to cover the floor?

Since each row has the same number of tiles, we multiply.

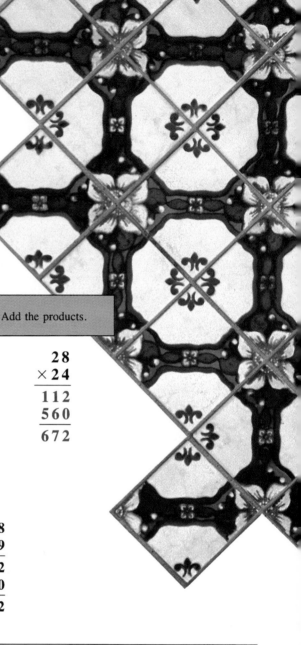

Multiply by the ones.

```
  28
× 24
 112
```

Multiply by the tens.

```
  28
× 24
 112
 560
```

Add the products.

```
  28
× 24
 112
 560
 672
```

It will take 672 tiles to cover the floor.

Other Examples

```
   257          523         4,078
 ×  38        ×  70        ×   59
  2056        36,610        36702
  7710                     203900
  9,766                    240,602
```

Warm Up

Multiply.

1. 72 × 46
2. 67 × 76
3. 94 × 73
4. 59 × 60
5. 80 × 52
6. 163 × 42
7. 309 × 75
8. 1,824 × 56
9. 5,227 × 43
10. 9,561 × 84

Multiply.

1. 25 × 43
2. 63 × 82
3. 79 × 37
4. 68 × 87
5. 94 × 69

6. 772 × 37
7. 306 × 48
8. 529 × 87
9. 623 × 92
10. 375 × 64

11. 990 × 81
12. 403 × 94
13. 924 × 98
14. 965 × 19
15. 665 × 38

16. 2,137 × 58
17. 5,074 × 83
18. 9,544 × 97
19. 6,626 × 49
20. 7,189 × 81

21. 8,064 × 53
22. 7,009 × 47
23. 1,398 × 45
24. 7,799 × 86
25. 5,238 × 42

26. 6,078 × 39
27. 3,009 × 74
28. 9,849 × 76
29. 4,396 × 17
30. 7,468 × 84

31. 17 × 12
32. 42 × 38
33. 24 × 18
34. 56 × 62

35. 129 × 31
36. 248 × 72
37. 365 × 98
38. 279 × 84

39. 824 × 76
40. 747 × 39
41. 9,055 × 99
42. 8,630 × 70

43. 8,296 × 57
44. 56,296 × 18
45. 20,304 × 56
46. 36,068 × 77

47. Lisa is covering a rectangular table with small square tiles. It will take 32 rows of tiles with 48 tiles in each row. How many tiles are needed to cover the table?

48. A classroom floor is covered with square tiles. There are 14 blue tiles and 15 white tiles in each row. There are 23 rows of tiles. How many tiles are on the floor?

More Practice, page 421, Set B

Multiplying by Larger Factors

Carroll County has 798 people per square mile (mi^2) and an area of 572 mi^2. What is the total population of Carroll County?

Since the number of people is given for each square mile, we multiply to find the total number of people.

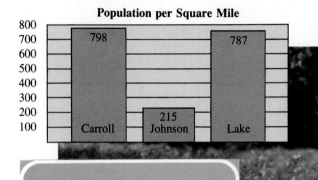

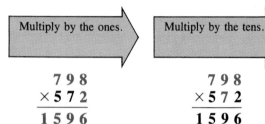

| Multiply by the ones. | Multiply by the tens. | Multiply by the hundreds. | Add the products. |

```
    7 9 8         7 9 8         7 9 8         7 9 8
  × 5 7 2       × 5 7 2       × 5 7 2       × 5 7 2
    1 5 9 6       1 5 9 6       1 5 9 6       1 5 9 6
                5 5 8 6 0     5 5 8 6 0     5 5 8 6 0
                            3 9 9 0 0 0   3 9 9 0 0 0
                                          4 5 6,4 5 6
```

The total population of Carroll County is 456,456.

Other Examples

```
      6 2 4              5 1 9            7,0 4 6
    × 3 0 7            × 6 0 0          ×   1 3 0
    4 3 6 8            3 1 1,4 0 0       2 1 1 3 8 0
  1 8 7 2 0 0  ← We can omit              7 0 4 6
  1 9 1,5 6 8     these zeros.            9 1 5,9 8 0
```

Warm Up

Multiply.

1. 225 × 164
2. 818 × 376
3. 807 × 209
4. 1,255 × 800

5. 628 × 350
6. 1,829 × 607
7. 3,009 × 850
8. 9,132 × 278

Multiply.

1.	375 × 267	**2.**	432 × 357	**3.**	842 × 109	**4.**	828 × 839	**5.**	783 × 497
6.	260 × 195	**7.**	907 × 226	**8.**	834 × 506	**9.**	777 × 296	**10.**	509 × 386
11.	419 × 700	**12.**	923 × 850	**13.**	622 × 266	**14.**	8,139 × 116	**15.**	9,067 × 340

16. 213 × 819   **17.** 297 × 664   **18.** 1,365 × 745

19. 8,281 × 706   **20.** 2,684 × 804   **21.** 52,609 × 905

Multiply.

22. 23 × 37 × 84

23. 27,469 × 129

24. (9 × 8 × 7) − (6 × 5 × 4)

25. (120 × 119 × 118) − (117 × 116)

26. (100 × 100) − (99 × 99)

27. (1,000 × 1,000) − (999 × 999)

28. Jackson County has an area of 510 mi^2. The population density is 1,164 people per square mile. What is the population of the county?

29. Oakvale County has an area of 1,147 mi^2. The population density is 780 people per square mile. The population of the county is how many less than 1 million?

THINK MATH

Finding Patterns

Palindromic numbers are numbers that remain unchanged when their digits are written in reverse order.

These are examples of palindromic numbers.

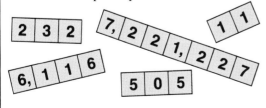

Find the smallest multiple of 13 that is a palindromic number. You may want to use a calculator.

What is the second smallest multiple of 13 that is a palindromic number?

List all the palindromic multiples of 13 that are less than 1,000. Can you see a pattern in these numbers?

More Practice, page 421, Set C

Exponential Notation

Crystals grow by adding new layers of their own substance on exposed geometric surfaces. The early growth is **exponential**.

On a unit with 5 surfaces for growth, 5 new units will form. How many surfaces for growth will there be in 4 generations of growth?

When the same factor is repeated, we can use **exponential notation**.

$$5 \times 5 \times 5 \times 5 = 5^4 \quad \begin{array}{l}\leftarrow \text{Exponent} \\ \leftarrow \text{Base}\end{array}$$

Repeated Factors

Repeated factors	Exponential notation	We read	Standard numeral
$5 \times 5 \times 5 \times 5$	5^4	"five to the fourth power"	625
$5 \times 5 \times 5$	5^3	"five cubed" or "five to the third power"	125
5×5	5^2	"five squared" or "five to the second power"	25

There will be 625 surfaces for growth.

The exponents 0 and 1 have special meanings. $5^0 = 1$ and $5^1 = 5$

Other Examples

A centered dot can be used instead of an $\times$ for multiplication.

$7 \cdot 7 \cdot 7 = 7^3 = 343$ $10 \cdot 10 \cdot 10 \cdot 10 = 10^4 = 10{,}000$ $4^1 = 4$
$4 \cdot 10^2 = 4 \cdot (10 \cdot 10) = 400$ $3 \cdot 2^5 = 3 \cdot (2 \cdot 2 \cdot 2 \cdot 2 \cdot 2) = 3 \cdot 32 = 96$ $6^0 = 1$

Warm Up

Read. State which number is the **base** and which is the **exponent**.

1. 2^5 **2.** 4^7 **3.** 5^3 **4.** 9^2 **5.** 6^4

Give in exponential notation.

6. $8 \cdot 8 \cdot 8$ **7.** 3 squared   **8.** 10 cubed   **9.** $5 \cdot 5 \cdot 5 \cdot 5 \cdot 5$ **10.** $3 \cdot 3$

Give the standard numeral.

11. 6^2 **12.** 2^4 **13.** 3^3 **14.** $8 \cdot 10^3$ **15.** $5 \cdot 2^2$

Write in exponential notation.

1. $2 \cdot 2 \cdot 2 \cdot 2$
2. $8 \cdot 8 \cdot 8 \cdot 8$
3. $10 \cdot 10$
4. $3 \cdot 3 \cdot 3 \cdot 3 \cdot 3$
5. $9 \cdot 9 \cdot 9 \cdot 9 \cdot 9 \cdot 9$
6. $5 \cdot 5 \cdot 5$
7. 4 squared
8. 3 to the fourth power
9. 12 cubed
10. 2 to the tenth power
11. 25 squared
12. 7 to the seventh power
13. 100 cubed
14. 10 to the sixth power
15. 3 to the first power

Write the standard numeral.

16. 6^2
17. 2^3
18. 9^2
19. 5^1
20. 10^2
21. 4^3
22. 5^4
23. 10^3
24. 2^5
25. 12^2
26. 3^0
27. 2^7
28. 3^5
29. 8^2
30. 10^6

Find the products.

31. $5 \cdot 10^2$
32. $4 \cdot 5^2$
33. $3 \cdot 2^3$
34. $2 \cdot 7^2$
35. $9 \cdot 10^4$
36. $6 \cdot 10^6$

37. What is 7^3 written as a standard numeral?

38. What is the missing exponent in $3^{\text{▊}} = 43{,}046{,}721$?
Hint: 3 is the repeated factor.

THINK MATH

Computer Exponents

Many computer languages use an upward arrow (↑) to show exponents.

```
2 ↑ 3 MEANS 2³ OR 8.
3 ↑ 2 MEANS 3² OR 9.
```

Write in the usual exponential notation. Then write the standard numeral.

1. 3 ↑ 4
2. 2 ↑ 5
3. 5 ↑ 2
4. 10 ↑ 3
5. 7 ↑ 2
6. 5 ↑ 3
7. 10 ↑ 4
8. 12 ↑ 2
9. 8 ↑ 3
10. 4 ↑ 4
11. 100 ↑ 2
12. 2 ↑ 10

More Practice, page 422, Set A

PROBLEM SOLVING: Practice

QUESTION
DATA
PLAN
ANSWER
CHECK

Solve.

1. Amber spends about $3 a day to drive to work. Three people want to ride with her, and the 3 passengers will divide the driving costs equally. What will it cost each of the 3 passengers to ride with Amber for 5 working days?

2. Carl Fleetwing pays $645 a year for car insurance. If he joins a car pool, his insurance will be reduced by $78 a year. How much will Carl then pay a year for insurance?

3. Sara James has a new car. She estimates that it will cost about 25¢ per kilometer to drive her car. How much will it cost to drive 20,000 km?

4. Scott Whitson has a car that costs 15¢ per kilometer to drive. How much will it cost Scott if he drives 20,000 km the first year he has the car?

5. Connie pays about $2,000 a year for gasoline, oil, and tolls when she drives to work alone. If she rides in a company van, she will pay $1,200 a year. How much can she save a year by riding in the van?

6. Joe spent $56 for gasoline in November. In December he cut his gasoline bill by $15. How much did Joe spend for gasoline for the two months?

7. Leslie Farmer pays $6 per day to ride a commuter train. What will it cost her to ride the train for 23 days?

8. Kaye Werner's budget for car expense is $25.00 per week. One week she buys gasoline twice. The two tanks cost $11.46 and $9.29. She also buys one can of oil for $1.75. How much is left in her budget for car expense that week?

9. Donna McCormick pays $387 for automobile insurance every six months. If she pays for insurance once a year, she will save $24. What is the total cost of the insurance if she pays once a year?

10. Susan Martinez budgeted $1,500 for car expense during one year. Her expenses were as follows: January through March, $387; April through June, $411; July through September, $435; October through December, $328. How much more or less than $1,500 were her expenses?

11. The cost of crossing a toll bridge is 75¢. Jill Fox paid $12 for a book of 20 tickets to cross the bridge. How much will she save?

12. A book of 20 tickets for crossing a bridge costs 60¢ a ticket. When Walt Swanson turned in a book with 7 unused tickets, he got a refund of $1.75. How much did he lose on the 7 unused tickets?

13. When Mr. Banks drives alone, he leaves home at 8:15 a.m. and gets to work at 8:55 a.m. If he joins a car pool, it will take him 20 minutes longer to get to work. What time must he leave home to get to work at 8:55 a.m.?

★ **14.** Mr. Kent has savings between $8,500 and $9,500. He wants to buy a car that costs between $6,900 and $7,300. If he pays cash for the car, what is the least amount and the greatest amount that could be left in his savings account?

15. Try This Four cars—one red, one yellow, one green, and one black—arrive at a toll booth. How many different ways can they line up to pass through the toll area? Hint: Make an organized list.

Special Quotients: Mental Math

When we know special products, we can use them to find some special quotients.

$300 \div 10 = 30$ because $30 \times 10 = 300$

$1{,}200 \div 100 = 12$ because $12 \times 100 = 1{,}200$

$6{,}000 \div 1{,}000 = 6$ because $6 \times 1{,}000 = 6{,}000$

$800 \div 40 = 20$ because $20 \times 40 = 800$

$4{,}000 \div 50 = 80$ because $80 \times 50 = 4{,}000$

We can use a shortcut to find some quotients.

$200 \div 50 = \dfrac{20\cancel{0}}{5\cancel{0}} = \dfrac{20}{5} = 4$

(Divide each number by 10.) (This bar means divide.)

$15{,}000 \div 300 = \dfrac{15{,}0\cancel{00}}{3\cancel{00}} = \dfrac{150}{3} = 50$

(Divide each number by 100.)

Check: $4 \times 50 = 200$

Check: $50 \times 300 = 15{,}000$

Other Examples

$1{,}800 \div 100 = \dfrac{1{,}8\cancel{00}}{1\cancel{00}} = 18$

$30{,}000 \div 50 = \dfrac{30{,}00\cancel{0}}{5\cancel{0}} = 600$

Find the quotients mentally.

1. $\dfrac{2{,}000}{500}$
2. $\dfrac{900}{300}$
3. $\dfrac{36{,}000}{4{,}000}$
4. $\dfrac{21{,}000}{700}$

5. $1{,}600 \div 40$
6. $8{,}000 \div 400$
7. $2{,}800 \div 70$
8. $15{,}000 \div 3{,}000$

9. $60{,}000 \div 20$
10. $1{,}200 \div 300$
11. $2{,}800 \div 70$
12. $1{,}300 \div 100$

13. $29{,}000 \div 100$
14. $1{,}900 \div 10$
15. $14{,}000 \div 20$
16. $900{,}000 \div 300$

17. $240 \div 60$
18. $1{,}500 \div 300$
19. $800 \div 40$
20. $350 \div 70$

21. $600 \div 10$
22. $900 \div 30$
23. $1{,}200 \div 600$
24. $4{,}200 \div 600$

25. $4{,}000 \div 50$
26. $6{,}300 \div 900$
27. $7{,}200 \div 800$
28. $60{,}000 \div 2{,}000$

Estimating Quotients

We can use special quotients for estimation.

To estimate quotients, we round each number to 1-digit accuracy.

Estimate $2{,}163 \div 39$.

$$\begin{array}{r}2{,}163 \to \\ 39 \to\end{array} \frac{2{,}0\cancel{00}}{4\cancel{0}} = \frac{200}{4} = 50$$

The estimated quotient is 50.

Other Examples

$39{,}419 \div 523$

$$\begin{array}{r}39{,}419 \to \\ 523 \to\end{array} \frac{40{,}0\cancel{00}}{5\cancel{00}} = \frac{400}{5} = 80$$

$321\overline{)936}$

$$\begin{array}{r}936 \to \\ 321 \to\end{array} \frac{9\cancel{00}}{3\cancel{00}} = \frac{9}{3} = 3$$

Estimate each quotient by rounding to 1-digit accuracy.

1. $1{,}927 \div 41$
2. $629 \div 19$
3. $4{,}214 \div 53$
4. $31{,}427 \div 620$

5. $1{,}999 \div 49$
6. $390 \div 81$
7. $33{,}747 \div 6{,}419$
8. $78{,}109 \div 427$

9. $409 \div 21$
10. $622 \div 32$
11. $1{,}877 \div 37$
12. $899 \div 28$

13. $2{,}774 \div 58$
14. $788 \div 38$
15. $3{,}919 \div 83$
16. $6{,}077 \div 29$

17. $\dfrac{5{,}166}{493}$
18. $\dfrac{3{,}792}{77}$
19. $\dfrac{2{,}919}{309}$
20. $\dfrac{7{,}935}{412}$

21. $\dfrac{8{,}283}{396}$
22. $\dfrac{5{,}877}{1{,}939}$
23. $\dfrac{38{,}717}{5{,}062}$
24. $\dfrac{9{,}273}{319}$

25. $71\overline{)719}$
26. $607\overline{)2{,}888}$
27. $795\overline{)41{,}266}$
28. $5{,}776\overline{)27{,}509}$

29. $25\overline{)857}$
30. $432\overline{)7{,}754}$
31. $519\overline{)24{,}562}$
32. $8{,}252\overline{)81{,}652}$

More Practice, page 422, Set B

Dividing by a 1-digit Divisor

Four students are collecting aluminum cans. They collected 138 kg of aluminum. If the weight is divided equally, how many kilograms is this for each person?

To find the number of kilograms for each person, we divide.

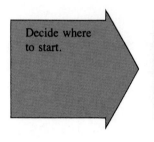

Decide where to start.

$4\overline{)138}$

4 > 1 Not enough hundreds.
4 < 13 Divide the tens.

Dividing Tens
- Divide
- Multiply
- Subtract
- Compare

$$\begin{array}{r} 3 \\ 4\overline{)138} \\ \underline{12} \\ 1 \end{array}$$

Dividing Ones
- Bring down
- Divide
- Multiply
- Subtract
- Compare

$$\begin{array}{r} 34\text{ R2} \\ 4\overline{)138} \\ \underline{12}\downarrow \\ 18 \\ \underline{16} \\ 2 \end{array}$$

Check:
```
    34  ← quotient
  ×  4  ← divisor
   136
  +  2  ← remainder
   138  ← dividend
```

The students collected 34 kg of aluminum per person and 2 kg extra.

Other Examples

$$\begin{array}{r} 163 \\ 3\overline{)489} \\ \underline{3} \\ 18 \\ \underline{18} \\ 09 \\ \underline{9} \\ 0 \end{array} \qquad \begin{array}{r} 240\text{ R2} \\ 9\overline{)2{,}162} \\ \underline{18} \\ 36 \\ \underline{36} \\ 02 \\ \underline{0} \\ 2 \end{array} \qquad \begin{array}{r} 308\text{ R5} \\ 6\overline{)1{,}853} \\ \underline{18} \\ 053 \\ \underline{48} \\ 5 \end{array}$$

Find the quotients and remainders.

1. $6\overline{)882}$
2. $4\overline{)806}$
3. $6\overline{)1{,}762}$
4. $2\overline{)7{,}045}$

5. $7\overline{)2{,}817}$
6. $3\overline{)1{,}773}$
7. $4\overline{)2{,}435}$
8. $8\overline{)4{,}538}$

9. James, Diane, and Sandy collected 74 kg of aluminum cans. How many kilograms is this per person?

10. Laurie collected 147 bottles for a refund of 60¢ a carton. How many cartons of 6 bottles were there?

More Practice, page 422, Set C

Short Division

Five classes collected 1,378 kg of glass for the Martin Luther King, Jr. Middle School. How many kilograms was this per class?

Since we want to find the number of kilograms per class, we divide.

Short division can be used when the divisor is a 1-digit number. Most of the work must be done mentally.

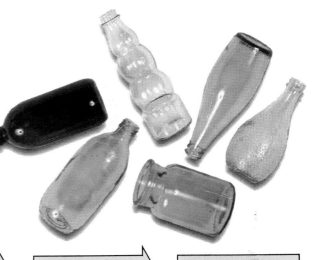

| Decide where to start. | Divide the hundreds. Write the remainder by the tens. | Divide the tens. Write the remainder by the ones. | Divide the ones. Write the remainder if necessary. |

$5\overline{)1,378}$ 　　 $5\overline{)1,3^37\,8}$ 　　 $5\overline{)1,3^37^2 8}$ 　　 $5\overline{)1,3^37^2 8}$ = 275 R 3

(top: 2) 　　 (top: 2 7) 　　 (top: 2 7 5 R 3)

(13 ÷ 5 = 2 with R3)　(37 ÷ 5 = 7 with R2)　(28 ÷ 5 = 5 with R3)

Each class collected about 276 kg of glass.

Find the quotients and remainders. Use the short division method.

1. 6)316　　2. 4)2,055　　3. 3)155　　4. 8)393　　5. 6)902

6. 4)842　　7. 8)9,728　　8. 2)3,825　　9. 5)8,215　　10. 3)1,520

11. 9)4,623　　12. 6)3,290　　13. 7)2,828　　14. 3)9,531　　15. 5)1,023

SKILLKEEPER

1. 3×7	2. 3×70	3. 3×700	4. 30×70	5. 30×700
6. 4×80	7. 40×80	8. 400×8	9. $4,000 \times 8$	10. 400×80
11. 9×6	12. 9×60	13. 90×60	14. 900×6	15. $9,000 \times 6$
16. 7×8	17. 70×8	18. 7×80	19. 70×80	20. 700×8

More Practice, page 423, Set A

Dividing by a 2-digit Divisor

Bob Feller pitched 428 major league baseball games over a period of 18 years. About how many games did he pitch each year?

Divide the number of games pitched by the number of years.

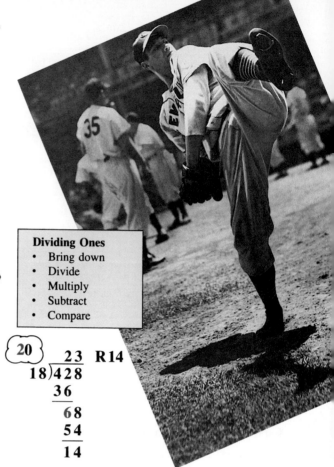

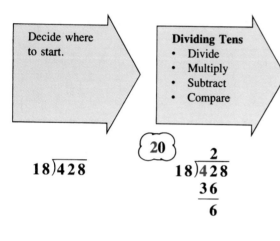

Bob Feller pitched over 23 games each year.

Other Examples

$$\begin{array}{r}7\text{ R }40\\63\overline{)481}\\441\\\hline 40\end{array}\quad(60)$$

$$\begin{array}{r}32\text{ R }4\\89\overline{)2{,}852}\\267\\\hline 182\\178\\\hline 4\end{array}\quad(90)$$

$$\begin{array}{r}402\text{ R }11\\57\overline{)22{,}925}\\228\\\hline 12\\0\\\hline 125\\114\\\hline 11\end{array}\quad(60)$$

Warm Up

Divide. Check your answers.

1. 32)796
2. 48)984
3. 54)683
4. 35)850
5. 35)507
6. 45)2,288
7. 51)3,423
8. 68)1,878
9. 75)2,580
10. 24)1,448

Find the quotients and remainders.

1. 36)876
2. 52)970
3. 34)673
4. 83)3,735
5. 47)1,268
6. 74)4,626
7. 25)5,049
8. 72)3,603
9. 84)5,265
10. 25)2,274
11. 16)5,401
12. 63)2,520
13. 49)2,819
14. 81)4,006
15. 88)929
16. 65)1,581
17. 25,740 ÷ 45
18. 21,633 ÷ 93
19. 13,398 ÷ 29
20. 7,255 ÷ 15
21. 17,360 ÷ 57
22. 38,019 ÷ 70
23. 27,676 ÷ 68
24. 43,800 ÷ 72

25. Cy Young pitched in 826 major league games over a period of 22 years. About how many games did Cy Young pitch each year?

26.

Pitcher	Winning Games	Years
Bob Feller	266	18

How many games did he win per year?

★ 27. Red Ruffing pitched for 22 years. If he had pitched 14 more games, he would have pitched 23 games a year. How many games did Ruffing pitch?

28. **DATA BANK** Who pitched more games per year, Warren Spahn or Grover Alexander? (See page 414.)

THINK MATH

Computer Operations

The tables show the symbols and order of operations for many computer languages.

Symbol	Operation
+	Addition
−	Subtraction
*	Multiplication
/	Division
↑	Exponentiation

Order of operations
1. Operations within parentheses
2. Exponentiation
3. Multiplication and division
4. Addition and subtraction

What number would a computer print for each of these?

1. 9 * 2 − 5
2. 20/4 + 7
3. 6 * 3 − (8 + 4)
4. 5 ↑ 2 * 4
5. 10 + 5 * 2
6. (10 + 5) * 3
7. 18/6 + 42/7
8. 2 ↑ 3 + 3 ↑ 2
9. 9 ↑ 2/9 − 9
10. 15 − (24 − 16)
11. 4 * 4 − 4 ↑ 2
12. (9 * 4)/3 − 8

More Practice, page 423, Set B

Using Larger Divisors

A restaurant serves butter in small pats. There are about 204 servings of butter in 1 kg of butter. Each kilogram of butter contains about 7,142 calories. How many calories are in each serving of butter?

Since we want to find the number of calories in each serving, we divide.

Decide where to start. → **Dividing Tens**
- Divide
- Multiply
- Subtract
- Compare

→ **Dividing Ones**
- Bring down
- Divide
- Multiply
- Subtract
- Compare

```
                 (200)              (200)
        3                  35 R 2
204)7,142    204)7,142         204)7,142
              612                612
              102               1022
                                1020
                                   2
```

Each serving of butter has about 35 calories.

Other Examples

```
(100)                (400)                  (500)
      56 R 60              80 R 35                 207 R 83
129)7,284            392)31,395             482)99,857
    645                  3136                   964
    834                    35                   345
    774                     0                     0
     60                    35                   3457
                                                3374
                                                  83
```

Warm Up

Divide.

1. 316)8,286
2. 547)21,898
3. 793)27,467
4. 987)641,228

Find the quotients and remainders.

1. 294)6,428
2. 727)12,955
3. 509)8,143
4. 419)9,207
5. 633)53,229
6. 711)27,542
7. 385)36,397
8. 528)19,636
9. 166)6,569
10. 977)297,008
11. 402)153,929
12. 849)648,342
13. 26,145 ÷ 421
14. 10,267 ÷ 295
15. 43,324 ÷ 682
16. 43,932 ÷ 523
17. 335,345 ÷ 775
18. 139,461 ÷ 198
19. 350,208 ÷ 912
20. 516,883 ÷ 824

Estimate each quotient by rounding to 1-digit accuracy.
Then find the exact quotients.

21. 6,358,842 ÷ 8,334
22. 4,598,979 ÷ 5,093
23. 947,947 ÷ 1,001
24. 34,864,545 ÷ 62,819
25. 49,946,395 ÷ 50,707
26. 39,978,001 ÷ 19,999

27. A loaf of bread has 1,680 calories. There are 24 slices of bread in the loaf. How many calories are in each slice?

28. A slice of wheat bread has 80 calories. One pat of butter has 35 calories. If 2 pats of butter are spread on each slice of bread, how many slices of buttered bread equal 1,500 calories?

THINK MATH

Calculator Division

You can use a calculator to find the whole-number quotient and the remainder for 377,809 ÷ 689.

Study the method below. Follow the steps and use a calculator to do exercises 13–20 on this page.

Enter	Press	Display	Comments
377809 ÷ 689	=	548.34397	Decimal quotient
548	M+	M 548.	Enters whole number part of the quotient in the memory.
377809 − (689 × MR)	=	237.	Remainder

Solution: 377,809 ÷ 689 = 548 R237

More Practice, page 423, Set C

PROBLEM SOLVING: Choosing the Operation

Wayne Burgess owns a glider school and trains people to fly gliders. Gliders are aircraft that have no engines. They are used mostly for recreational flying.

One of the most important features of a glider is its glide ratio. This is the distance the glider will fly forward while it is losing 1 unit in altitude.

What is the glide ratio of a glider that can fly 2,500 feet (ft) forward while losing 100 ft in altitude? The ratio will help us choose the operation to solve the problem.

$$\frac{2,500}{100} = 2,500 \div 100 = 25$$

The glide ratio is 25. The glider can fly 25 ft while losing 1 ft in altitude. Some gliders have glide ratios of more than 40.

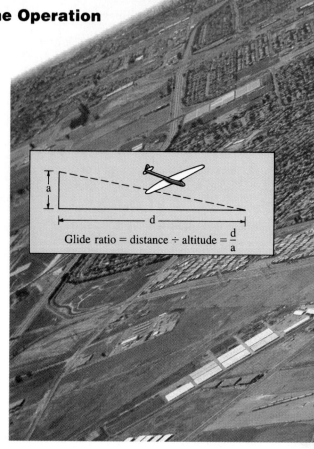

Glide ratio = distance ÷ altitude = $\frac{d}{a}$

Choose the operation to solve these problems.

1. A glider moves forward 12,600 ft while losing 360 ft in altitude. What is the glide ratio for this glider?

2. A glider has a glide ratio of 35. How far forward will it fly while losing 150 ft in altitude?

3. The glide ratio of a glider is 25. It is at an altitude of 3,200 ft. How far will it glide forward while it drops in altitude to 2,500 ft?

4. A glider has a glide ratio of 28. How much altitude will it lose in a glide that is 2,520 ft long?

5. A glider has long, narrow wings. One glider has wings 810 in. long. The wings are 30 times as long as they are wide. How wide are the wings?

6. A glider flew 159 mi in 3 hours. How many miles per hour (mph) did the glider fly?

7. A record altitude of 46,266 ft for a glider was set by Paul Bikle in 1961. Commercial jet airliners fly at about 35,250 ft. How much higher is the glider's record altitude?

8. In 1972 Hans Werner Grosse made a long-distance flight of 907.7 mi in a glider. In 1977 Karl Striedieck made a flight of 1,016 mi. How much farther was Striedieck's flight?

9. The record speed for a short glider race was 102.74 mph. The record speed for a race ten times as long was 54.78 mph. How much faster was the speed in the shorter race?

10. **Try This** When you travel by air, you select your seating by making different choices. How many different selections of seating are possible using the four choices below?

> **Universal Airlines Seating**
> 1. Aisle, middle, or window (A,M,W)
> 2. First class or coach (F,C)
> 3. Smoking or nonsmoking (S,NS)
> 4. Right or left aisle (R,L)

Skills Practice

Multiply.

1. 376 × 4
2. 8,097 × 9
3. 71,549 × 8
4. 60,526 × 7
5. 21,009 × 5

6. 598 × 43
7. 614 × 57
8. 72,885 × 68
9. 94,301 × 92
10. 10,653 × 48

11. 924 × 375
12. 618 × 593
13. 23,110 × 406
14. 57,291 × 813
15. 28,962 × 489

16. 312 × 784
17. 2,197 × 356
18. 3,091 × 424
19. 87,135 × 1,099
20. 26,290 × 1,847

Divide.

21. 5)9,173
22. 7)2,682
23. 8)510,329
24. 6)749,141

25. 73)45,016
26. 92)305,729
27. 56)9,410,321
28. 81)7,648,113

29. 113)75,043
30. 702)94,680
31. 255)107,332
32. 648)386,190

33. 811)63,172
34. 304)40,879
35. 582)951,632
36. 375)154,875

Multiply or divide.

37. 6,821 × 754
38. 9,763 × 1,340
39. 491)60,152
40. 387)92,610

41. 8,762 ÷ 523
42. 112,764 ÷ 109
43. 2,413 × 8,760
44. 6,243 × 7,095

45. 4,991 × 3,714
46. 5,261 × 2,788
47. 294,344 ÷ 660
48. 829,726 ÷ 741

PROBLEM SOLVING: Using a Formula

QUESTION
DATA
PLAN
ANSWER
CHECK

Ronald typed 285 words in 5 minutes. He made 6 errors. For each error made, 10 words are subtracted from the total. What is his typing speed?

The formula for finding typing speed is

$$S = \frac{W - (10 \cdot e)}{n}$$

Each letter represents a number.

S = typing speed e = number of errors
W = number of words n = number of minutes

Step 1: List the value for each letter in the formula.

$W = 285$ $e = 6$ $n = 5$

Step 2: Substitute the numbers for the letters in the formula.

$$S = \frac{285 - (10 \cdot 6)}{5}$$

Remember, this bar means divide.

Step 3: Perform the operations in the formula.

$$S = \frac{285 - 60}{5} = \frac{225}{5} = 45$$

Ronald's typing speed is 45 words per minute.

Use the formula to find these typing speeds.

1. Dave typed 380 words in 5 minutes with 7 errors.

2. Lamar typed 886 words in 9 minutes. He made 13 errors.

3. A legal secretary typed 2,450 words in 20 minutes with 17 errors.

4. Vicky typed 554 words in 12 minutes and made 11 errors.

5. **DATA HUNT** What is your typing speed? Have another person watch the time while you type for 5 minutes. Use the formula to find your speed.

6. **Try This** Ann typed 5 more words per minute than Jeff. Together they typed 73 words in one minute. How many words did each person type in one minute?

81

PROBLEM SOLVING: Find a Pattern

QUESTION
DATA
PLAN
ANSWER
CHECK

Try This

On Sunday, Bill decides to start a soccer team fan club. On Monday, Bill gets 2 friends to join the fan club. Each of the 2 friends gets 2 more people to join the club on Tuesday. New members each get 2 more people to join the club the day after they join. This goes on all week. How many people, including Bill, are members of the fan club by the end of Saturday?

Some problems seem difficult because they are long and complicated. To solve these problems, you may need to use the strategy **Find a Pattern.** To solve the above problem, find the pattern.

What is the pattern for the new members?

Is there a pattern for the total members?

The solution for the problem is given below.

There are 127 members of the club by the end of Saturday.

	New members	Total members
Sunday	1	1
Monday	2	3
Tuesday	4	7
Wednesday	8	15
Thursday	16	
Friday		
Saturday		

Solve.

1. Kristy's parents put $100 in a savings account on Kristy's first birthday. Each year on her birthday they put in $200 more than on her last birthday. What will the total be when she is 12 years old? What will the total be when she is 18 years old?

Year	1	2	3
Amount	$100	$300	$500
Total	$100	$400	$

2. A bamboo shoot is 2 cm high on the first day, 5 cm high on the second day, 9 cm high on the third day, 14 cm high on the fourth day, and so on. If this growth pattern continues, how high will the bamboo shoot be at the end of two weeks?

Day	1	2	3	4	5
Height in centimeters	2	5	9	14	

CHAPTER REVIEW/TEST

Estimate each product by rounding to 1-digit accuracy.

1. 42×68 **2.** 53×79 **3.** 216×61 **4.** 45×96

Find the products mentally.

5. 4×90 **6.** $8 \times 3,000$ **7.** 700×80 **8.** 60×30

Find the products.

9. 67×3 **10.** 39×6 **11.** 283×9 **12.** $1{,}406 \times 4$

13. 27×58 **14.** 682×93 **15.** 428×336 **16.** 237×496

Write in exponential notation.

17. 7 squared   **18.** 2 to the sixth power   **19.** $4 \cdot 4 \cdot 4 \cdot 4 \cdot 4$

Write the standard numeral.

20. 3^3 **21.** 5^2 **22.** $3^2 \cdot 2^3$

Estimate each quotient by rounding to 1-digit accuracy.

23. $4{,}177 \div 439$ **24.** $5{,}972 \div 298$ **25.** $3{,}879 \div 78$ **26.** $4{,}392 \div 52$

Find the quotients and remainders.

27. $6 \overline{)259}$ **28.** $82 \overline{)1{,}095}$ **29.** $47 \overline{)3{,}166}$ **30.** $91 \overline{)5{,}880}$

31. $66 \overline{)12{,}445}$ **32.** $29 \overline{)82{,}821}$ **33.** $187 \overline{)37{,}629}$ **34.** $433 \overline{)18{,}109}$

35. There are 24 hours in one day. How many hours are there in one year of 365 days?

36. A loaf of raisin bread has 3,630 calories. There are 22 slices in the loaf. How many calories are there in each slice?

ANOTHER LOOK

```
        2 zeros
  ↓  ↓    ↓↓
30 × 40 = 1,200
  ↓  ↓   ✓
  3 × 4 = 12
```

Find the products mentally.

1. 7×30
2. 80×20
3. 400×3
4. $2,000 \times 9$
5. 60×50
6. 700×60

```
    2 7 6
  ×   3 4
  1 1 0 4  ← 4 × 276
  8 2 8 0  ← 30 × 276
  9,3 8 4  ← 34 × 276
```

Multiply.

7. 27×8
8. 126×7
9. $4,073 \times 5$
10. 34×29
11. 82×67
12. 259×83
13. 213×175
14. $1,918 \times 207$
15. $5,217 \times 468$

```
2 as a factor 4 times
2 · 2 · 2 · 2 = 2⁴ = 16
2 to the fourth power
```

Write in exponential notation.

16. $3 \cdot 3 \cdot 3 \cdot 3 \cdot 3$
17. 8 squared
18. 10 to the fifth power
19. 7 cubed

```
1,600 ÷ 80 = 20
because 20 × 80 = 1,600
3,500 ÷ 7 = 500
because 7 × 500 = 3,500
```

Find the quotients mentally.

20. $800 \div 4$
21. $2,500 \div 50$
22. $9,000 \div 3$
23. $6,400 \div 8$
24. $20,000 \div 40$
25. $5,600 \div 70$

```
       60
       3 5 R 7       6 2
   62)2,1 7 7      ×   3
       1 8 6       1 8 6
         3 1 7
         3 1 0       6 2
             7    ×    5
                    3 1 0
```

Find the quotients and remainders.

26. $7\overline{)52}$
27. $9\overline{)237}$
28. $8\overline{)1,026}$
29. $21\overline{)584}$
30. $69\overline{)1,609}$
31. $53\overline{)2,418}$
32. $70\overline{)15,247}$
33. $13\overline{)4,294}$
34. $89\overline{)12,649}$

ENRICHMENT

History of Mathematics

A simple calculator that could be used to multiply any number by a 1-digit number was invented by the Scottish mathematician John Napier (1550–1617). Napier's calculator is called **Napier's Rods** or **Napier's Bones.** It may have been given the name Napier's Bones because some were made of bone.

Here is the way to use Napier's Bones to find 849×7.

Place the strips headed 8, 4, and 9 side by side. Place the index beside the three strips. Use the numbers that are opposite 7 on the index. Start at the right and add diagonally to find the product.

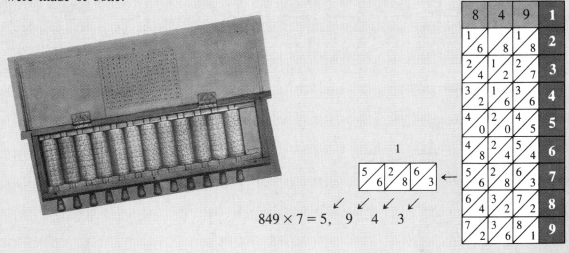

Make a copy of the ten strips of Napier's Bones. Use them to find these products.

1. 56×3
2. 87×5
3. 498×7
4. 128×9
5. 635×8
6. 298×7
7. 726×3
8. $2{,}675 \times 9$
9. $9{,}167 \times 4$

CUMULATIVE REVIEW

1. Which number is largest?

   A 0.48 B 0.40
   C 0.4   D not given

2. Round 0.9567 to the nearest thousandth.

   A 0.956 B 0.957
   C 0.96   D not given

3. Find the sum.

   27.2   A 18.78   B 187.8
   61.6   C 188.70   D not given
   + 99

4. Find the difference.

   123.4   A 36.77   B 35.71
   − 87.63   C 35.77   D not given

5. Multiply.

   722   A 28,158   B 8,564
   × 39   C 28,058   D not given

6. Find the quotient.

   88)9,255 A 15 R10   B 105 R15
             C 105   D not given

7. Estimate the difference.

   $92.68   A $70   B $66
   − 29.05   C $60   D not given

8. Estimate the sum by rounding each number to 1-digit accuracy.

   59¢ + 83¢ + 39¢ + 42¢

   A $2.20   B $25.00
   C $2.00   D not given

9. Divide.

   72,000 ÷ 9   A 800   B 80
                C 900   D not given

10. Find the quotient.

    9)3,847   A 427 R4   B 427
              C 427 R3   D not given

11. Choose the best estimate.

    276   A 10,000   B 15,000
    × 54   C 1,500   D not given

12. Find the product.

    $3^3 \cdot 8^2$ A 576   B 432
                      C 1,296 D not given

13. One year, the cost of automobile insurance was $279.14. The next year the cost was $323.70. How much was the increase?

    A $44.56   B $54.56
    C $44.66   D not given

14. Clara typed 340 words in 5 minutes with 12 errors. What was her typing speed? Use this formula.

    $$S = \frac{W - (10 \cdot e)}{n}$$

    A 49 words per minute
    B 57 words per minute
    C 44 words per minute
    D not given

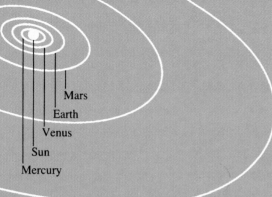

4 Multiplication of Decimals

Astronomers began a special search of the skies in the 1870s. They believed that a large unknown planet was affecting the orbits of Uranus and Neptune.

In 1930 the young astronomer Clyde Tombaugh discovered Planet X by using a process called "blinking." At least two pictures of each section of the sky were taken several days apart. When the pictures were compared in a blink microscope, all the stars were in the same places. Planet X, later named Pluto, was noticed because its position had changed.

Pluto is too small to affect the orbits of larger planets. Astronomers were looking for a planet with at least 6.6 times the mass of Earth. Some astronomers continue to search for a tenth and larger planet.

January 23, 1930

January 29, 1930

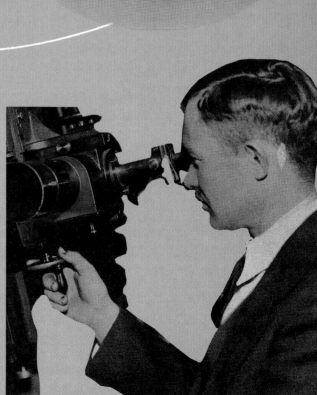

Decimal Multiplication

Carole left an electric lamp on for 4 hours. The bulb in the lamp uses 0.2 kilowatt (kW) of electricity each hour. How many kilowatts did the bulb use in 4 hours?

We can find the answer by adding 0.2 four times.

0.2 + 0.2 + 0.2 + 0.2 = 0.8

Since the same addend is repeated, we can also use multiplication.

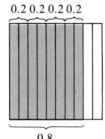

4 × 2 tenths = 8 tenths
4 × 0.2 = 0.8

The bulb used 0.8 kW in 4 hours.

Here is how we can think about multiplying a decimal times a decimal.

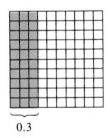
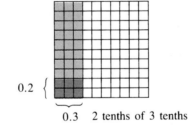
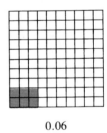

2 tenths × 3 tenths = 6 hundredths
0.2 × 0.3 = 0.06

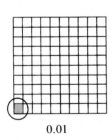
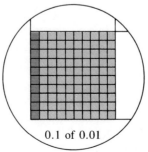
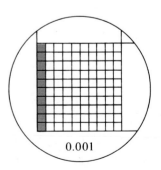

1 tenth × 1 hundredth = 1 thousandth
0.1 × 0.01 = 0.001

Find the sums and the products.

1. $0.3 + 0.3 + 0.3$
   3×0.3

2. $0.6 + 0.6 + 0.6 + 0.6$
   4×0.6

3. $0.07 + 0.07$
   2×0.07

4. $0.012 + 0.012 + 0.012$
   3×0.012

5. $0.5 + 0.5 + 0.5 + 0.5$
   4×0.5

6. $0.09 + 0.09 + 0.09 + 0.09 + 0.09$
   5×0.09

Write a multiplication statement for each figure.

7. 0.6 by 0.4

8. 0.8 by 0.5

9. 0.9 by 0.9

10. 0.2 by 0.4

11. 0.5 by 0.1

12. 0.2 by 0.5

13. 0.2 by 0.2

14. 0.7 by 0.6

15. 0.9 by 0.3

Multiply.

16. 2×0.7

17. 3×0.4

18. 4×0.04

19. 4×0.07

20. 6×0.009

21. 8×0.07

22. 0.3×0.02

23. 0.5×0.09

24. 0.1×0.01

Finding Products Using Estimation

Charles has 8 quarters. He puts 1 quarter on a balance scale and finds that it weighs 5.6 grams (g). What is the total weight of 8 quarters?

To find the total weight, we can multiply.

Use an estimate to help place the decimal point in the product.

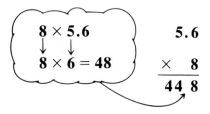

```
  5.6
×   8
 44.8
```

The 8 quarters have a total weight of 44.8 g.

Other Examples

64 × 1.8
Estimate: 60 × 2 = 120

```
    1.8
  × 6 4
     7 2
   1 0 8
  1 1 5.2
```

5 × $3.96
Estimate: 5 × $4 = $20

```
  $ 3.9 6
  ×     5
  $ 1 9.8 0
```

23 × 6.025
Estimate: 20 × 6 = 120

```
    6.0 2 5
  ×     2 3
    1 8 0 7 5
  1 2 0 5 0
  1 3 8.5 7 5
```

Warm Up

Use estimation to place the decimal point in each product.

1. 6.3 × 9 = 567
2. 7 × 7.186 = 50302
3. 6 × $19.75 = $11850

4.
```
  $2.35
 ×   29
 $6815
```

5.
```
  32.41
 ×    7
  22687
```

6.
```
  5.784
 ×   31
 179304
```

Find the products. Use estimation to place the decimal point.

7.
```
  6.19
 ×   8
```

8.
```
  $2.04
 ×   57
```

9.
```
  9.306
 ×   12
```

10.
```
  87.3
 ×  44
```

90

Use estimation to choose each product.

1. 5×6.28
   A 3.140
   B 31.40
   C 314.0

2. 8×92.7
   A 741.6
   B 74.16
   C 7,416

3. $28 \times \$6.35$
   A $17.78
   B $177.80
   C $1,778.00

4. $8 \times \$27.95$
   A $2,236.00
   B $22,360.00
   C $223.60

5. 72×9.44
   A 67,968
   B 67.968
   C 679.68

6. 39×1.897
   A 7.3983
   B 739.83
   C 73.983

7. 11×119.4
   A 1,313.4
   B 131.34
   C 13,134

8. 42×0.996
   A 0.41832
   B 4.1832
   C 41.832

9. 9.7×140
   A 1,358.0
   B 13.580
   C 135.80

Find the products. Use estimation to place the decimal point.

10. 3.4×8
11. 1.9×5
12. 7.2×7
13. 5.6×9
14. 8.8×6

15. $\$1.43 \times 7$
16. $\$6.09 \times 13$
17. $\$12.79 \times 42$
18. $\$63.18 \times 28$
19. $\$83.41 \times 76$

20. 2.9×34
21. 6.8×90
22. 8.5×54
23. 6.7×75
24. 9.8×47

25. 59×32.4
26. 86×8.08
27. 63×72.5
28. 19×3.447
29. 68×3.108

30. One nickel weighs about 5.2 g. What is the weight of a roll of 40 nickels?

31. One quarter weighs about 5.6 g. One nickel weighs 5.2 g. What is the total weight of 5 nickels and 5 quarters?

SKILLKEEPER

1. $320 \div 10$
2. $600 \div 100$
3. $10,000 \div 1,000$
4. $56,000 \div 10$

5. $850,000 \div 1,000$
6. $25,000 \div 100$
7. $9,000 \div 10$
8. $7,200 \div 100$

9. $41 \overline{)25,720}$
10. $93 \overline{)21,633}$
11. $29 \overline{)13,398}$
12. $15 \overline{)7,255}$

13. $57 \overline{)17,360}$
14. $70 \overline{)38,019}$
15. $68 \overline{)27,676}$
16. $72 \overline{)43,800}$

Multiplying Decimals

Myra Bailey is a cartographer, or map maker. She is working on a map that uses 1 cm to represent an actual distance of 25.4 km. What distance would 3.7 cm represent?

Since each centimeter on the map represents the same distance, we multiply.

| Multiply as with whole numbers. | Write the product so it has the same number of decimal places as the sum of the decimal places in the factors. |

```
    2 5.4                2 5.4   ← 1 decimal place
  ×   3.7              ×   3.7   ← 1 decimal place
   1 7 7 8              1 7 7 8    Tenths × tenths
   7 6 2                7 6 2      equals hundredths.
   9 3 9 8              9 3.9 8  ← 2 decimal places
```

The distance represented is 93.98 km.
Use an estimate to check. $25 \times 4 = 100$
The estimate and product are close.

Other Examples

```
    0.3 3   ← 2 decimal places       4 1.3 7   ← 2 decimal places       0.6 5 9   ← 3 decimal places
  ×   6.5   ← 1 decimal place      ×     1 2   ← 0 decimal places     ×     0.8   ← 1 decimal place
    1 6 5                            8 2 7 4                           0.5 2 7 2   ← 4 decimal places
    1 9 8                            4 1 3 7
    2.1 4 5 ← 3 decimal places       4 9 6.4 4 ← 2 decimal places
```

Warm Up

Find the products.

1. 2.9 × 6.4
2. 38.7 × 0.26
3. 5.123 × 0.85
4. 27.8 × 0.07
5. 0.86 × 0.74
6. 3.125 × 9
7. 125 × 0.008
8. 5.2^2

Find the products.

1. 3.9 × 0.8
2. 6.8 × 0.37
3. 0.715 × 3.8
4. 0.89 × 0.89
5. 42.8 × 0.06

6. 72.5 × 4.7
7. 0.813 × 6.9
8. 8,000 × 0.52
9. 3.14 × 19
10. 9.075 × 3.7

11. 8.54 × 6.9
12. 9.033 × 0.8
13. 257.2 × 0.45
14. 0.33 × 8.4
15. 10.09 × 8.6

16. $23.75 × 6
17. $415.95 × 3
18. $0.75 × 15
19. $18.95 × 5
20. $1.33 × 128

21. 34.90 × 1.05
22. 87.50 × 1.25
23. 6,740 × 0.16
24. 172.75 × 0.09
25. 62.44 × 12

26. 7.1 × 19.76
27. 0.67 × 600
28. 56 × 0.938
29. 0.93 × 2.31

30. 0.5 × 3.51
31. 14 × 0.55
32. 0.99 × 4.56
33. 1.06 × 2.66

34. 8.4^2
35. 0.68^2
36. 36.8^2
37. 3.03^2

38. On a map, 1 cm represents 38.5 km. What distance would 6.3 cm represent?

39. On a map, 3 cm represents 282 km. What distance would 0.7 cm represent?

THINK MATH

Number Puzzle

Which rings on the dart board must be hit to score exactly 99 points with 6 darts?

Which rings on the dart board must be hit to score 100 points with 6 darts?

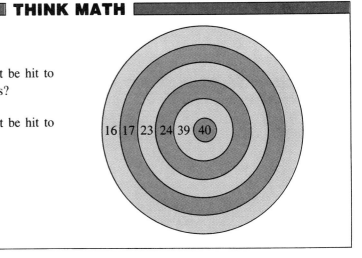

More Practice, page 424, Set A

Zeros in Products

Maria wants to buy a notebook that costs $0.79. The sales tax is 0.06 of the cost. To the nearest cent, what is the sales tax on the notebook?

Since we want to find 0.06 of the cost, we multiply.

Multiply as with whole numbers.

Write zeros to make the product have the same number of decimal places as the sum of the decimal places in the factors.

$ 0.7 9 ← 2 decimal places
× 0.0 6 ← 2 decimal places
———
4 7 4 ← Must have 4 decimal places.

$ 0.7 9
× 0.0 6
———
$ 0.0 4 7 4   Zero is written to make 4 decimal places.

$0.0474 rounded to the nearest cent, or hundredth of a dollar, is $0.05. The sales tax is 5 cents.

Other Examples

```
   0.0 6   ← 2 places           3 8 4    ← 0 places           0.0 1 3   ← 3 places
 × 0.0 9   ← 2 places        × 0.0 0 0 2 4 ← 5 places       ×   0.2 5   ← 2 places
 ———                            ———                          ———
 0.0 0 5 4 ← 4 places           1 5 3 6                         6 5
                                7 6 8                           2 6
                                ———                          ———
                                0.0 9 2 1 6 ← 5 places       0.0 0 3 2 5 ← 5 places
```

Warm Up

Give the number of decimal places in each product. Find the products.

1.   0.76
    × 0.09

2.   0.033
    × 0.17

3.   0.028
    × 0.035

4.   0.019
    ×   0.6

5.   0.35
    × 0.041

Find the products.

1. 0.05 × 0.7
2. 1.7 × 0.0008
3. 0.0024 × 0.04
4. 68 × 0.0006
5. 0.47 × 0.001

6. 0.71 × 0.09
7. 0.028 × 0.05
8. 3.2 × 0.07
9. 0.0086 × 0.8
10. 0.27 × 0.04

11. 0.36 × 0.13
12. 2.7 × 0.025
13. 0.51 × 0.028
14. 14.6 × 0.04
15. 0.079 × 0.034

16. 1.025 × 0.066
17. 3.94 × 0.059
18. 6.01 × 0.78
19. 0.034 × 0.062
20. 463 × 0.0096

21. 0.0742 × 67
22. 8.72 × 0.053
23. 0.918 × 0.422
24. 12.15 × 0.013
25. 9.49 × 0.0038

Find the products. Round the answers to the nearest cent.

26. $2.84 × 0.08
27. $45.99 × 0.06
28. $0.95 × 1.04
29. $235.50 × 0.04
30. $80.00 × 0.035

31. $59.95 × 0.05
32. $147.90 × 0.13
33. $4.95 × 0.03
34. $0.67 × 0.50
35. $9.99 × 0.45

36. A package of notebook paper costs $1.33. The sales tax is 0.06 of the cost. What is the amount of sales tax, rounded to the nearest cent?

37. A ballpoint pen costs $0.69. The sales tax is 0.06 of the cost. What is the sales tax, rounded to the nearest cent? What is the total cost of the pen, including sales tax?

More Practice, page 424, Set B

PROBLEM SOLVING: Using Data from a Table

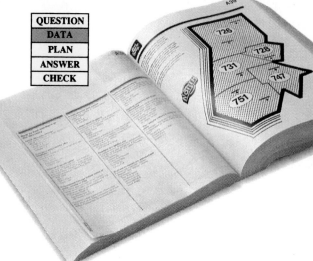

QUESTION
DATA
PLAN
ANSWER
CHECK

Telephone Rates

Prefix	First minute	Each additional minute
726	$0.13	$0.08
728	0.19	0.12
731	0.22	0.13
747	0.10	0.06
751	0.22	0.13
755	0.19	0.12
756	0.19	0.12
761	0.19	0.12
762	0.22	0.13
768	0.22	0.13
769	0.22	0.13
771	0.22	0.13
772	0.22	0.13
775	0.22	0.13
781	0.22	0.13

Telephone numbers are usually 10-digit numbers.

$$311 - 555 - 2368$$
↑ area code   ↑ prefix   ↑ local number

A telephone directory gives tables of rates for making calls to nearby areas.

What would be the cost of making a 4-minute call to a number with a 755 prefix?

The rate for the first minute is $0.19. Each additional minute costs $0.12.

```
  $ 0.1 2   ← Rate for additional minutes
×       3   ← Additional minutes
  0.3 6
+ 0.1 9     ← Rate for first minute
  $ 0.5 5   ← Cost of a 4-minute call
```

Find the cost of each call.

1. a 5-minute call to 756 prefix
2. an 8-minute call to 726 prefix
3. a 10-minute call to 772 prefix
4. a 3-minute call to 751 prefix
5. a 20-minute call to 747 prefix
6. a 30-minute call to 771 prefix
7. a 9-minute call to 761 prefix
8. a 11-minute call to 781 prefix
9. a 43-minute call to 775 prefix
10. Which telephone call costs more, a 7-minute call to a number with a 756 prefix, or a 6-minute call to a number with a 731 prefix? How much more does it cost?

Long distance calls may be dial-direct or operator-assisted.
The table below shows long distance rates from a city in California.

Cost of a 6-minute dial-direct call to Phoenix:

$0.50 + (3 \times 0.34) = \$0.50 + 1.02 = \$1.52$

Cost of a 4-minute operator-assisted call to Phoenix:

$\$2.05 + 0.34 = \$2.39$

Weekday Long Distance Rates

To	Dial-direct		Operator-assisted	
	First 3 minutes	Each additional minute	First 3 minutes	Each additional minute
Phoenix	$0.50	$0.34	$2.05	$0.34
Denver	0.52	0.36	2.15	0.36
Washington, D.C.	0.54	0.38	2.25	0.38
Miami	0.54	0.38	2.25	0.38
Honolulu	0.65	0.48	2.55	0.52
Chicago	0.52	0.36	2.15	0.36
Reno	0.44	0.30	1.85	0.30
New York	0.54	0.38	2.25	0.38
Seattle	0.50	0.34	2.05	0.34

Find the cost of each long distance call.

1. a 5-minute dial-direct call to Washington, D.C.

2. a 6-minute operator-assisted call to Denver

3. an 8-minute dial-direct call to Seattle

4. a 7-minute operator-assisted call to Reno

5. a 12-minute dial-direct call to Honolulu

6. a 15-minute operator-assisted call to Miami

7. How much more does a 5-minute operator-assisted call to Denver cost than a dial-direct call of the same length of time?

8. How much would it cost for a 1-hour call to Chicago using the direct-dial method?

9. **Try This** A long distance call costs $3.77. The costs for the first 4 minutes are shown in the table. How long was the telephone call? Hint: Find a pattern to complete the table.

Minutes	1	2	3	4
Total cost	$0.53	$0.89	$1.25	$1.61

Multiplying Decimals: Mental Math

A machinist is making two holes in a machine part. The blueprint shows that the centers of the holes are to be 0.613 m apart. What is this distance in decimeters, centimeters, and millimeters?

To change the number of meters to decimeters (dm), centimeters (cm), and millimeters (mm), we multiply by 10, 100, and 1,000.

Study the shortcuts below.

$0.613 \times 10 = 6.13$ $0.613 \times 100 = 61.3$ $0.613 \times 1,000 = 613.$
shift 1 place right   shift 2 places right   shift 3 places right

The centers of the holes should be 6.13 dm, 61.3 cm, or 613 mm apart.

We can also multiply decimals by 0.1, 0.01, and 0.001 mentally.

$543.8 \times 0.1 = 54.38$ $543.8 \times 0.01 = 5.438$ $543.8 \times 0.001 = 0.5438$
shift 1 place left   shift 2 places left   shift 3 places left

Other Examples

$3.5 \times 100 = 350$ $0.075 \times 1,000 = 75$ $18.26 \times 1,000 = 18,260$

$0.87 \times 0.1 = 0.087$ $65 \times 0.01 = 0.65$ $12.7 \times 0.001 = 0.0127$

Warm Up

Multiply each number by 10, 100, and 1,000.

1. 6.25   **2.** 0.8   **3.** 0.235   **4.** 0.047   **5.** 9.8

Multiply each number by 0.1, 0.01, and 0.001.

6. 31.4   **7.** 190   **8.** 5.2   **9.** 200   **10.** 0.88

Multiply each number by 10.

1. 2.71   2. 0.86   3. 31.42   4. 0.081   5. 9.4
6. 0.006   7. 21.79   8. 0.728   9. 746.2   10. 1.014

Multiply each number by 100.

11. 7.476   12. 0.913   13. 82.47   14. 0.0077   15. 9.2
16. 0.083   17. 0.0009   18. 1.5   19. 16.56   20. 8.728

Multiply each number by 1,000.

21. 3.142   22. 0.9163   23. 12.75   24. 0.0048   25. 17.8
26. 0.72   27. 3.14   28. 0.0046   29. 0.255   30. 1.427

Multiply each number by 0.1.

31. 2.87   32. 34.5   33. 726   34. 0.9   35. 62.75
36. 0.04   37. 142.5   38. 6,276   39. 0.007   40. 8.044

Multiply each number by 0.01.

41. 175.6   42. 9.7   43. 23.6   44. 0.5   45. 276.97
46. 0.78   47. 0.043   48. 200   49. 2,746.1   50. 847.3

Multiply each number by 0.001.

51. 7,468.3   52. 29,744   53. 59.7   54. 6.28   55. 0.97
56. 47.63   57. 5,094   58. 927.5   59. 3,000   60. 679.24

61. A machine part is made from a piece of metal 0.76 cm long. What is the length of metal needed to make 10 parts?

62. A steel bar 2.38 cm thick had 0.01 of the thickness ground off. How thick is the bar now?

THINK MATH

Logical Reasoning

How can a 3-minute egg timer and a 5-minute egg timer be used together to time exactly 7 minutes of cooking?

Estimating Products Using Decimals

Melissa sees a flash of lightning. She hears the thunder from the lightning 4.8 seconds later. If sound travels 0.33 km in 1 second, about how far away is the flash of lightning?

Since the distance is the same each second, we can multiply.

To make an estimate of the product, we round each number to 1-digit accuracy.

```
  0.3 3   →     0.3
× 4.8     →   ×   5
                1.5
```

The lightning is about 1.5 km away.

Other Examples

```
  6.8 2   →       7        $ 2 9.4 3  →   $ 3 0        0.5 9   →   0.6
× 2.3 5   →     × 2        × 0.0 8 2  →   × 0.0 8    × 0.7 2   → × 0.7
                1 4                        $ 2.4 0                  0.4 2
```

Estimate each product.

1.	9.1 × 3.8	**2.**	6.27 × 0.73	**3.**	0.413 × 0.609	**4.**	29.3 × 8.8	**5.**	7.26 × 0.05
6.	0.36 × 5.2	**7.**	31.6 × 57.4	**8.**	9.95 × 3.87	**9.**	0.772 × 0.619	**10.**	42.7 × 8.5
11.	1.74 × 6.6	**12.**	34.5 × 0.19	**13.**	2.7 × 3.2	**14.**	209 × 3.1	**15.**	28.6 × 0.77

Estimate each amount of money.

16.	$2.98 × 3.75	**17.**	$10.25 × 0.58	**18.**	$43.45 × 0.18	**19.**	$19.09 × 0.04	**20.**	$0.92 × 0.75
21.	$39.95 × 0.06	**22.**	$62.42 × 8.16	**23.**	$295.98 × 0.028	**24.**	$22.75 × 0.05	**25.**	$548.73 × 0.67

PROBLEM SOLVING: Using Estimation

| QUESTION |
| DATA |
| PLAN |
| ANSWER |
| CHECK |

1. Jackie's paycheck for a week of work is $289.67. Estimate how much Jackie will earn in a year of 52 weeks.

2. Armando found gasoline that costs $0.399 per liter (L). He bought 29.9 L of gasoline. What is an estimate of the cost of the gasoline?

3. A warehouse worker is checking an order for 10,000 advertising brochures which are packed in boxes with 335 to a box. What is an estimate of the number of boxes of brochures?

4. There are 1,287 registered voters in precinct 23. In the last election, 0.52 of those registered voted. Estimate the number of people who voted in the last election.

5. Harry took 314 strides to cross the courtyard. Each of his strides is 0.69 m long. What is Harry's estimate of the distance across the courtyard?

6. Brian bought 2 shirts that cost $18.95 each, and 3 pairs of socks for $2.25 each. What is an estimate of the amount that Brian spent?

7. Leslie has $831.66 in her checking account. She must write a check for $89.75 and another check for $62.50. What is an estimate of the amount Leslie will have left in her account?

8. Shirley Pierson wants to buy a new car that costs $6,893. The sales tax on the car is 0.0625 of the cost of the car. What is an estimate of the sales tax?

9. At Willow Junior High School, the enrollment is 595 students. The secretary used a calculator to compute that 0.0303 of the students were absent one day. Estimate the number of students that were absent.

10. **Try This** One month Hannah and Larry Brown earned a total of $4,000. Hannah earned $600 more than Larry. What did each person earn?

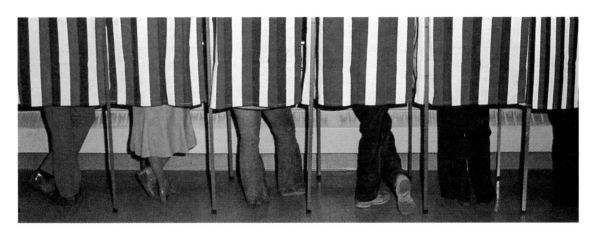

101

Scientific Notation

Light from the sun, traveling 300,000 km each second, takes 500 seconds to reach the earth. How many kilometers does the light travel?

Since the distance is the same for each second, we can multiply using a calculator or a pencil and paper.

scientific notation
1.5   08 means 1.5×10^8
↑   ↑
a number   a power of 10
from 1 to 10

$1.5 \times 10^8 = 150{,}000{,}000$
shift decimal 8 places

The light travels 150,000,000 km.

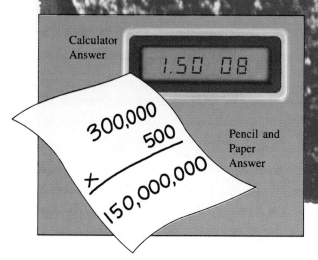

Calculator Answer

Pencil and Paper Answer

Other Examples

$2.4 \times 10^6 = 2{,}400{,}000$
6 places

$8.06 \times 10^4 = 80{,}600$
4 places

$7 \times 10^7 = 70{,}000{,}000$
7 places

$5{,}040{,}000 = 5.04 \times 10^6$
6 places

$300 = 3 \times 10^2$
2 places

$639{,}000 = 6.39 \times 10^5$
5 places

Warm Up

Give the missing exponent.

1. $5{,}000 = 5.0 \times 10^{\square}$

2. $45 = 4.5 \times 10^{\square}$

3. $27{,}500{,}000{,}000 = 2.75 \times 10^{\square}$

Give the missing factor.

4. $9{,}600 = \square \times 10^3$

5. $271{,}000{,}000 = \square \times 10^8$

6. $68{,}000{,}000 = \square \times 10^7$

Write the standard numeral.

7. 6.3×10^6

8. 1.9×10^{12}

9. 3.11×10^9

Write in scientific notation.

1. 25,000
2. 580,000,000
3. 610,000,000,000
4. 300,000
5. 11,000,000,000
6. 2,400

7.
8.
9.

Write the standard numeral.

10. 3.2×10^5
11. 6.5×10^8
12. 8.9×10^3
13. 3.09×10^6
14. 1.86×10^7
15. 6.75×10^{10}

16. [display: 3.60 07]
17. [display: 4.66 09]
18. [display: 1.37 10]

19. The planet Saturn is about 1.43×10^9 km from the sun. Write the standard numeral for this number.

20. Jupiter is about 778,000,000 km from the sun. Write the distance in scientific notation.

21. Light from the sun takes about 9,600 seconds to reach the planet Uranus. About how far does light travel from the sun to Uranus? Give both the standard numeral and the scientific notation.

22. **DATA BANK** Find the distance from the sun to the planet Neptune. About how many seconds would it take light from the sun to reach Neptune? (See page 415.)

SKILLKEEPER

1. 5.6×0.1
2. 98×0.1
3. $8,769 \times 0.1$
4. 4.78×10
5. 0.359×10
6. 7.254×100
7. 6.50×0.01
8. 45.9×0.01
9. 0.005×0.01
10. 2.729×100
11. 0.77×0.001
12. 7.007×0.001
13. $956.4 \times 1,000$
14. 5.6×0.001
15. $99.99 \times 1,000$
16. 0.0947×0.001
17. 0.52×10
18. 62.51×0.01
19. $2.3 \times 1,000$
20. $9,675 \times 0.001$
21. 0.257×0.1
22. 585.6×10
23. 0.06×0.001
24. $7,952 \times 0.01$

More Practice, page 424, Set D

Skills Practice

Multiply.

1. 3.4 × 7.8
2. 5.7 × 0.03
3. 4.2 × 9.1
4. 0.65 × 1.6
5. 0.95 × 0.08

6. $1.03 × 29
7. $8.46 × 13
8. $3.52 × 40
9. $6.93 × 57
10. $4.75 × 16

11. 11.83 × 0.02
12. 26.72 × 0.57
13. 72.04 × 0.36
14. 31.90 × 0.25
15. 68.53 × 0.18

16. 0.09 × 0.03
17. 0.013 × 0.07
18. 0.006 × 0.104
19. 0.092 × 0.017
20. 3.4 × 0.0052

21. 1.092 × 0.57
22. 22.01 × 1.1
23. 65.1 × 0.09
24. 83.45 × 0.13
25. 70.01 × 0.045

26. 6.58 × 0.46
27. 88.9 × 0.62
28. 4.06 × 0.39
29. 25.3 × 5.7
30. 0.689 × 4.2

31. 0.622 × 0.23
32. 39.4 × 3.11
33. 4.27 × 0.381
34. 0.575 × 4.14
35. 0.361 × 0.084

Multiply. Round to the nearest cent.

36. $23.04 × 1.15
37. $62.15 × 3.39
38. $48.25 × 6.20
39. $39.31 × 7.45
40. $81.24 × 10.05

Find the products.

41. 36.01 × 23.5
42. 0.654 × 3.91
43. 783.1 × 0.0072
44. 92.3 × 40.01
45. 86.33 × 0.9271
46. 0.018 × 0.907

PROBLEM SOLVING: Using Data from a Table

| QUESTION |
| DATA |
| PLAN |
| ANSWER |
| CHECK |

Gloria Rodriguez works in the classified advertising department of a daily newspaper. She must compute the cost of advertisements.

The Daily News

Classified Advertisement Rates (Minimum 2 lines)	
Consecutive day insertions	Charge per line per day
12 or more days	$1.08
9 days	1.34
7 days	1.66
6 days	1.93
4–5 days	1.96
1–3 days	2.61

What is the cost of running a 4-line advertisement for 5 days?

```
   $ 1.9 6   ← Cost per line per day
 ×       4   ← Number of lines
  $ 7.8 4
 ×       5   ← Number of days
  $ 3 9.2 0  ← Total cost
```

Find the total cost for each advertisement.

1. **Cat Lost** Reward. Long hair, white with orange polka-dots. Declawed but dangerous. Answers to the name of Gwendolyn or Sid.

   Insert for 3 days

2. **Labrador Lost** Female, with 3 puppies and 2 kittens pretending to be puppies. Reward: 50¢ plus 2 puppies and 1 kitten.

   Insert for 9 days

3. **Terrier Found** Looking for owner of funny-looking pink dog found underneath bed. Will not come out and refuses to eat. Owner please call Sherlock Bones to claim.

   Insert for 7 days

4. **Goofy Dog** found. Small collie with limp; missing upper fang. Call Pearl at 999-9999.

   Insert for 2 days

5. **Black Cat** found. At the top of my tulip tree; won't come down. Owner please claim. Bring ladder, fire truck, or trampoline.

   Insert for 15 days

6. **Goldfish Found** in supermarket parking lot. Owner please claim before my cat does.

   Insert for 12 days

7. Which costs more, a 3-line advertisement for 3 days or a 2-line advertisement for 5 days?

8. Mr. Craney wrote a 5-line advertisement to run for 4 days. Gloria helped him shorten his advertisement to 3 lines. How much will Mr. Craney save?

9. **Try This** The total cost of two advertisements was $113. One advertisement cost $18 less than the other. What was the cost of each advertisement?

PROBLEM SOLVING: Practice

	Dive	Difficulty	Ratings	Score
QUESTION / DATA / PLAN / ANSWER / CHECK	1	1.6	8.5; 9.0; 8.5; 9.0; 7.5	68.00

Mimi is on the school diving team. She enters the springboard diving competition which is judged by experts. There are from three to seven judges who rate the dives from 0 to 10. Each dive is assigned a number called the **degree of difficulty.**

Mimi completed her first dive. Here is the way her score was computed.

Score = degree of difficulty × (sum of judges' ratings)
= 1.6 × (8.5 + 9.0 + 8.5 + 9.0 + 7.5)
= 1.6 × 42.5 = 68.00

Mimi's score was 68.00.

Find the score for each dive.

1. Forward $1\frac{1}{2}$-somersault, tuck position
   Degree of difficulty: 1.5
   Judges' ratings: 7.0; 7.5; 7.5; 7.0; 8.0

2. Back $2\frac{1}{2}$-somersault, pike position
   Degree of difficulty: 3.0
   Judges' ratings: 6.5; 6.5; 7.0; 6.5; 6.0

3. Reverse, layout position
   Degree of difficulty: 2.0
   Judges' ratings: 8.0; 8.5; 8.5; 8.0; 9.0

4. Reverse $2\frac{1}{2}$-somersault, tuck position
   Degree of difficulty: 2.8
   Judges' ratings: 7.5; 7.5; 7.0; 7.0; 7.0

5. Inward double somersault, pike position
   Degree of difficulty: 2.6
   Judges' ratings: 7.5; 7.5; 8.5; 8.0; 8.0

6. Forward $2\frac{1}{2}$-somersault, 2 twists, free position
   Degree of difficulty: 3.2
   Judges' ratings: 6.5; 7.0; 6.0; 7.0; 7.5

Find the score for each dive and the total for each diver.

7.

Dive	Difficulty	Ratings	Score
1	2.2	7.0; 7.0; 6.5; 7.5	
2	1.7	8.5; 8.0; 8.5; 7.5	
3	2.5	6.0; 6.0; 6.0; 6.5	
4	1.9	7.0; 7.5; 8.5; 8.0	
		Total	

Diver: C. Williams

8.

Dive	Difficulty	Ratings	Score
1	1.8	9.0; 9.0; 8.5; 8.5	
2	1.7	7.0; 7.0; 7.0; 7.0	
3	2.4	6.5; 7.0; 6.5; 7.0	
4	2.8	5.5; 6.0; 6.0; 6.5	
		Total	

Diver: J. Montez

9.

Dive	Difficulty	Ratings	Score
1	1.4	8.0; 8.5; 8.0; 9.0	
2	2.1	6.5; 6.5; 7.0; 6.5	
3	2.3	7.0; 6.5; 8.0; 7.5	
4	3.3	5.5; 6.0; 6.0; 5.5	
		Total	

Diver: L. Freel

10.

Dive	Difficulty	Ratings	Score
1	2.5	6.0; 6.0; 5.5; 6.5	
2	2.4	9.0; 9.5; 8.5; 9.0	
3	2.8	8.0; 7.5; 7.0; 7.0	
4	3.0	6.5; 6.5; 6.5; 6.0	
		Total	

Diver: M. Kennard

11. **Try This** After completing three dives, Nadine has a total score of 201 points. The score for her second dive is 2 points more than for her first dive. The score for her third dive is 2 points more than for her second dive. What is her score for each of the three dives?

Forward dive (layout position)

Forward $1\frac{1}{2}$-somersault dive (tuck position)

Inward dive (pike position)

Half-twist dive (free position)

PROBLEM SOLVING: Draw a Picture

| QUESTION |
| DATA |
| PLAN |
| ANSWER |
| CHECK |

Try This

Three towns—Alton, Hull, and York—are all on the same straight road. It is 123 km from Alton to Hull. It is 64 km from York to Alton. The town of Alton lies between Hull and York. What is the distance between Hull and York?

Some problems contain groups of information. It may be difficult to remember how the information is related. A good strategy to try on such problems is **Draw a Picture**.

Here is how a picture might be made for the problem above.

It is 123 km from Alton (A) to Hull (H).

Label your pictures to show the data.

It is 64 km from York (Y) to Alton.

This tells us the distance, but not where York is located.

Alton lies between York and Hull.

Now we know where York lies.

Use the information given in the picture to answer the question.

$64 + 123 = 187$

The distance between Hull and York is 187 km.

Solve.

1. Three towns—Adding, Plus, and Minus—are on one straight road. It is 186 km from Adding to Plus. It is 57 km from Minus to Plus. Minus is between Adding and Plus. What is the distance between Adding and Minus?

2. The combined height of Jack and Bob is 340 cm. This is 158 cm more than Al's height. Al is 8 cm taller than Jack. How tall is Bob?

CHAPTER REVIEW/TEST

Find the products.

1. 3.8
   × 0.7

2. 0.92
   ×   8

3. 6.3
   × 7.5

4. 0.007
   ×   6.8

5. 0.722
   × 0.04

6. $21.24
   ×   24

7. $18.50
   × 0.06

8. 3.125
   × 10.4

9. 0.05
   × 0.07

10. 0.012
   × 0.13

11. 420
   × .015

12. 0.059
   × 0.027

13. 10 × 7.295
14. 1,000 × 8.34
15. 100 × 12.729
16. 0.1 × 62.8
17. 0.01 × 534
18. 0.001 × 72.9
19. 0.001 × 8,647.7
20. 0.001 × 38.9
21. 0.001 × 0.883

Estimate each product.

22. 19.42
   ×   0.7

23. 3.842
   × 5.41

24. 0.72
   × 0.96

Write the standard numeral.

25. 6.4×10^5
26. 8.1×10^8
27. 3.03×10^4

28. A local call costs $0.13 for the first minute, plus $0.08 for each additional minute. What is the cost of a 10-minute call?

29. A long distance telephone call costs $2.04 for the first 3 minutes, plus $0.34 for each additional minute. What is the cost of a 12-minute call?

30. Oranges cost $0.44 each. How much would 10 oranges cost?

31. For a 4-line advertisement, a newspaper charges $1.46 per line for each day. What is the cost to run the advertisement for 5 days?

ANOTHER LOOK

```
  2.7 4   ← 2 decimal places
×   0.8   ← 1 decimal place
  2.1 9 2 ← 3 decimal places

    0.1 3 ← 2 decimal places
×  0.0 6  ← 2 decimal places
  0.0 0 7 8 ← 4 decimal places
         Add 2 zeros to make
         4 decimal places.
```

Find the products.

1. 12.8 × 6
2. 0.72 × 9
3. 2.038 × 25
4. 4.7 × 0.5
5. 0.84 × 0.9
6. 0.825 × 0.03
7. 0.42 × 0.03
8. 1.09 × 0.07
9. 0.043 × 0.15
10. 0.234 × 0.008
11. 0.0006 × 0.28
12. 10.45 × 6.28

```
10 × 0.46 = 4.6
100 × 3.142 = 314.2
0.01 × 6,285 = 62.85
0.001 × 9,000 = 9.000
```

Find the products mentally.

13. 100 × 6.428
14. 10 × 0.762
15. 1,000 × 3.1416
16. 0.1 × 22.7
17. 0.001 × 7,200
18. 0.01 × 62.57
19. 100 × 0.67
20. 0.001 × 16.5

```
Estimate.
0.92 × 0.48
Round to 1-digit accuracy.
0.9 × 0.5 = 0.45
```

Estimate the products.

21. 9 × 8.89
22. 0.72 × 0.88
23. 59.25 × 8.09
24. 17 × $2.12
25. 8 × $0.92
26. 115 × $3.09
27. 19.4 × 5.7
28. 0.27 × 0.04

ENRICHMENT

Number Theory

Casting Out Nines is an old method of checking computation. To cast out nines from a number, you can use either of these methods.

Add the digits of the number. Add the digits of the sum until you get a single digit. If the single digit is 9, cast it out to get 0.

2,749 → 2 + 7 + 4 + 9 = 22 → 2 + 2 = 4
30,213 → 3 + 0 + 2 + 1 + 3 = 9 → 0

Look for the digits in the number that have a total of 9. Cast out these digits. Add the other digits to get a single-digit sum.

2̸ 7̸ 4 9̸ → 4
3̸ 0 2̸ 1̸ 3̸ → 0

You can use the digits you get by casting out nines to check your computation.

Multiplication

```
  2 5 9  →  7        7
×   6 5  →  2       ×2
  1 2 9 5          14 → 5
1 5 5 4
1 6,8 3 5  →  5     5 and 5
                    It checks.
```

Division

```
                  1 3  →  4       8  ← divisor
  8 ← 2 9 6)3,8 4 9 → 6         ×4  ← quotient
            2 9 6                32
            ─────                +1  ← remainder
              8 8 9              ──
              8 8 8              33 → 6 dividend
              ───                     It checks.
                1  →  1
```

Which answers are correct? Check by casting out nines.

1. 728
    × 86
    ────
    4368
    5824
    ─────
    62,508

2. 819
    × 69
    ────
    7371
    4914
    ─────
    56,511

3. 32)6,909 215 R29
       64
       ──
       50
       32
       ──
       189
       160
       ───
        29

4. 619)84,443 139 R402
       61 9
       ────
       23 54
       18 57
       ─────
        5 973
        5 571
        ─────
          402

Find the products and quotients. Check by casting out nines.

5. 237
    × 29

6. 564
    × 57

7. 2,618
    ×   93

8. 9,356
    × 247

9. 57)2,949

10. 86)38,081

11. 566)20,944

12. 777)93,746

TECHNOLOGY

Computer Programs and Flowcharts

Computers must be given a precise set of instructions called a **program** to accomplish any task. These instructions must be written in a special language that the computer understands. We are going to use a language called **BASIC** (Beginners' All-Purpose Symbolic Instruction Code).

Flowcharts are sometimes used to plan computer programs. Study the example below.

Mathematical Symbol	BASIC
3 + 4	3 + 4
12 − 9	12 − 9
25 × 17	25 * 17
63 ÷ 7	63 / 7
3(9 + 4)	3 * (9 + 4)
2^5	2 ↑ 5 or 2**5
0 (zero)	∅

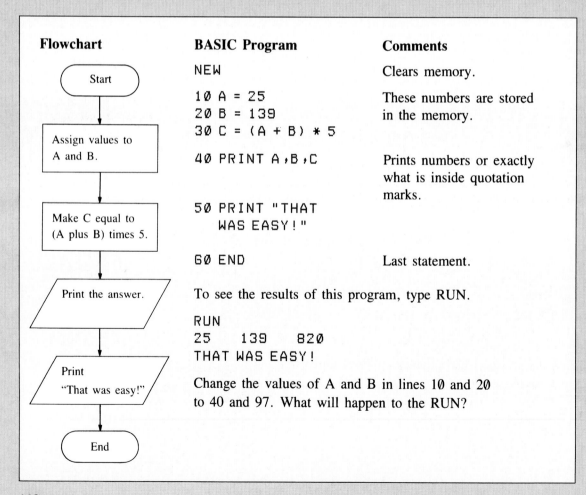

Flowchart

Start → Assign values to A and B. → Make C equal to (A plus B) times 5. → Print the answer. → Print "That was easy!" → End

BASIC Program

```
NEW
10 A = 25
20 B = 139
30 C = (A + B) * 5
40 PRINT A,B,C
50 PRINT "THAT
   WAS EASY!"
60 END
```

Comments

NEW — Clears memory.

These numbers are stored in the memory.

Prints numbers or exactly what is inside quotation marks.

Last statement.

To see the results of this program, type RUN.

```
RUN
25    139    820
THAT WAS EASY!
```

Change the values of A and B in lines 10 and 20 to 40 and 97. What will happen to the RUN?

Write the RUN for each program.

1. ```
   10 M = 100
   20 T = 4
   30 S = M/T
   40 PRINT M,T,S
   50 END
   ```

2. ```
   10 A = 48
   20 B = 72
   30 C = (A + B)/12
   40 D = A + B/12
   50 PRINT A,B,C,D
   60 END
   ```

3. ```
   10 K = 5
   20 Z = 2
   30 M = K↑Z
   40 Q = Z↑K
   50 PRINT K,Z,M,Q
   60 END
   ```

4. ```
   10 A = 10
   20 B = 70
   30 PRINT A + B,B - A
   40 PRINT A * B,B/A
   50 PRINT A↑3,B↑2
   60 END
   ```

5. ```
   10 A = 10
   20 PRINT A;" TO THE SECOND
      POWER IS ";A↑2
   30 PRINT A;" TO THE THIRD
      POWER IS ";A↑3
   40 PRINT A;" TO THE FOURTH
      POWER IS ";A↑4
   50 END
   ```

6. ```
   10 A = 35
   20 B = 12
   30 PRINT A;" TIMES ";B;
      " EQUALS ";A*B
   40 END
   ```

7. Assign different values to the variables in exercises 1–6. Give each new RUN.

8. Make a flow chart for your own program. Then write the program and show the RUN.

Rick Rasay checks a computer program.

CUMULATIVE REVIEW

1. Multiply.

    529
    × 705

    A 372,845　**B** 372,945
    C 39,675　**D** not given

2. Find the quotient.

    56)9,124

    A 162 R20　**B** 16 R52
    C 162 R52　**D** not given

3. Estimate the quotient by rounding each number to 1-digit accuracy.

    98,759 ÷ 988

    A 90　**B** 1,000
    C 100　**D** not given

4. Choose the best estimate.

    609 × 25

    A 18,000　**B** 1,800
    C 180,000　**D** not given

5. Find the quotient.

    64,000 ÷ 800

    A 9,000　**B** 8,000
    C 800　**D** not given

Multiply.

6.　2.76
    × 9.3

    A 25.668　**B** 286.68
    C 28.668　**D** not given

7.　0.057
    × 0.95

    A 0.05515　**B** 0.05415
    C 0.5415　**D** not given

8. $7^2 \cdot 9^5$

    A 35,721　**B** 2,893,401
    C 413,349　**D** not given

9. Estimate the product by rounding each number to 1-digit accuracy.

    0.129 × 0.62

    A 0.6　**B** 0.06
    C 6.0　**D** not given

10. Find the product.

    912.47 × 100

    A 91.247　**B** 9,124.7
    C 91,247　**D** not given

11. Multiply.

    0.545 × 0.001

    A 0.000545　**B** 545
    C 0.00545　**D** not given

12. Find the product.

    6.59
    × 0.1

    A 6,590　**B** 65.90
    C 0.659　**D** not given

13. Each row of chairs in a large room has 27 chairs. There are 18 rows of chairs. How many chairs are there all together?

    A 45　**B** 486
    C 288　**D** not given

14. A 7-minute telephone call costs $0.54 for the first minute and $0.38 for each additional minute. What is the total cost of the call?

    A $0.92　**B** $3.62
    C $2.82　**D** not given

114

5
Division of Decimals

When you see a large bronze sculpture, you might not realize that a sculptor starts by making a small clay model. Larger models that may be 0.9 m high and 2.70 m high are then made.

The models are sent to a foundry where the founder uses them to make a full-size hollow wax model. The sculptor works with the wax model before it is filled and coated with plaster. This lump of plaster is baked for three days. The wax melts at 63°C and runs out. Bronze, heated to 1,100°C, is poured into the space where the wax used to be. When the plaster is chipped away, only the hardened bronze sculpture is left. The sculptor applies the final finishing touches to the surface.

Dividing a Decimal by a Whole Number

The Alaskan brown bear at the Cumberland Zoo is fed 28.25 kilograms (kg) of special biscuits in 5 days. How many kilograms of biscuits is the bear fed each day?

Since we want to find the amount of kilograms fed each day, we divide.

Divide the whole number part. → Place the decimal point. Divide the tenths. → Divide the hundredths.

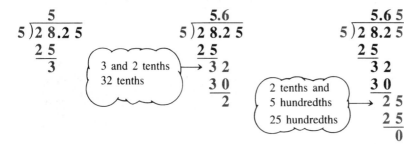

The bear is fed 5.65 kg of biscuits each day.

Other Examples

When dividing a decimal by a whole number, place the decimal point in the quotient directly above the decimal point in the dividend.

```
      2.0 6
24)4 9.4 4
    4 8
    1 4 4
    1 4 4
          0
```

Zeros show not enough ones or tenths. Divide the hundredths.

```
      0.0 5 4
  4)0.2 1 6
      2 0
        1 6
        1 6
            0
```

Warm Up

Find the quotients. Check by multiplying.

1. 32)18.24
2. 9)0.504
3. 7)26.74
4. 19)77.33

Find the quotients.

1. $2\overline{)6.54}$
2. $7\overline{)20.3}$
3. $8\overline{)42.4}$
4. $9\overline{)66.6}$
5. $8\overline{)27.2}$
6. $5\overline{)17.05}$
7. $6\overline{)343.2}$
8. $4\overline{)29.12}$
9. $7\overline{)227.43}$
10. $5\overline{)32.65}$
11. $9\overline{)10.71}$
12. $6\overline{)0.54}$
13. $8\overline{)2.72}$
14. $4\overline{)0.148}$
15. $5\overline{)1.65}$
16. $3\overline{)1.458}$
17. $42\overline{)268.8}$
18. $71\overline{)248.5}$
19. $78\overline{)252.72}$
20. $52\overline{)254.8}$
21. $85\overline{)276.25}$
22. $29\overline{)89.03}$
23. $91\overline{)518.7}$
24. $18\overline{)0.702}$
25. $73\overline{)1.168}$
26. $15.96 \div 57$
27. $76.14 \div 81$
28. $34.65 \div 45$
29. $2.07 \div 9$
30. $197.6 \div 52$

31. An animal trainer at the zoo feeds 6 sea lions a total of 59.76 kg of mackerel each day. How many kilograms does each sea lion get?

32. Three pairs of hornbills are fed 1.32 kg of soaked raisins as part of their daily diet. How much is fed to each hornbill?

33. **DATA BANK** If 6 giraffes were fed a total of 3.0 kg of grain per day, would this be about the right amount for each giraffe? (See page 416.)

THINK MATH

Finding Patterns

How many different squares can you count in each figure?

Continue the pattern with a 4×4 square and a 5×5 square.

How many different squares are in an 8×8 checkerboard?

1^2

$1^2 + 2^2 = 5$

$1^2 + 2^2 + 3^2 = 14$

?

More Practice, page 425, Set A

Rounding Decimal Quotients

James Tong buys a roll of camera film with 36 exposures for $4.89. What is the cost (to the nearest cent) of each exposure?

To find the cost (to the nearest cent) of each exposure, we divide and round the quotient to the nearest hundredth of a dollar.

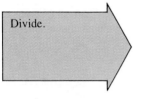

```
    $ 0.1 3
3 6 )$ 4.8 9
     3 6
     1 2 9
     1 0 8
         2 1
```

```
    $ 0.1 3 5
36 )$ 4.8 9 0
     3 6
     1 2 9
     1 0 8
         2 1 0
         1 8 0
             3 0
```
Annex zeros as needed.

$ 0.1 3 5 rounded to the nearest cent is **$0.14** or **14¢**.

Each exposure will cost 14¢.

Other Examples

```
          7.4 4  →  7.4  (nearest tenth)
2 9 )2 1 6.0 0
     2 0 3
       1 3 0
       1 1 6
           1 4 0
           1 1 6
               2 4
```
Divide to the hundredths place.

```
        0.2 1 7 3  →  0.2 1 7  (nearest thousandth)
2 3 )5.0 0 0
     4 6
       4 0
       2 3
       1 7 0
       1 6 1
           9 0
           6 9
           2 1
```
Divide to the ten-thousandths place.

Warm Up

Find the quotients to the nearest thousandth.

1. 9 ÷ 23   **2.** 43.00 ÷ 17   **3.** 28.6 ÷ 7

Find the quotients to the nearest cent.

4. 7)$13.80   **5.** 15)$33.71   **6.** 34)$52.86

Find the quotients to the nearest tenth.

1. 6)20
2. 9)14
3. 3)14
4. 7)12

5. 14)26
6. 12)75
7. 15)32
8. 13)37.2

9. 23)62.8
10. 56)375
11. 42)74
12. 35)37

Find the quotients to the nearest thousandth.

13. 7)9.2
14. 3)7.1
15. 6)8
16. 9)17

17. 12)4.1
18. 17)13
19. 18)19.2
20. 19)21

21. 34)21.4
22. 28)35.4
23. 47)26
24. 52)38

Find the quotients to the nearest cent.

25. 6)$24.80
26. 18)$15.75
27. 64)$172.96
28. 36)$53.88

29. 72)$50.00
30. 120)$97.75
31. 54)$19.63
32. 250)$688.76

33. Film with 24 exposures costs $3.89. What is the cost (to the nearest cent) of each exposure?

34. One dozen rolls of film costs $42.65. What is the cost (to the nearest cent) of each roll?

35. A roll of home movie film costs $7.10. The roll has a total of 3,240 frames. What is the cost (to the nearest cent) of 10 frames?

36. **DATA HUNT** Find the cost of developing a roll of color print film. What is the cost (to the nearest cent) of each print?

SKILLKEEPER

1. 9.7×10
2. 0.35×100
3. 12.67×100
4. 0.004×10
5. 5.5×100

6. 0.056×100
7. $15.7 \times 1{,}000$
8. 0.0009×10
9. 10.73×10
10. $0.47 \times 1{,}000$

11. 0.423×0.1
12. 123×0.001
13. 5.98×0.1
14. 2.3×0.001
15. 2.734×0.1

16. 593.7×0.01
17. 12.9×0.001
18. 0.062×0.1
19. 5.6×0.01
20. 3.35×0.01

More Practice, page 425, Set B

Dividing Decimals: Mental Math

Lele ran a 10-kilometer race in 56.4 minutes (min). At this rate, how long did it take her to run 1 km?

Since we know the time for 10 km, we divide to find the time for 1 km.

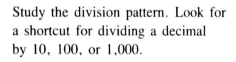

It took Lele 5.64 min to run 1 km.

Study the division pattern. Look for a shortcut for dividing a decimal by 10, 100, or 1,000.

56.4 ÷ 10	= 5.64	shift 1 place left
56.4 ÷ 100	= 0.564	shift 2 places left
56.4 ÷ 1,000	= 0.0564	shift 3 places left

Other Examples

1,950 ÷ 10 = 195.0	32.6 ÷ 10 = 3.26	0.25 ÷ 10 = 0.025
1,950 ÷ 100 = 19.50	32.6 ÷ 100 = 0.326	0.25 ÷ 100 = 0.0025
1,950 ÷ 1,000 = 1.950	32.6 ÷ 1,000 = 0.0326	0.25 ÷ 1,000 = 0.00025

Find the quotients mentally. Write only the answers.

Divide each number by 100.

1. 0.7   2. 52.95   3. 2,372   4. 0.045   5. 1,875

Divide each number by 1,000.

6. 31,682   7. 5,746   8. 397   9. 4,280   10. 127

PROBLEM SOLVING: Practice

QUESTION
DATA
PLAN
ANSWER
CHECK

Solve.

1. Ned can walk 10 km in 95 min. How many minutes will it take him to walk 1 km?

2. The winner of a 10-kilometer roller-skating race had a time of 21.87 min. About how many minutes did it take the winner to skate 1 km?

3. Karen ran 100 m in 13.2 seconds (s). Her time was 1.3 s slower for the second 100 m. What was her total time for running the 200 m?

4. In a 15-kilometer cross-country ski race, one skier skied the first 5 km in 17.4 min. At this rate, how long would it take that person to ski the 15 km?

5. In a 1,000-meter speed-skating race, the winning time was 1 min 19.32 s. How many seconds did it take the winner to skate 1 m? Round your answer to the nearest hundredth of a second.

6. An airplane flew 1,000 km in 62 min. At this rate, how many seconds did it take the airplane to fly 1 km?

7. In a 100-meter butterfly-stroke swimming race, the winning time was 60.56 s. The second place time was 61.03 s. How much slower was the second place time?

8. In a 1,500-meter race, a runner runs the first 1,000 m in 2.9 min. At this rate, how long will it take the runner to run the 1,500 m?

9. A car traveled 100 km in 68 min. How long did it take the car to travel 1 km? How many seconds is this time?

10. **Try This** May, Carrie, Wilma, and Leslie are the first four finishers in a 10-kilometer run. May ran the race in 43.56 min. Carrie was 1.09 min faster than Leslie. Wilma was 2.35 min slower than Leslie. May was 0.65 min behind Leslie. What was the order of the finish? Hint: Draw a picture.

PROBLEM SOLVING: Using Data from a Table

To encourage people to conserve water, the Sunnyside City Council approved new water rates.

Sunnyside Water Rates

Cubic feet (ft³) of water	Rate per 100 ft³
0–600	$0.300
601–2,000	0.350
2,001–5,000	0.365
5,001–50,000	0.375
50,001–125,000	0.385
125,001–250,000	0.395
250,001 and above	0.405

The more water a customer uses, the higher the water rate is.

During the month of May, the Jessel family used 6,400 cubic feet (ft³) of water. What is the cost of this amount of water?

To solve the problem, you must use several steps.

Find the correct rate from the table.	6,400 falls between 5,001 and 50,000. The rate for this amount is $0.375 per 100 ft³.
Find the number of hundreds of cubic feet of water used.	6,400 ÷ 100 = 64
Multiply this amount by the rate.	64 × $0.375 = $24.00

The cost of the water is $24.00.

Use the rate table to find the cost of each amount of water. Round to the nearest cent.

1. 500 ft³

2. 1,500 ft³

3. 3,750 ft³

4. 63,880 ft³

Use the water rate table on page 122 to solve these problems. Round your answers to the nearest cent when necessary.

1. The Ahmeds used 4,950 ft^3 of water one month. What is the cost of this amount of water?

2. Mrs. Wagner used only 780 ft^3 of water in a month. What was her water bill for the month?

3. The Custom Manufacturing Company uses water to make its products. During the month of June, the company used 329,000 ft^3 of water. What was the cost of this amount of water?

4. The Cardillas used 4,560 ft^3 of water in July. During August, they watered the lawn and garden heavily and used 7,380 ft^3. How much more was their water bill in August than in July?

5. The Chesters used 2,100 ft^3 of water in May, 3,200 ft^3 in June, and 4,800 ft^3 in July. What rate did they pay for the water each month? What is the total number of cubic feet used?

6. What was the Chester's total water bill for the months of May, June, and July? (See problem 5.)

7. The Rockwell household paid a total of $306.00 for water in one year. What was the cost of water per month?

8. The Morgans were averaging 7,600 ft^3 of water use per month. They started being more careful and cut their use of water by 0.35 of this amount. How much water did they now use a month? What was the cost of this amount of water?

9. The Universal Company was using 450,000 ft^3 of water per month. By using conservation methods, they reduced the amount of water used per month to 375,000 ft^3. How much money did the company save on water per month?

10. **Try This** Kelly MacGregor paid $6.30 for water for one month. What amount of water did she use during the month? Hint: Use the rate table to guess and check.

Dividing by a Decimal

Paper cups that will hold 0.25 liters (L) of liquid will be used for a club party. How many servings this size are there in 6 L of fruit punch?

Since each serving will be the same amount, we divide the total amount of punch by the amount in one serving.

| Multiply the divisor by a power of 10 to make it a whole number. | Multiply the dividend by the same power of 10. | Divide. |

```
          0.25)6̅
```
0.25 × 100 = 25

```
         0.25)6.00
```
6 × 100 = 600

```
              2 4.
         0.25)6.0 0
              5 0
              1 0 0
              1 0 0
                  0
```

Check:
```
      24
   × 0.25
     120
      48
    6.00
```

There are 24 servings of fruit punch.

Other Examples

```
           2.8 5    → 2.9 (nearest tenth)
      0.7)2.0 0 0
          1 4
            6 0
            5 6
              4 0
              3 5
                5
```
Multiply divisor and dividend by 10.

```
            0.0 3 7   → 0.04 (nearest hundredth)
       4.8)0.1 7 8 0
           1 4 4
             3 4 0
             3 3 6
                 4
```
Multiply divisor and dividend by 10.

Warm Up

Find the quotients to the nearest tenth.

1. 3.8)53.7
2. 0.42)1.848
3. 0.039)0.814

Find the quotients to the nearest hundredth.

4. 0.47)1.92
5. 7.6)913
6. 0.082)0.1487

Find the quotients to the nearest tenth.

1. $0.6 \overline{)1.7}$
2. $0.9 \overline{)7}$
3. $0.4 \overline{)1.9}$
4. $0.7 \overline{)1.35}$
5. $3.6 \overline{)23.74}$
6. $1.6 \overline{)130}$
7. $0.82 \overline{)2.795}$
8. $1.5 \overline{)50}$
9. $2.8 \overline{)130}$
10. $0.75 \overline{)3.882}$
11. $0.94 \overline{)1.875}$
12. $6.9 \overline{)123.4}$

Find the quotients to the nearest hundredth.

13. $6.1 \overline{)32.46}$
14. $6.8 \overline{)0.75}$
15. $7.3 \overline{)38.42}$
16. $0.54 \overline{)2.793}$
17. $0.048 \overline{)0.702}$
18. $9.8 \overline{)103.6}$
19. $0.17 \overline{)0.132}$
20. $0.9 \overline{)7}$
21. $2.6 \overline{)5.84}$
22. $0.59 \overline{)0.3954}$
23. $7.2 \overline{)26.48}$
24. $0.11 \overline{)0.4184}$

Find the quotients to the nearest thousandth.

25. $0.36 \overline{)5}$
26. $4.5 \overline{)8.924}$
27. $8.1 \overline{)580}$
28. $0.9 \overline{)1.4}$
29. $0.67 \overline{)0.384}$
30. $0.052 \overline{)0.159}$
31. $2.8 \overline{)120}$
32. $1.6 \overline{)0.9284}$

33. How many 0.27-liter servings are there in 3.78 L of milk?

34. How much money would be earned by selling 7.5 L of apple juice if each serving of 0.3 L costs 25¢?

35. Which division problem at the right has a quotient different from the others? Try to decide without dividing. Then check your answer.

A $0.68 \overline{)17.68}$
B $6.8 \overline{)176.8}$
C $0.068 \overline{)1.768}$
D $68 \overline{)0.1768}$

SKILLKEEPER

1. $0.09 \div 10$
2. $56,600 \div 1,000$
3. $29.62 \div 100$
4. $576.3 \div 10$
5. $2.65 \div 100$
6. $0.596 \div 10$
7. $634.22 \div 100$
8. $9,270 \div 1,000$
9. $7,954 \div 1,000$
10. $0.6 \div 100$
11. $93.54 \div 10$
12. $256 \div 100$

More Practice, page 425, Set C

PROBLEM SOLVING: Finding Averages

QUESTION
DATA
PLAN
ANSWER
CHECK

What is the average daily attendance at Jefferson School?

Jefferson School Daily Attendance Form	
Day	Attendance
Monday	372
Tuesday	384
Wednesday	378
Thursday	390
Friday	388

The **average** of a list of numbers is the sum of the numbers divided by how many numbers there are in the list.

```
  3 7 2
  3 8 4
  3 7 8
  3 9 0
+ 3 8 8
-------
1,9 1 2   ← total attendance
```

```
          3 8 2.4   ← average daily
       ┌─────────      attendance
    5 )1,9 1 2.0
       1 5
       ───
         4 1
         4 0
         ───
           1 2
           1 0
           ───
             2 0
             2 0
             ───
               0
```

The average daily attendance was 382.4, or about 382 students per day.

What score must Katy make on the fifth test in order to raise her test score average to 90?

Katy's scores must total 450 to average 90 on the five tests ($5 \times 90 = 450$). The total for the first four tests is 351. Therefore, Katy must score 99 ($450 - 351$) on the fifth test.

Katy's Test Scores	
Test	Score
1	84
2	92
3	86
4	89
5	?

Solve.

1. The daily attendance for one week at Creekside School was 273, 252, 269, 271, and 265. What was the average daily attendance?

2. Reggie's weekly math test scores for one month were 85, 82, 90, and 67. What was his average score for the month?

3. What is the average price of soap powder?

4. Find the average height to the nearest whole inch (in.).

Name	Height (in.)
Carla	60
Charlie	58
Ross	64
Laverne	63
Mark	62
Jose	60
Ken	59

5. Find the average gasoline price to the nearest tenth of a cent.

Station	Price per Gallon
A	140.9¢
B	139.9¢
C	138.0¢
D	136.7¢
E	135.4¢

6. Find the test score averages to the nearest whole number. Who has the higher average?

Name	Test Scores				Average
Leona K.	78	87	84	90	
Pete R.	80	80	85	86	

7. Find the average weight to the nearest pound (lb).

Name	Weight (lb)
Michel	115
Nancy	94
Polly	120
Matt	98
Bev	95
Dave	109
Kathy	101

8. Raoul had scores of 93, 88, and 92 on his first three tests. What is his average test score?

9. Raoul is going to take a fourth test. What is the lowest score he could make on the fourth test and have an average of 90? (Use the data in problem 8.)

10. **DATA HUNT** Find the height of eight of your classmates. What is the average height?

11. **Try This**

    Lennie: "What were your last two test scores?"
    Minnie: "I raised my test score on the last test by 12. That made a total of 170 for the two tests."

    What was Minnie's score on each test?

Skills Practice

Find the quotients to the nearest tenth.

1. $0.9 \overline{)33.1}$
2. $0.6 \overline{)8}$
3. $0.5 \overline{)2.8}$
4. $0.3 \overline{)0.43}$

5. $3.3 \overline{)83.46}$
6. $0.97 \overline{)1.261}$
7. $6.4 \overline{)85.03}$
8. $7.3 \overline{)4.594}$

9. $0.28 \overline{)0.9832}$
10. $3.11 \overline{)40.126}$
11. $1.01 \overline{)63.781}$
12. $4.51 \overline{)0.0532}$

13. $6.43 \overline{)80.137}$
14. $1.03 \overline{)866.02}$
15. $0.055 \overline{)10.936}$
16. $0.247 \overline{)0.0226}$

Find the quotients to the nearest hundredth.

17. $0.2 \overline{)87.653}$
18. $6.7 \overline{)3.2951}$
19. $8.1 \overline{)670.13}$
20. $0.01 \overline{)0.37855}$

21. $0.056 \overline{)0.8725}$
22. $39.1 \overline{)0.2643}$
23. $4.55 \overline{)0.0952}$
24. $86.1 \overline{)29.031}$

25. $1.6 \overline{)\$118.37}$
26. $3.2 \overline{)\$27.55}$
27. $64.1 \overline{)\$109.20}$
28. $55.3 \overline{)\$268.43}$

29. $15.8 \overline{)\$456.02}$
30. $3.72 \overline{)\$11.66}$
31. $0.09 \overline{)\$8.47}$
32. $43.5 \overline{)\$58.78}$

Find the quotients to the nearest thousandth.

33. $2.48 \overline{)0.0389}$
34. $0.091 \overline{)0.0982}$
35. $65.3 \overline{)5.451}$
36. $1.01 \overline{)32.085}$

37. $82 \div 0.815$
38. $11.01 \div 3.06$
39. $37 \div 22.9$
40. $0.042 \div 65.1$

41. $0.001 \div 0.093$
42. $8.6921 \div 47.3$
43. $0.1825 \div 0.071$
44. $0.6 \div 63.72$

45. $0.55 \div 33.49$
46. $8.6392 \div 0.092$
47. $4.71 \div 0.4713$
48. $0.009 \div 3.9$

Estimating Quotients with Decimals

It takes 21.9 L of purified water to fill Helen's aquarium. The water costs $3.69. What is the estimated cost of 1 L of water?

To find the estimated cost of 1 L, we round each number to 1-digit accuracy and divide the cost by the number of liters.

Problem	Estimate
21.9)$3.69	$0.20 20)$4.00

The estimated cost of the water is $0.20 per liter.

Other Examples

Problem	Estimate		Problem	Estimate
0.838)2.755	→	3. 0.8)3.0	4.12)0.866	→	0.2 4)0.9

Estimate each quotient.

1. 3.56)16.86
2. 0.818)1.731
3. 6.8)0.5214

4. 0.077)0.1913
5. 37.4)5.817
6. 9.93)68.09

7. 0.188)4.26
8. 28.144)609.85
9. 0.54)10.37

10. 6.5)0.1954
11. 0.68)0.2546
12. 0.032)9.1

13. Robert filled his aquarium with 38.6 L of purified water. He paid $7.56 for the water. Estimate the cost of 1 L of water. What was the actual cost per liter?

14. Mary bought 21.0 L of water and a filter for her aquarium. The total cost was $12.25. The filter cost $1.95. Estimate the cost of the water per liter. Find the actual cost.

More Practice, page 425, Set D

PROBLEM SOLVING: Work Backward

QUESTION
DATA
PLAN
ANSWER
CHECK

Try This

A girl in Colonial Williamsburg said, "I divided the number for my birth year by 2, subtracted 365, and divided that answer by 9. The result was 59. In what year was I born?"

Some problems start with an unknown number. Several operations are performed, and you are told the final result. To find the starting number you can use the strategy called **Work Backward**. Here is a way to use this strategy to solve the problem above.

Show the steps of the problem in the order given.

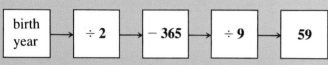

Work backward by using the numbers in opposite order and use opposite or **inverse** operations.

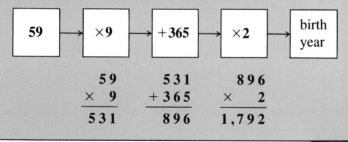

The girl was born in 1792.

Solve.

1. Ted's grandfather said, "If you multiply my age by 3, then add 17 to that answer, then divide by 2 you will get exactly 100." How old is Ted's grandfather?

2. Mavis Dean baked 6 batches of corn muffins. She gave 5 muffins to a friend, she ate 3 muffins, and she put 14 muffins in the freezer. There were 128 muffins left. How many muffins were in each batch?

CHAPTER REVIEW/TEST

Find the quotients.

1. $6\overline{)15.6}$
2. $4\overline{)15.04}$
3. $21\overline{)134.4}$
4. $56\overline{)2{,}027.2}$

Find the quotients to the nearest tenth.

5. $4\overline{)17.1}$
6. $5\overline{)12.3}$
7. $20\overline{)43}$
8. $16\overline{)34.02}$

Divide.

9. $26.4 \div 10$
10. $648.3 \div 1{,}000$
11. $2.73 \div 100$
12. $47 \div 1{,}000$

Find the quotients to the nearest hundredth.

13. $0.31\overline{)0.6}$
14. $4.25\overline{)18.52}$
15. $14.2\overline{)351}$
16. $29.5\overline{)226.8}$

17. $25\overline{)\$181.28}$
18. $5\overline{)\$34.92}$
19. $251\overline{)\$1{,}396.88}$
20. $14\overline{)\$0.19}$

Find the quotients to the nearest thousandth.

21. $17\overline{)44.82}$
22. $3.6\overline{)46.7}$
23. $0.06\overline{)4.451}$
24. $0.23\overline{)251}$

Estimate each quotient.

25. $0.4\overline{)18.32}$
26. $3.1\overline{)0.58}$
27. $2.381\overline{)18.592}$
28. $0.63\overline{)0.3441}$

29. Clifford bought 15 oranges for $1.38. What was the cost (to the nearest cent) of each orange?

30. Jan bought a can of 3 tennis balls for $2.59. What was the cost (to the nearest cent) of each ball?

31. Find the average test score.

Name	Test Scores
Kaye	79
Reed	85
Hector	96
Dotty	95
Jeff	89
Liz	90

ANOTHER LOOK

```
  0.18
21)3↑78   whole
          number
          divisor
```

Find the quotients.

1. 25)60
2. 8)7.2
3. 7)0.126

```
To round to the
nearest tenth,
find the quotient
to the hundredths
place.
   0.16 → 0.2
21)3.48
```

Find the quotients to the nearest tenth.

4. 8)23.0
5. 16)27.8
6. 25)42.4

7. 12)54.9
8. 29)1.972
9. 54)95.8

```
12.8 ÷ 10 = 1.28
      ↑      ↶
   1 zero  shift
          1 place
```

Divide each number by 10, 100, and 1,000.

10. 7.2
11. 0.16
12. 0.48

```
Make the divisor
a whole number.

6.2)17.4 81
  ↷   ↷
 ×10  ×10
```

Find the quotients to the nearest hundredth.

13. 0.4)0.3519
14. 5.6)29.34
15. 2.3)4.0

16. 0.86)1.208
17. 0.69)0.748
18. 0.079)0.334

Find the quotients to the nearest thousandth.

19. 7.4)5
20. 3.6)46.7
21. 0.03)0.1723

22. 0.024)0.76
23. 57)9.385
24. 0.26)0.755

```
0.43)0.76

       2.   ← estimate
0.4)0.8
  ↶  ↶
```

Estimate each quotient.

25. 0.2)5.61
26. 0.29)0.0893
27. 1.5)0.8

ENRICHMENT

History of Mathematics

Karl Friedrich Gauss was a famous German mathematician. When Gauss was a young student, his mathematics teacher gave him the task of adding all the whole numbers from 1 to 100. Gauss surprised his teacher by using a shortcut to quickly find the sum.

Karl Friedrich Gauss
German, 1777–1855

To see how Gauss might have discovered a shortcut, examine this simpler problem.

What is the sum of all the whole numbers from 1 to 10?

Suppose we reverse the order of the addends and add the pairs of numbers vertically.

$$\begin{array}{c}1 + 2 + 3 + 4 + 5 + 6 + 7 + 8 + 9 + 10\\ 10 + 9 + 8 + 7 + 6 + 5 + 4 + 3 + 2 + 1\\ \hline 11 + 11 + 11 + 11 + 11 + 11 + 11 + 11 + 11 + 11 = 10 \times 11\end{array}$$

The product of 10×11 is twice the sum we are seeking, because each number from 1 to 10 has been added twice. Therefore, the correct sum is

$$\frac{10 \times 11}{2} = \frac{110}{2} = 55$$

Use the shortcut to find these sums.

1. The sum of the whole numbers from 1–15.
2. The sum of the whole numbers from 1–20.
3. The sum of the odd numbers from 1–19.
4. The sum of the whole numbers from 1–100.
5. The sum of the odd numbers from 1–99.
6. The sum of the even numbers from 2–100.

CUMULATIVE REVIEW

1. Find the product.

    218
    × 59

    A 15,142   B 15,042
    C 12,862   D not given

2. Multiply.

    $6^2 \cdot 5^3$

    A 4,500   B 3,750
    C 7,500   D not given

3. Find the quotient.

    37)35,742

    A 966   B 836
    C 976   D not given

4. Find the quotient and remainder.

    76)2,749

    A 36 R13   B 37 R37
    C 37 R47   D not given

5. Estimate the product by rounding to 1-digit accuracy.

    427 × 85

    A 32,000   B 37,800
    C 36,000   D not given

6. Divide.

    24)4.212

    A 0.1756   B 0.2755
    C 0.1766   D not given

7. Estimate the quotient by rounding to 1-digit accuracy.

    3,200 ÷ 262

    A 16   B 10
    C 15   D not given

8. Find the quotient.

    0.18)4.212

    A 0.234   B 23.4
    C 2.34   D not given

9. Divide. Round to the nearest hundredth.

    9.5)0.654

    A 0.068   B 0.060
    C 0.069   D not given

10. Divide.

    7.596 ÷ 100

    A 75.96   B 759.6
    C 7,596   D not given

11. Find the quotient.

    5,956 ÷ 1,000

    A 5.956   B 59.56
    C 0.5956   D not given

12. Find the product.

    2,597 × 808

    A 2,098,376   B 209,837
    C 2,020,000   D not given

13. Alex collects 884 golf balls at the driving range. He places the same number of golf balls in 17 baskets. How many golf balls are in each basket?

    A 52   B 42
    C 53   D not given

14. Five students in Miss Marshall's class have scores of 154, 162, 152, 149, and 158. What is the average score?

    A 152   B 155
    C 158   D not given

134

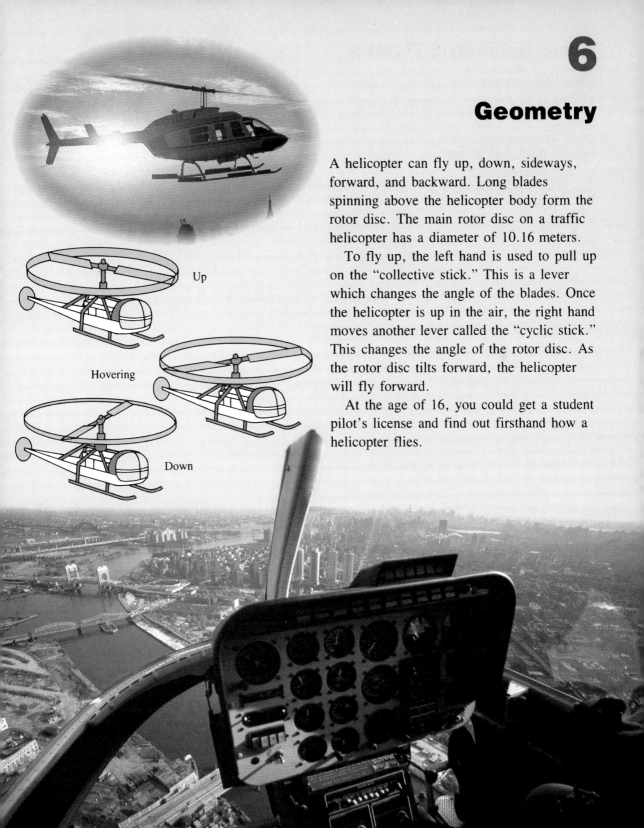

6

Geometry

A helicopter can fly up, down, sideways, forward, and backward. Long blades spinning above the helicopter body form the rotor disc. The main rotor disc on a traffic helicopter has a diameter of 10.16 meters.

To fly up, the left hand is used to pull up on the "collective stick." This is a lever which changes the angle of the blades. Once the helicopter is up in the air, the right hand moves another lever called the "cyclic stick." This changes the angle of the rotor disc. As the rotor disc tilts forward, the helicopter will fly forward.

At the age of 16, you could get a student pilot's license and find out firsthand how a helicopter flies.

Basic Geometric Figures

Everywhere we look we see objects that remind us of different geometric figures or shapes.

Points, **lines**, and **planes** are the simplest geometric figures.

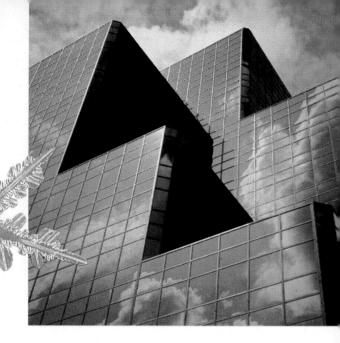

We think of a **point** as a position in space with no length, width, or thickness.

P.

Point *P*: *P*

We think of a **line** as a straight path of points.

Line *RS*: $\overleftrightarrow{RS}$ or $\overleftrightarrow{SR}$

We think of a **plane** as a flat surface.

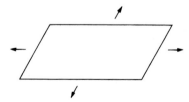

A **segment** is two points on a line and all the points on the line between them.

Segment *AB*: $\overline{AB}$ or $\overline{BA}$

A **ray** is the part of a line that has one endpoint and extends endlessly in one direction.

Ray *PQ*: $\overrightarrow{PQ}$

In the same plane, two lines are either **parallel** or they **intersect**. Parallel lines do not meet.

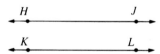

Line *HJ* is parallel to line *KL*. $\overleftrightarrow{HJ} \parallel \overleftrightarrow{KL}$

Intersecting lines meet at one point.

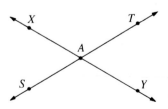

Lines *XY* and *ST* intersect at point *A*.

Draw a picture of each geometric figure. Write the symbol for the picture.

1. Segment *MN*
2. Point *Z*
3. Ray *OP*
4. Line *TR*
5. Use the three points to name three different segments.

   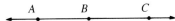

6. Use the four points to name six different segments.

7. Name all the segments shown in the figure.

   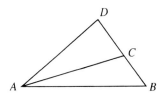

8. Name ten different segments shown in the figure.

   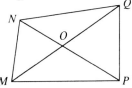

9. Name all the rays shown in the figure.

   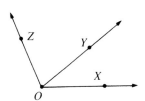

10. Use points *A* and *B* to name two rays.

11. Which two lines appear to be parallel lines?

    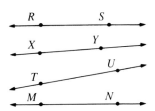

12. Draw a picture of a line intersecting two parallel lines.

137

Angles

An angle is two rays from the same endpoint or **vertex**. The rays are the sides of the angle.

We can use a **protractor** to find the **measure** of an angle in **degrees**.

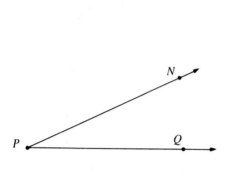

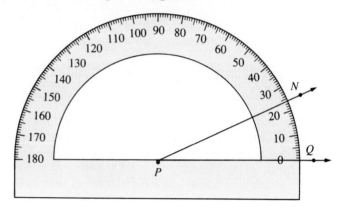

Angle *NPQ* or angle *QPN* or angle *P*
∠*NPQ*   ∠*QPN*   ∠*P*

The measure of ∠*NPQ* is 24 degrees.
m ∠*NPQ* = 24°

Angles are named according to their measures.

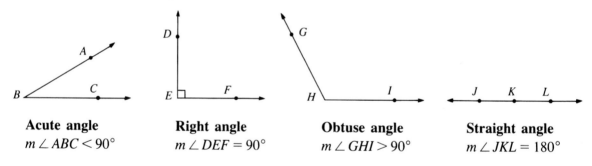

Acute angle
m ∠ *ABC* < 90°

Right angle
m ∠ *DEF* = 90°

Obtuse angle
m ∠ *GHI* > 90°

Straight angle
m ∠ *JKL* = 180°

Two angles are **complementary** if their measures have a sum of 90°.

Two angles are **supplementary** if their measures have a sum of 180°.

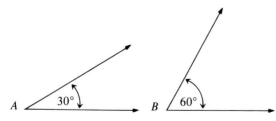

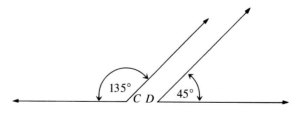

30° + 60° = 90°
∠*A* and ∠*B* are complementary angles.

135° + 45° = 180°
∠*C* and ∠*D* are supplementary angles.

Give the measure of each angle. State if the angle is **right**, **acute**, **obtuse**, or **straight**.

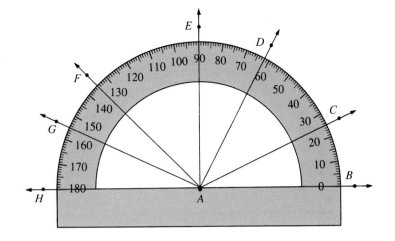

1. ∠BAC
2. ∠BAD
3. ∠BAE
4. ∠BAF
5. ∠BAG
6. ∠BAH
7. ∠EAF
8. ∠GAH

Estimate the measure of each angle. Then measure the angle with a protractor.

9.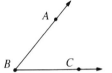
10.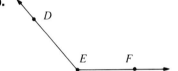
11.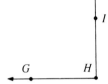

Give the measure of angles that are complementary to angles having the measures below.

12. 40°
13. 75°
14. 12°
15. 32°
16. 80°
17. 29°
18. 66°
19. 45°
20. 88°
21. 11°

Give the measure of angles that are supplementary to angles having the measures below.

22. 50°
23. 120°
24. 80°
25. 135°
26. 90°
27. 65°
28. 148°
29. 106°
30. 177°
31. 33°

SKILLKEEPER

Give the place value of each underlined digit.

1. 7<u>5</u>9
2. 6,<u>7</u>78
3. 1<u>5</u>,952
4. <u>7</u>5,912
5. 9,62<u>7</u>
6. 6.9<u>2</u>
7. 0.5<u>1</u>9
8. 7.<u>9</u>65
9. 15.6<u>2</u>
10. 0.21<u>5</u>
11. 3.596<u>7</u>
12. 72.32
13. <u>7</u>92.8
14. 0.6<u>2</u>1
15. 7.95<u>8</u>

Triangles

Some diatoms are shaped like **triangles**. Triangle *ABC* (△ *ABC*) has three **sides**, three **vertices**, and three **angles**. The sides are $\overline{AB}$, $\overline{AC}$, and $\overline{BC}$. The vertices are points *A*, *B*, and *C*. The angles are ∠*A*, ∠*B*, and ∠*C*.

Triangles are named according to the measures of their angles or the lengths of their sides.

Light micrograph of diatoms

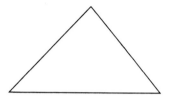

Acute triangle
All angles measure less than 90°.

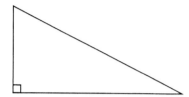

Right triangle
One angle measures 90°.

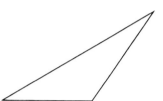

Obtuse triangle
The measure of one angle is greater than 90°.

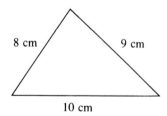

Scalene triangle
All three sides have different lengths.

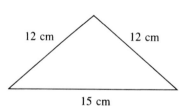

Isosceles triangle
At least two sides have the same length.

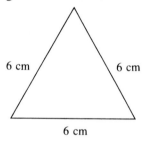

Equilateral triangle
All three sides have the same length.

The sum of the measure of the angles of any triangle is 180°.

In △ *ABC*, *m* ∠ *A* = 58° and *m* ∠ *C* = 30°. What is the measure of the third angle?

m ∠ *A* + *m* ∠ *C* = 58° + 30° = 88°

m ∠ *B* = 180° − 88° = 92°

m ∠ *A* + *m* ∠ *B* + *m* ∠ *C* = 180°

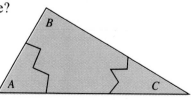

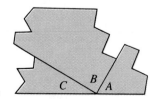

Name each triangle according to the measure of its angles.

1.
2.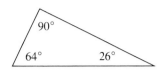
3.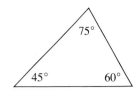

Name each triangle according to the length of its sides.

4.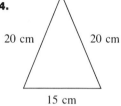
5.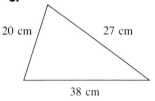
6.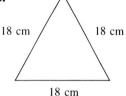

Give two names for each triangle.

7.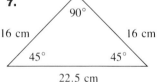
8.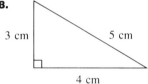
9.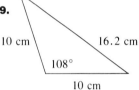

Find the measure of the third angle in each triangle.

10.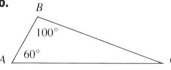
11.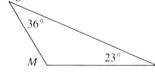
12.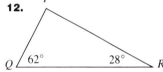

13.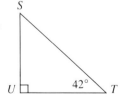
14.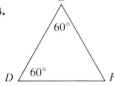
15.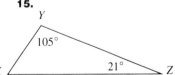

16. The measures of the angles of a triangle are 28°, 37°, and 115°. Is the triangle acute, right, or obtuse?

17. The measures of two angles of a triangle are 67° and 72°. Is the triangle acute, right, or obtuse?

18. Can a triangle have two obtuse angles? Explain your answer.

19. Can a triangle have one right angle and one obtuse angle? Explain your answer.

141

Quadrilaterals

Quadrilaterals have four sides and four angles.

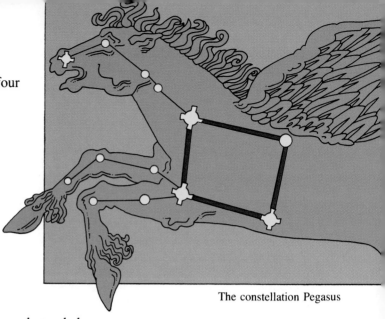

The constellation Pegasus

Quadrilateral *RSTU*
Sides: $\overline{RS}$, $\overline{ST}$, $\overline{TU}$, $\overline{UR}$
Angles: $\angle R$, $\angle S$, $\angle T$, $\angle U$

Some special kinds of quadrilaterals are shown below.

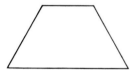

Trapezoid
A quadrilateral with a pair of parallel sides

Parallelogram
A quadrilateral with both pairs of opposite sides parallel

Rhombus
A parallelogram with all sides the same length

Rectangle
A parallelogram with four right angles

Square
A rectangle with all sides the same length

$\overline{AC}$ is a **diagonal** of quadrilateral *ABCD*. This diagonal divides the quadrilateral into two triangles.

The sum of the measures of the angles in each triangle is 180°.

Therefore, the sum of the measures of the four angles of any quadrilateral is 360°.

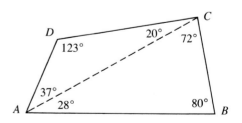

$m\angle A + m\angle B + m\angle C + m\angle D = 360°$

Name the kind of quadrilateral shown.

1.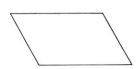

2.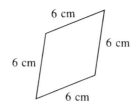

3.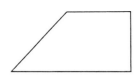

Find the measure of the fourth angle in each quadrilateral.

4.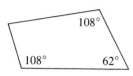

5.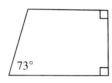

6.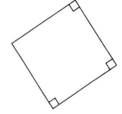

7.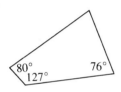

8.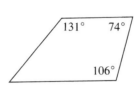

9.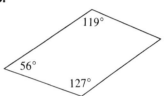

10. What is the measure of ∠C of a quadrilateral ABCD when m∠A = 70°, m∠B = 107°, and m∠D = 80°?

11. Draw a parallelogram that has a right angle. What kind of figure will result?

THINK MATH

Shape Perception

Draw a large quadrilateral ABCD. Measure each side and mark the middle points M, N, O, and P.

Draw a quadrilateral MNOP. It should be a parallelogram.

Cut off the four triangles in the figure. Try to place the four triangles so that they will exactly cover quadrilateral MNOP.

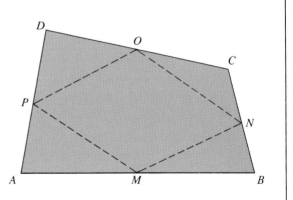

143

Polygons

Honeybees build six-sided cells for storage. A geometric figure having many sides that are segments is a **polygon**.

Polygons are named according to the number of sides they have.

Polygon	Number of sides	Name
△	3	triangle
▱	4	quadrilateral
⬠	5	pentagon
⬡	6	hexagon

Polygon	Number of sides	Name
⬡	7	heptagon
⬢	8	octagon
⬠	9	nonagon
⬠	10	decagon

Polygons with all sides the same length and all angles having the same measure are called **regular polygons**.

Equilateral triangle   **Square**   **Regular pentagon**   **Regular hexagon**   **Regular octagon**

Name each polygon.

1.
2.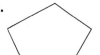
3.
4.

5.
6.
7.
8.

Use the drawing at the right for exercises 1–4.

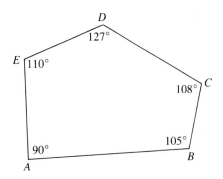

1. Name polygon ABCDE.

2. Name each side of the polygon.

3. Give the measure of each angle of the polygon.

4. What is the sum of the angles of the polygon?

Name each polygon. State if it is regular.

5.

6.

7.

8.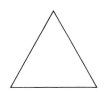

9.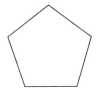

10.
90°	88°
90°	92°

★ 11. Can a quadrilateral have two right angles and no parallel sides? Draw a picture.

★ 12. What is the greatest number of right angles possible in a pentagon? Draw a picture.

THINK MATH

Shape Perception

Trace the figure and the dotted lines. Cut out your tracing and then cut along the dotted lines. Fit the four pieces together to make a square.

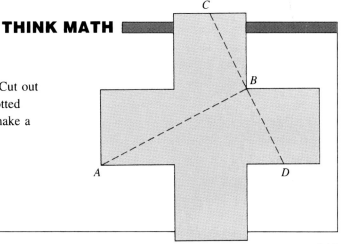

Circles

In the Chinese jade disk, called a *bi*, there are shapes that remind us of circles.

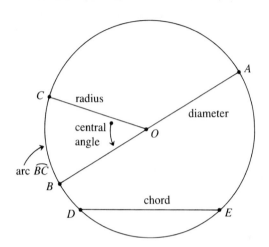

A **circle** is all the points in a plane that are the same distance from one point called the **center**. Point *O* is the center of circle *O*.

A **chord** is a segment with its endpoints on the circle. $\overline{AB}$ and $\overline{DE}$ are chords.

A **diameter** is a chord that passes through the center of the circle. $\overline{AB}$ is a diameter.

A **radius** is a segment from the center of the circle to a point on a circle. $\overline{OC}$, $\overline{OB}$, and $\overline{OA}$ are radii.

A **central angle** has its vertex at the center of a circle. $\angle BOC$ and $\angle AOC$ are central angles.

An **arc** is a part of a circle. $\overset{\frown}{BC}$ or $\overset{\frown}{CB}$ is the shorter arc with endpoints *B* and *C*.

Use circle *Q* for exercises 1–8.

1. Name the center of the circle.
2. Name a diameter of the circle.
3. Name two chords of the circle.
4. Name three radii of the circle.
5. Name two central angles of the circle.
6. Name four arcs of the circle.
7. If the length of $\overline{QZ}$ is 1.8 cm, what is the length of $\overline{WY}$?
8. If $m\angle ZQY = 85°$, what is $m\angle ZQW$?

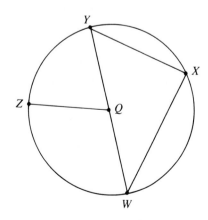

Follow the directions and answer the questions.

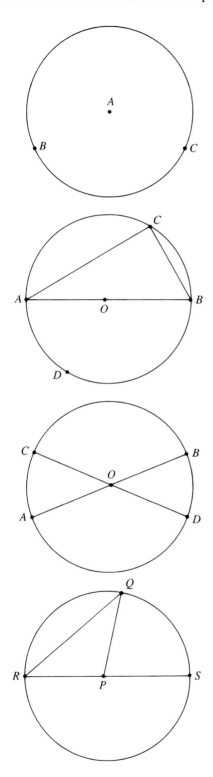

9. Draw a circle with center point A. Draw any two radii, $\overline{AB}$ and $\overline{AC}$. Draw chord $\overline{BC}$. What kind of triangle is △ABC?

10. Draw a circle with center O. Draw diameter $\overline{AB}$ and mark any point C on the circle. Draw $\overline{AC}$ and $\overline{BC}$. What is the measure of ∠ACB?

11. Use the same circle as in exercise 10. Mark another point D on the circle. Draw $\overline{AD}$ and $\overline{BD}$. What is the measure of ∠ADB?

12. What is the measure of an angle whose vertex is on a circle and whose sides pass through the endpoints of a diameter of the circle?

13. Draw a circle with center O. Draw diameters $\overline{AB}$ and $\overline{CD}$. Draw chords $\overline{AD}$, $\overline{DB}$, $\overline{BC}$, and $\overline{AC}$. What kind of quadrilateral is polygon ADBC?

14. Draw a circle with center P. Draw a diameter $\overline{RS}$. Mark a point Q on the circle. Draw $\overline{PQ}$ and $\overline{RQ}$. What is the measure of the central ∠QPS?

What is the measure of ∠QRS?

How do the measures of these two angles compare in size?

15. Name the angle that is supplementary to ∠RPQ.

16. Give the measure of ∠RPS.

147

Congruent Figures

Two geometric figures are **congruent** to each other if they have the same size and shape.

Two segments are congruent to each other if they have the same length.

$\overline{AB}$ is **congruent** to $\overline{XY}$.
$\overline{AB} \cong \overline{XY}$

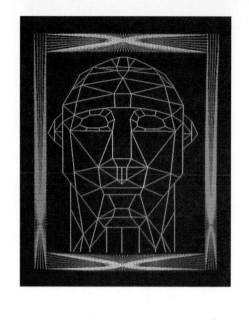

Two angles are congruent if they have the same measure.

$m \angle P = 37°$ $m \angle Q = 37°$
$\angle P$ is congruent to $\angle Q$.
$\angle P \cong \angle Q$

Two polygons are congruent if their vertices can be matched so that the matching sides are congruent and the matching angles are congruent.

Vertices A, B, and C can be matched to vertices E, D, and F so that

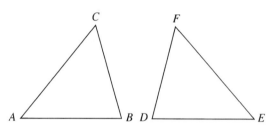

$\angle A \cong \angle E$ $\overline{AB} \cong \overline{ED}$
$\angle B \cong \angle D$ $\overline{BC} \cong \overline{DF}$
$\angle C \cong \angle F$ $\overline{CA} \cong \overline{FE}$

$\triangle ABC$ is congruent to $\triangle EDF$.
$\triangle ABC \cong \triangle EDF$

1. Which pair of segments is congruent?

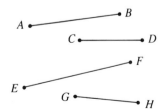

2. Which pair of angles is congruent?

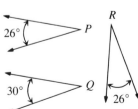

3. Which pair of triangles is congruent?

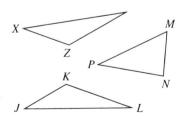

Each pair of triangles is congruent. List the pairs of congruent angles and pairs of congruent sides.

4.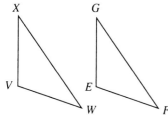

5.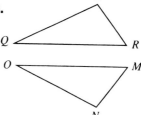

6.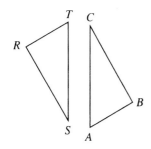

7.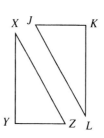

Use △ABC and △TRS for exercises 8-16.   △ABC ≅ △TRS

8. ∠A ≅ ▓   **9.** ∠B ≅ ▓   **10.** ∠C ≅ ▓

11. $\overline{AB}$ ≅ ▓   **12.** $\overline{AC}$ ≅ ▓   **13.** $\overline{BC}$ ≅ ▓

14. What is the length of $\overline{RT}$?

15. What is the measure of ∠S?

16. What is the length of $\overline{ST}$?

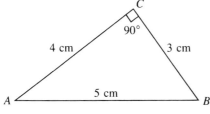
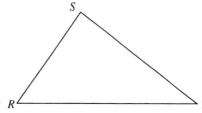

THINK MATH

Shape Perception

How many pairs of congruent triangles can you find in this picture?

Name the pairs you find.

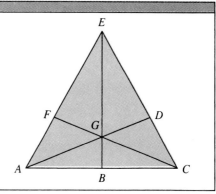

149

Lines of Symmetry

Think of folding the picture of the hockey rink on the red line or centerline. The two halves of the rink will match.

The red line is a **line of symmetry** of the rink.

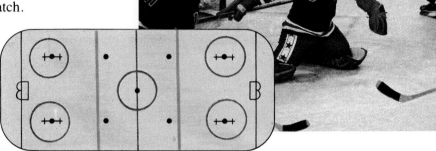

Some geometric figures have lines of symmetry.

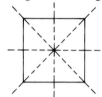

Square
four lines of symmetry

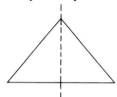

Isosceles triangle
one line of symmetry

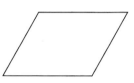

Parallelogram
no lines of symmetry

Think of folding each figure on the dotted line.
Is the dotted line a line of symmetry?

1.

2.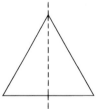

3.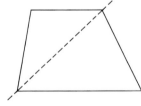

4.

5.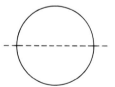

6.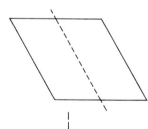

7.

8.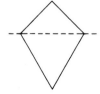

9.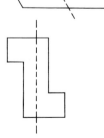

Draw each figure on graph paper. Draw all lines of symmetry.

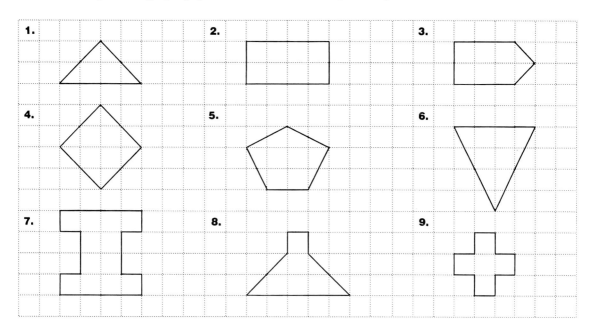

How many lines of symmetry does each polygon have?

10.

Equilateral triangle

11.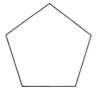

Regular pentagon

12.

Regular hexagon

13.

Regular octagon

14. How many lines of symmetry does a regular decagon have?

15. What is true of any one of the lines of symmetry of a circle?

SKILLKEEPER

1. 9.7 × 7	2. 0.35 × 3	3. 12.67 × 19	4. 0.004 × 67	5. 5.05 × 92
6. 0.56 × 0.5	7. 5.7 × 0.20	8. 0.0009 × 0.8	9. 10.73 × 0.5	10. 0.47 × 0.7
11. 0.423 × 0.015	12. 0.598 × 5.6	13. 3.8 × 0.05	14. 12.34 × 6.2	15. 2.724 × 0.16

Constructing Parallel Lines and Perpendicular Lines

Two lines in the same plane that do not intersect are **parallel lines**.

Two lines that intersect to form right angles are **perpendicular lines**.

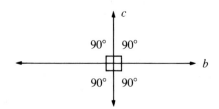

Line r is parallel to line s.
We write: $r \parallel s$

Lines b and c are perpendicular lines.
We write: $b \perp c$

Construct a line perpendicular to a line through a point on the line.

Given: **Step 1**   **Step 2**   **Step 3**

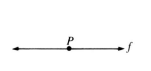 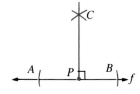

Draw arcs with center at P. Label points A and B.

Open compass wider and draw arcs with centers A and B. Label the point of intersection C.

Draw $\overleftrightarrow{CP}$.
$\overleftrightarrow{CP} \perp f$ at point P.

Construct a line parallel to a given line.

Given: **Step 1**   **Step 2**

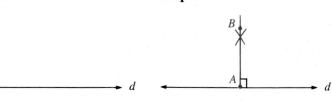 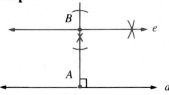

Mark a point A on line d. Construct a line through A perpendicular to line d. Mark a point B on the constructed line.

Construct line e through point B so that $e \perp \overleftrightarrow{AB}$.
$d \parallel e$

1. Draw a line *t* and mark a point *R* on the line. Construct a line *m* through *R* perpendicular to *t*.

2. Draw any line *m*. Construct a line *a* parallel to *m*.

3. Draw two lines, each perpendicular to $\overleftrightarrow{AB}$, one through *A* and the other through *B*. How are the two constructed lines related?

4. Draw any line *s*. Construct two lines each parallel to *s* and on opposite sides of *s*.

5. Use a ruler and a compass to construct a square.

6. Construct a trapezoid that has two right angles.

7. Lines *a* and *b* are parallel. Line *t* intersects *a* and *b*. Which numbered angles appear to be congruent angles?

8. Using the figure for exercise 7, suppose the measure of ∠1 = 50°. What is the measure of each of the other numbered angles in the figure?

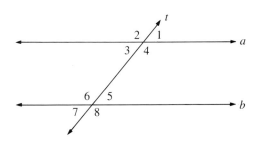

Bisecting Segments and Angles

A geometric figure is **bisected** if it is divided into two congruent parts.

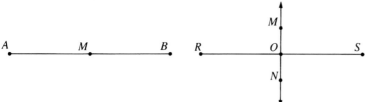
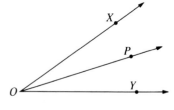

$\overline{AM} \cong \overline{MB}$
Point M bisects $\overline{AB}$.

$\overleftrightarrow{MN}$ is the **perpendicular bisector** (⊥ bis) of $\overline{RS}$.
Point O is the **midpoint**.
$\overline{RO} \cong \overline{OS}$ and $\overleftrightarrow{MN} \perp \overleftrightarrow{RS}$.

$\overrightarrow{OP}$ bisects $\angle XOY$.
$\angle XOP \cong \angle YOP$

Construct the perpendicular bisector of a segment.

Given: | **Step 1** | **Step 2** | **Step 3**

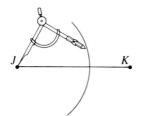
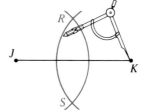
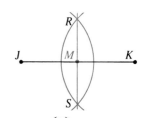

Open the compass to more than one half the length of $\overline{JK}$ and draw an arc with center at J.

With the same opening, draw an arc with center K. Label the intersections of the arcs R and S.

Draw $\overleftrightarrow{RS}$. Label point M.
M bisects $\overline{JK}$.
$\overleftrightarrow{RS} \perp$ bis $\overline{JK}$.

Construct the bisector of an angle.

Given: | **Step 1** | **Step 2** | **Step 3**

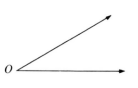
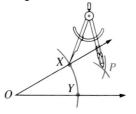
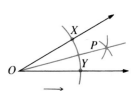

Draw any arc with center O. Label points X and Y.

Draw arcs from points X and Y. Label the intersection P.

Draw $\overrightarrow{OP}$.
$\overrightarrow{OP}$ bisects $\angle XOY$.

1. Draw $\overline{XY}$. Then bisect $\overline{XY}$ using a ruler and a compass.

2. Draw $\overline{MN}$. Then construct the perpendicular bisector of $\overline{MN}$.

3. Draw acute $\angle G$. Then bisect $\angle G$ with a ruler and a compass.

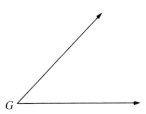

4. Draw obtuse $\angle H$. Then bisect it.

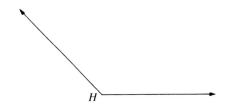

5. Draw any segment. Divide the segment into four congruent parts by bisecting the segment, then bisecting each half.

6. Draw any angle. Divide the angle into four congruent angles by bisecting the angle repeatedly.

7. Draw $\triangle XYZ$. Construct the perpendicular bisectors of each of the three sides of the triangle.

8. Draw $\triangle RST$. Bisect each angle of the triangle.

9. Point K is the midpoint of $\overline{HG}$. $\overline{HG}$ is 3.2 cm long. What is the length of $\overline{HK}$?

10. The measure of $\angle BAC$ is 78°. $\overrightarrow{AP}$ is the bisector of the angle. What is the measure of $\angle BAP$?

THINK MATH

Shape Perception

Follow the steps below to draw the shape of an egg.

1. Draw a circle with $\overline{AB} \perp \overline{CD}$.
2. Extend chords $\overline{AC}$ and $\overline{BC}$ as shown.
3. Draw $\widehat{BF}$ with center at A. Draw $\widehat{AE}$ with center at B.
4. Draw $\widehat{EF}$ with center at C.
5. Shade in the egg-shaped region.

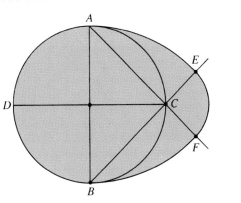

155

Space Figures

Geometric figures whose points do not all lie in the same plane are called **space figures**.

A **prism** has two bases which lie in parallel planes. The bases are polygonal regions. The other faces are regions formed by parallelograms.

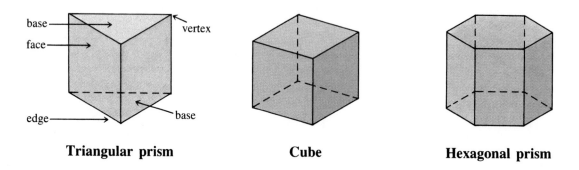

Triangular prism   **Cube**   **Hexagonal prism**

A **pyramid** has one base which is a polygonal region. The other faces of a pyramid are triangular regions.

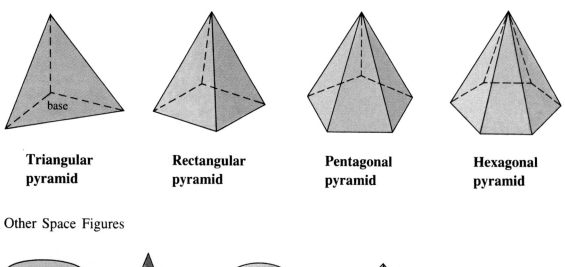

Triangular pyramid   **Rectangular pyramid**   **Pentagonal pyramid**   **Hexagonal pyramid**

Other Space Figures

Cylinder   **Cone**   **Sphere**   **Octahedron**   **Icosahedron**

Give the number of vertices (V), faces (F), and edges (E) of each figure.

1. 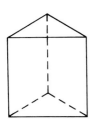 **2.** 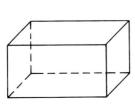 **3.** 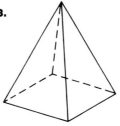 **4.**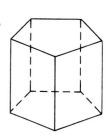

5. 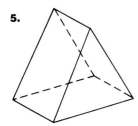 **6.** 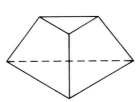 **7.** 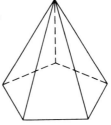 **8.**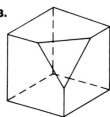

9. Use your answers for the numbers of vertices, faces, and edges in exercises 1–8 to show that $V + F - E = 2$.

10. DATA HUNT How many vertices, faces, and edges does a dodecahedron have?

Name the space figure that would be formed if each of the patterns were folded on the dashed lines.

11. 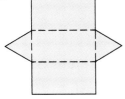 **12.** 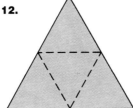 **13.** 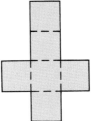 **14.**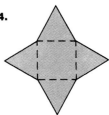

THINK MATH

Shape Perception

In what ways is this odd-shaped space figure like a cube?

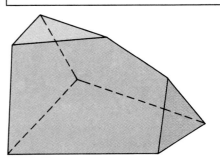

PROBLEM SOLVING: Solve a Simpler Problem

QUESTION
DATA
PLAN
ANSWER
CHECK

Try This

Sully and Howard are railroad engineers. Sully averages 388 km a day and Howard averages 320 km a day. How far has Howard traveled when Sully has traveled 1,020 km farther than Howard?

Some of the data in problems may be large numbers. Several operations may be used. To solve such problems, it may help to use the strategy **Solve a Simpler Problem**. To use this strategy, we can think of a simpler problem with smaller numbers.

Sully averages 10 km a day and Howard averages 8 km a day. How far has Howard traveled when Sully has traveled 20 km farther than Howard?

$10 - 8 = 2$ Sully travels 2 km a day farther.
$20 \div 2 = 10$ In 10 days, he will have traveled 20 km more than Howard.
$10 \times 8 = 80$ Howard will have traveled 80 km in this number of days.

Now use the same steps to solve the harder problem.

$388 - 320 = 68$ Sully travels 68 km a day farther.
$1,020 \div 68 = 15$ In 15 days, Sully will have traveled 1,020 km more than Howard.
$15 \times 320 = 4,800$ Howard will have traveled 4,800 km during this time.

Solve.

1. In a city of 45,300, a survey found that 0.14 of the people listened to radio station KXYU. The survey also found that 1 out of every 6 listeners tuned in to both the morning and the evening news broadcast. How many people in the city listen to both morning and evening news from KXYU?

2. There were 16,780 cars parked in a stadium parking lot. The average number of people per car was 1.5. How many fewer cars would have been used if there were 2.5 people per car?

CHAPTER REVIEW/TEST

Write the symbol for each figure.

1. R———S   2. ←—T———R—→   3. P•——Q→   4.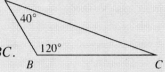

5. What kind of angle is ∠ABC?

6. What kind of angle is ∠BAC?

7. Give the measure of the angle that is supplementary to ∠ABC.

8. Give two names for △ABC.

9. Find the measure of ∠ACB.

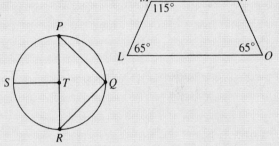

10. What kind of quadrilateral is figure LMNO?

11. Find the measure of ∠MNO.

12. Name the center of the circle.

13. Name two chords of the circle.

14. Name two central angles of the circle.

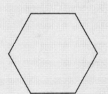

15. Which pair of triangles is congruent?

16. Name the polygon. Give the number of lines of symmetry.

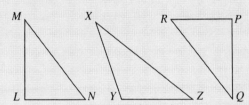

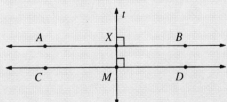

17. Name the line perpendicular to $\overleftrightarrow{AB}$.

18. Name the line parallel to $\overleftrightarrow{CD}$.

19. Give the measure of ∠XMD.

Name each space figure. Show that $V + F - E = 2$ for each.

20. 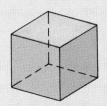   21. 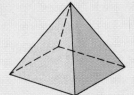   22.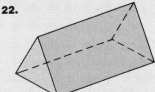

ANOTHER LOOK

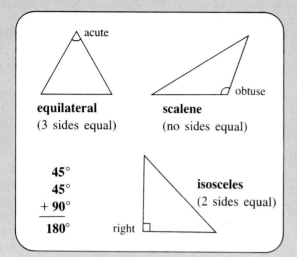

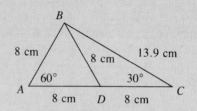

1. Is ∠BAD right, acute, or obtuse?
2. Is ∠BDC right, acute, or obtuse?
3. What kind of triangle is △ABD?
4. What kind of triangle is △BDC?
5. What kind of triangle is △ABC?
6. Give the measure of ∠ABC.
7. Give the measure of ∠ADC.

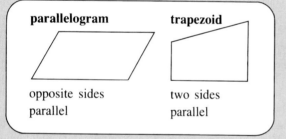

Name each quadrilateral.

8. 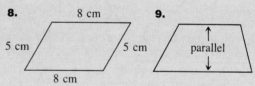 9.

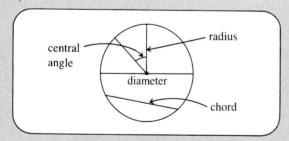

10. Name a radius of circle O.
11. Name a diameter.
12. Name a chord.
13. Name a central angle.

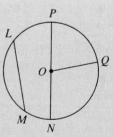

14. Name six vertices.
15. Name nine edges.

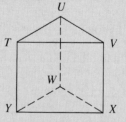

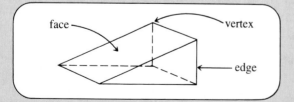

160

ENRICHMENT

Space Perception

There are five **regular polyhedrons** or solids. The faces of regular polyhedrons are congruent regular polygonal regions. These five regular polyhedrons are sometimes called the **Platonic solids**. They are named after Plato, one of the Greeks who studied them. The Greeks appear to have discovered that each regular polyhedron could be drawn inside a sphere.

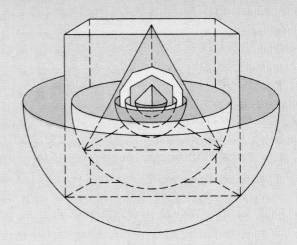

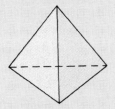
Tetrahedron

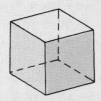
Cube

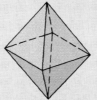
Octahedron

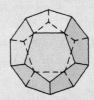
Dodecahedron

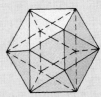
Icosahedron

Make a larger pattern for each polyhedron on lightweight posterboard. Cut out the patterns and tape the edges together to form models of the polyhedrons.

Use your models to complete the table for the number of vertices (V), faces (F), and edges (E).

Polyhedron	V	F	E	$V + F - E$
Tetrahedron				
Cube				
Octahedron				
Dodecahedron				
Icosahedron				

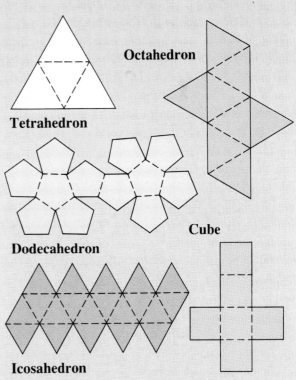

CUMULATIVE REVIEW

Find the products.

1.   7.29   A 45.927   B 459.27
      × 63   C 459.37   D not given

2.   0.56   A 4.604   B 46.04
      × 8.4   C 4.704   D not given

3. Estimate the product.

    $52.63 × 3.2

    A $150   B $200
    C $15.60   D not given

Multiply.

4. 0.627 × 100

    A 6.27   B 62.7
    C 627   D not given

5. 73.29 × 0.001

    A 7.329   B 7,329
    C 732.9   D not given

6. $954.62
     ×   8

    A $7,637.12   B $763.69
    C $7,636.96   D not given

7. Divide.

    $8)\overline{0.056}$

    A 0.07   B 7
    C 0.007   D not given

Find the quotients. Round to the nearest hundredth.

8. $12)\overline{\$15.75}$ A $1.31   B $1.312
                               C $1.32   D not given

9. $56)\overline{376}$ A 6   B 6.72
                              C 6.714   D not given

Find the quotients. Round to the nearest thousandth.

10. $0.59)\overline{17.3}$ A 0.293   B 29.322
                            C 2.932   D not given

11. $2.83)\overline{977.6}$ A 34.544   B 345.441
                             C 345.442   D not given

12. Estimate the quotient.

    $3.2)\overline{0.57}$ A 0.1   B 20
                       C 0.2   D not given

13. A record album costs $8.99. The sales tax is 0.06 of the cost. What is the total cost of the record album?

    A $9.05   B $9.53
    C $9.52   D not given

14. Purified water costs $9.35 for 21.5 L. What is the cost (to the nearest cent) per liter?

    A $0.43   B $0.40
    C $0.434   D not given

Number Theory and Equations

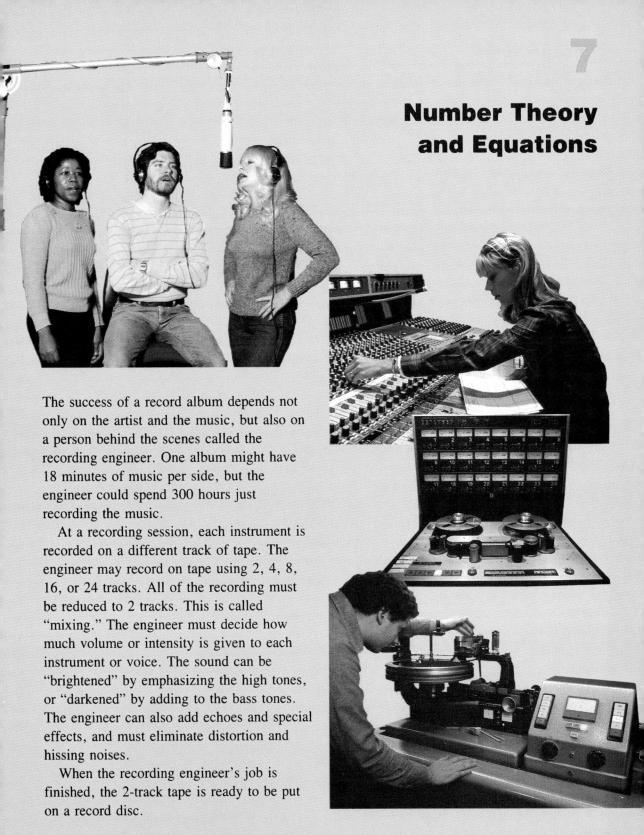

The success of a record album depends not only on the artist and the music, but also on a person behind the scenes called the recording engineer. One album might have 18 minutes of music per side, but the engineer could spend 300 hours just recording the music.

At a recording session, each instrument is recorded on a different track of tape. The engineer may record on tape using 2, 4, 8, 16, or 24 tracks. All of the recording must be reduced to 2 tracks. This is called "mixing." The engineer must decide how much volume or intensity is given to each instrument or voice. The sound can be "brightened" by emphasizing the high tones, or "darkened" by adding to the bass tones. The engineer can also add echoes and special effects, and must eliminate distortion and hissing noises.

When the recording engineer's job is finished, the 2-track tape is ready to be put on a record disc.

Factors

Which number has more factors, 36 or 40?

36 has 9 factors:
1, 2, 3, 4, 6, 9, 12, 18, and 36.

40 has 8 factors:
1, 2, 4, 5, 8, 10, 20, and 40.

36 has more factors.

Factors of 36	Factors of 40
1 × 36	1 × 40
2 × 18	2 × 20
3 × 12	4 × 10
4 × 9	5 × 8
6 × 6	

Any whole number, except 0, is divisible by each of its factors.

40 ÷ 1 = 40   40 ÷ 2 = 20   40 ÷ 4 = 10   40 ÷ 5 = 8
40 ÷ 8 = 5   40 ÷ 10 = 4   40 ÷ 20 = 2   40 ÷ 40 = 1

Zero is not a factor of any whole number except 0, but every whole number is a factor of 0.

0 × 0 = 0   1 × 0 = 0   2 × 0 = 0   3 × 0 = 0 . . .

We cannot divide any number by 0.

Other Examples

We can divide to find a factor of a number.

Is 8 a factor of 158?

$$\begin{array}{r} 19 \text{ R}6 \\ 8\overline{)158} \end{array}$$

The remainder is not 0.
8 is not a factor of 158.

Is 7 a factor of 105?

$$\begin{array}{r} 15 \text{ R}0 \\ 7\overline{)105} \end{array}$$

The remainder is 0.
7 and 15 are factors of 105.

Warm Up

List all the factors of each number.

1. 10   2. 12   3. 16   4. 11

Use division to decide if the first number is a factor of the second number.

5. 8 and 232   6. 16 and 500   7. 23 and 529   8. 54 and 324

List all the factors of each number.

1.
28
1 × 28
2 × 14
4 × 7

2.
51
1 × 51
3 × 17

3.
92
1 × 92
2 × 46
4 × 23

4.
54
1 × 54
2 × 27
3 × 18
6 × 9

5. 20   6. 24   7. 15   8. 17

9. 18   10. 33   11. 1   12. 80

13. 100   14. 56   15. 45   16. 68

17. 75   18. 84   19. 72   20. 120

Use division to decide if the first number is a factor of the second number.

21. 6 and 108   22. 7 and 162   23. 3 and 111   24. 4 and 138

25. 13 and 741   26. 18 and 666   27. 29 and 1,769   28. 25 and 525

29. 14 and 520   30. 31 and 967   31. 23 and 1,541   32. 42 and 2,142

33. 55 and 1,785   34. 120 and 600   35. 39 and 1,521   36. 231 and 28,413

37. What is the smallest whole number with exactly two different factors?

38. Which number is a factor of any even number?

39. What is the smallest whole number with exactly three different factors?

40. What is the smallest whole number with exactly four different factors?

Divide to decide if the first number is a factor of the second number.

41. 2,769 and 994,071   42. 3,427 and 23,396,129

43. 9,683 and 33,803,357   44. 47,287 and 13,902,378

45. 64,543 and 13,560,885   46. 15,625 and 1,000,000

Divisibility Rules

Beth Ann has 141 newspapers for a paper route. Can she divide the papers into 3 piles with the same number in each pile?

She could divide to find out. However, Beth Ann knows a rule that she can use without dividing.

$141 \rightarrow 1 + 4 + 1 = 6$
6 is divisible by 3,
so 141 is divisible by 3.

3 Add the digits of the number. If the sum is divisible by 3, so is the number.

$2,916 \rightarrow 2 + 9 + 1 + 6 = 18$
18 is divisible by 9,
so 2,916 is divisible by 9.

9 The divisibility rule for 9 is like the divisibility rule for 3.

You may already know the divisibility rules for 2, 5, and 10.

2 The ones digit must be **even**; 0, 2, 4, 6, 8.

5 The ones digit must be 5 or 0.

10 The ones digit must be 0.

Other Examples

$672 \rightarrow 6 + 7 + 2 = 15$
672 is divisible by 2.
672 is divisible by 3.
672 is not divisible by 5, 9, or 10.

$3,105 \rightarrow 3 + 1 + 0 + 5 = 9$
3,105 is divisible by 5.
3,105 is divisible by 3 and 9.
3,105 is not divisible by 2 or 10.

Warm Up

Use the divisibility rules to decide if each number is divisible by 2, 3, 5, 9, 10, or none of these numbers.

1. 912
2. 9,375
3. 5,550
4. 1,890
5. 2,331
6. 283
7. 8,577
8. 83,760

State if each number is divisible by 3.

1. 213
2. 624
3. 115
4. 810
5. 2,349
6. 6,227
7. 111
8. 1,011
9. 726
10. 555
11. 4,281
12. 57
13. 1,089
14. 4,431
15. 111,012

State if each number is divisible by 9.

16. 513
17. 784
18. 396
19. 405
20. 267
21. 1,236
22. 6,021
23. 108
24. 656
25. 2,070
26. 1,234
27. 6,669
28. 42,723
29. 56,725
30. 111,111,111

State if each number is divisible by 2, 3, 5, 9, 10, or none of these numbers.

31. 714
32. 420
33. 9,125
34. 123
35. 3,150
36. 5,310
37. 666
38. 4,277
39. 23,634
40. 88,524

41. Jeremy must deliver 112 Sunday newspapers. Can he divide that number of newspapers into 9 stacks with the same number in each stack?

42. Juan will deliver a number of newspapers that can be divided by 2, 3, 5, 9, and 10. What is the smallest number of papers he could deliver?

THINK MATH

Abundant Numbers

A number is **abundant** if the sum of its factors (other than the number itself) is greater than the given number.

Factors of 20: **1, 2, 4, 5, 10**

Sum of the factors: **1 + 2 + 4 + 5 + 10 = 22**

Since 22 > 20, 20 is an **abundant number**.

What is the smallest abundant number? Which numbers less than 50 are abundant numbers?

Prime and Composite Numbers

Jacqueline uses a computer to find the factors of some numbers. The computer screen shows the factors of the numbers. It also lists each number as **prime** or **composite**.

Prime numbers have exactly two factors.

Composite numbers have more than two factors.

The numbers 0 and 1 are neither prime nor composite.

1 has only one factor.
Every number is a factor of 0.

Warm Up

What would the computer screen show for numbers 11 through 20?

	Number	Factors	Prime or Composite
1.	11		?
2.	12		?
3.	13		?
4.	14		?
5.	15		?
6.	16		?
7.	17		?
8.	18		?
9.	19		?
10.	20		?

Write P (prime), C (composite), or N (neither) for each number.

1. 21
2. 31
3. 33
4. 47
5. 51
6. 0
7. 23
8. 99
9. 81
10. 91
11. 53
12. 63
13. 73
14. 83
15. 93

Find the prime number in each list.

16. 8, 9, 10, 11, 12
17. 0, 5, 10, 15, 20
18. 21, 23, 25, 27
19. 33, 35, 37, 39
20. 49, 50, 51, 52, 53
21. 81, 89, 91, 93

Find the composite number in each list. Find the factors of that number.

22. 11, 21, 31, 41
23. 23, 33, 43, 53
24. 17, 71, 16, 61
25. 0, 1, 3, 100
26. 89, 79, 69, 59
27. 7, 17, 49, 71

28. Make a list of all the prime numbers less than 100. You should find 25 prime numbers.

29. Choose any two prime numbers. Find their product. Is the product a prime number or a composite number?

30. What is the only even prime number?

31. How many composite numbers are less than 100?

32. What are the only two consecutive prime numbers? (Their difference is 1.)

33. Show that 413 is not a prime number. Find another factor of 413, other than 1 or 413, by dividing.

THINK MATH

Prime Number Patterns

The pattern at the right seems to always have sums that are prime numbers.

Write several more sums in the pattern.

Find the first sum in the pattern that is not a prime number.

$11 + 0 = 11$ ← prime
$11 + 2 = 13$ ← prime
$13 + 4 = 17$ ← prime
$17 + 6 = 23$ ← prime
$23 + ? = 31$ ← ?

More Practice, page 426, Set A

Prime Factorization

Each composite number can be expressed as a product of prime factors. This is the **prime factorization** of the composite number. Except for the order of the factors, there is only one prime factorization of a composite number.

Number		Prime Factorization
30	=	$2 \cdot 3 \cdot 5$
24	=	$2 \cdot 2 \cdot 2 \cdot 3$
99	=	$3 \cdot 3 \cdot 11$
143	=	$11 \cdot 13$

To find the prime factors of a number, find any two factors. Then find the prime factors of these factors.

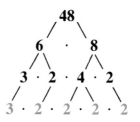

$$60 = 6 \cdot 10$$
$$60 = \underbrace{2 \cdot 3 \cdot 2 \cdot 5}_{\text{prime factorization}}$$

A **factor tree** may help to find the prime factors of a number. The lowest row of the tree contains only prime numbers.

```
         48
        /  \
       6    8
      /\   /\
     3·2  4·2
         /\
        2·2
```
$3 \cdot 2 \cdot 2 \cdot 2 \cdot 2$

We can write prime factorizations using exponents.

$48 = 2 \cdot 2 \cdot 2 \cdot 2 \cdot 3$
$48 = 2^4 \cdot 3$

Other Examples

$63 = 9 \cdot 7$
$63 = 3 \cdot 3 \cdot 7$

$72 = 8 \cdot 9$
$72 = 2 \cdot 2 \cdot 2 \cdot 3 \cdot 3$

$120 = 10 \cdot 12$
$120 = 2 \cdot 5 \cdot 2 \cdot 2 \cdot 3$

Warm Up

Complete each factor tree. Give the prime factorization of each number.

1. 42 = ▒ · 7; 42 = ▒ · ▒ · ▒

2. 36 = 9 · ▒; 36 = ▒ · ▒ · ▒ · ▒

3. 21 = ▒ · ▒

4. 27 = ▒ · ▒; 27 = ▒ · ▒ · ▒

Give the prime factorization of each number using exponents.

5. 8 **6.** 24 **7.** 63 **8.** 50

Complete each factor tree. Give the prime factorization of each number.

1.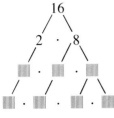

16 = ▨ · ▨ · ▨ · ▨

2.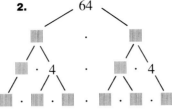

64 = ▨ · ▨ · ▨ · ▨ · ▨ · ▨

3.

28 = ▨ · ▨ · ▨

4.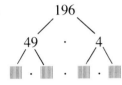

196 = ▨ · ▨ · ▨ · ▨

5.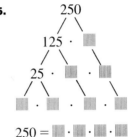

250 = ▨ · ▨ · ▨ · ▨

6.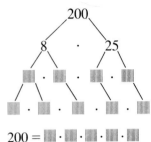

200 = ▨ · ▨ · ▨ · ▨ · ▨

Make a factor tree for each number. Give the prime factorization of the number.

7. 70 **8.** 84 **9.** 80 **10.** 81 **11.** 300

12. 120 **13.** 96 **14.** 105 **15.** 72 **16.** 1,000

Write the prime factorization of each number using exponents.

17. $968 = 2 \cdot 2 \cdot 2 \cdot 11 \cdot 11$

18. $1,125 = 3 \cdot 3 \cdot 5 \cdot 5 \cdot 5$

19. $162 = 2 \cdot 3 \cdot 3 \cdot 3 \cdot 3$

20. $1,372 = 2 \cdot 2 \cdot 7 \cdot 7 \cdot 7$

21. $18,865 = 5 \cdot 7 \cdot 7 \cdot 7 \cdot 11$

22. $396 = 2 \cdot 2 \cdot 3 \cdot 3 \cdot 11$

23. $900 = 2 \cdot 2 \cdot 3 \cdot 3 \cdot 5 \cdot 5$

24. $300 = 2 \cdot 2 \cdot 3 \cdot 5 \cdot 5$

25. Make a factor tree for 1 million. Give the prime factorization.

★ **26.** What are the next three numbers in this pattern? 2, 6, 30, 210, . . . Hint: What is the prime factorization of the numbers?

Give the prime factorizations. Each number below is the product of three of the prime numbers (one from each row in the table).

27. 27,178 **28.** 66,155

29. 42,051 **30.** 86,219

2	3	5	7
101	103	107	109
173	127	131	137

More Practice, page 426, Set B

Greatest Common Factor (GCF)

Pat has a piece of material 54 in. wide and 126 in. long. Squares must be cut out that are as large as possible without wasting any material. What is the greatest possible length of each side of the squares?

To solve the problem, we need to find the largest number that is a factor of both 54 and 126.

Factors of 54:   1, 2, 3, 6, 9, 18, 27, 54
Factors of 126: 1, 2, 3, 6, 7, 9, 14, 18, 21, 42, 63, 126

Common factors: 1, 2, 3, 6, 9, 18

The **greatest common factor** (**GCF**) is **18**.

Pat can cut the material into squares that are 18 in. on each side.

We can use the prime factorizations of two numbers to find the greatest common factor.

$54 = 2 \cdot 3 \cdot 3 \cdot 3$
$126 = 2 \cdot 3 \cdot 3 \cdot 7$
$GCF = 2 \cdot 3 \cdot 3 \quad = 18$

Two numbers are **relatively prime** if their GCF is 1.

Factors of 8: **1, 2, 4, 8**
Factors of 9: **1, 3, 9**
8 and 9 are relatively prime.

Other Examples

Find the GCF of 10 and 18.
  10 : 1, 2, 5, 10
  18 : 1, 2, 3, 6, 9, 18
GCF : 2

Find the GCF of 40 and 50.
  $40 = 2 \cdot 2 \cdot 2 \cdot 5$
  $50 = 2 \cdot 5 \cdot 5$
  $GCF = 2 \cdot 5 \ = 10$

Warm Up

List the common factors of each pair of numbers. Give the GCF.

1. 4, 10   **2.** 22, 33   **3.** 5, 15   **4.** 45, 75

Give the GCF of each pair of numbers. State if they are relatively prime.

5. 3, 5   **6.** 8, 18   **7.** 9, 13   **8.** 7, 14

List the common factors of each pair of numbers. Give the GCF.

1. 3, 18   **2.** 14, 28   **3.** 20, 50   **4.** 16, 32

Find the GCF of each pair of numbers.

5. $4 = 2 \cdot 2$
$6 = 2 \cdot 3$

6. $9 = 3 \cdot 3$
$45 = 3 \cdot 3 \cdot 5$

7. $42 = 2 \cdot 3 \cdot 7$
$105 = 3 \cdot 5 \cdot 7$

8. $90 = 2 \cdot 3 \cdot 3 \cdot 5$
$105 = 3 \cdot 5 \cdot 7$

9. $12 = 2 \cdot 2 \cdot 3$
$60 = 2 \cdot 2 \cdot 3 \cdot 5$

10. $42 = 2 \cdot 3 \cdot 7$
$28 = 2 \cdot 2 \cdot 7$

11. $150 = 2 \cdot 3 \cdot 5 \cdot 5$
$105 = 3 \cdot 5 \cdot 7$

12. $108 = 2 \cdot 2 \cdot 3 \cdot 3 \cdot 3$
$90 = 2 \cdot 3 \cdot 3 \cdot 5$

Use prime factorizations of each pair of numbers to find the GCF.

13. $28 = \blacksquare \cdot \blacksquare \cdot \blacksquare$
$35 = \blacksquare \cdot \blacksquare$

14. $30 = \blacksquare \cdot \blacksquare \cdot \blacksquare$
$24 = \blacksquare \cdot \blacksquare \cdot \blacksquare \cdot \blacksquare$

15. $18 = \blacksquare \cdot \blacksquare \cdot \blacksquare$
$45 = \blacksquare \cdot \blacksquare \cdot \blacksquare$

16. $36 = \blacksquare \cdot \blacksquare \cdot \blacksquare \cdot \blacksquare$
$40 = \blacksquare \cdot \blacksquare \cdot \blacksquare \cdot \blacksquare$

Find the GCF of each pair of numbers.

17. 27, 36   **18.** 18, 24   **19.** 36, 42   **20.** 20, 45

21. 60, 18   **22.** 30, 105   **23.** 42, 90   **24.** 36, 24

25. 44, 36   **26.** 14, 90   **27.** 18, 42   **28.** 8, 15

29. 24, 25   **30.** 39, 65   **31.** 63, 56   **32.** 100, 350

33. Hua has a piece of material that is 45 in. wide and 75 in. long. He wants to cut it into the largest possible squares without wasting any material. What is the length of the sides of the largest possible squares?

34. Janice is crocheting a tablecloth that will be 60 in. wide and 75 in. long. She wants to make the largest squares possible. How many squares will she have to make for the tablecloth?

SKILLKEEPER

1. $5\overline{)48}$ **2.** $23\overline{)5.934}$ **3.** $13\overline{)125.19}$ **4.** $8\overline{)60.4}$ **5.** $14\overline{)2.03}$

6. $673 \div 0.01$ **7.** $97.2 \div 0.1$ **8.** $9{,}247 \div 0.001$ **9.** $0.592 \div 0.01$ **10.** $9.12 \div 0.001$

11. $0.05\overline{)28.15}$ **12.** $2.3\overline{)14.697}$ **13.** $54\overline{)116.532}$ **14.** $0.36\overline{)53.28}$ **15.** $5.8\overline{)3.4162}$

More Practice, page 426, Set C

Least Common Multiple (LCM)

Two skaters start at the same point on an oval rink. One skater takes exactly 60 s to go once around the rink. A faster skater takes only 54 s. How many seconds will it be until the skaters are together at the starting point?

We need to find the smallest multiple of both 54 and 60. This number is the **least common multiple (LCM)** of 54 and 60.

We can use prime factorization to find the LCM of 54 and 60.

$$54 = 2 \cdot 3 \cdot 3 \cdot 3$$
$$60 = 2 \cdot 2 \cdot 3 \cdot 5$$
$$\text{LCM} = 2 \cdot 2 \cdot 3 \cdot 3 \cdot 3 \cdot 5 = 540$$

The skaters will be together at the starting point after 540 s, or 9 min.

Other Examples

Find the LCM of 4 and 15.

$$4 = 2 \cdot 2$$
$$15 = 3 \cdot 5$$
$$\text{LCM} = 2 \cdot 2 \cdot 3 \cdot 5 = 60$$

Find the LCM of 6 and 8 mentally.

> Multiples of 8: 8, 16, 24
> 24 is a multiple of 6.
> LCM = 24

Warm Up

Find the LCM of each pair of numbers. Use the prime factorization method.

1. $9 = 3 \cdot 3$
   $12 = 2 \cdot 2 \cdot 3$
2. $8 = 2 \cdot 2 \cdot 2$
   $10 = 2 \cdot 5$
3. $10 = 2 \cdot 5$
   $15 = 3 \cdot 5$
4. $7 = 1 \cdot 7$
   $21 = 3 \cdot 7$

Find the LCM of each pair of numbers mentally.

5. 4, 6
6. 6, 10
7. 4, 5
8. 8, 12

Use the prime factorizations of each pair of numbers to find the LCM.

1. $14 = 2 \cdot 7$
$35 = 5 \cdot 7$

2. $18 = 2 \cdot 3 \cdot 3$
$24 = 2 \cdot 2 \cdot 2 \cdot 3$

3. $30 = 2 \cdot 3 \cdot 5$
$12 = 2 \cdot 2 \cdot 3$

4. $36 = 2 \cdot 3 \cdot 2 \cdot 3$
$42 = 2 \cdot 3 \cdot 7$

5. $16 = 2 \cdot 2 \cdot 2 \cdot 2$
$32 = 2 \cdot 2 \cdot 2 \cdot 2 \cdot 2$

6. $14 = 2 \cdot 7$
$15 = 3 \cdot 5$

7. $15 = 3 \cdot 5$
$20 = 2 \cdot 2 \cdot 5$

8. $75 = 3 \cdot 5 \cdot 5$
$35 = 5 \cdot 7$

Find the LCM of each pair of numbers.

9. 4, 8
10. 10, 12
11. 9, 15
12. 6, 5
13. 8, 20
14. 18, 20
15. 21, 12
16. 6, 14
17. 6, 24
18. 15, 25
19. 24, 42
20. 21, 6
21. 11, 15
22. 20, 24
23. 48, 20
24. 12, 63

Copy and complete the table.

	Numbers	GCF	Product	Product ÷ GCF	LCM
	6, 10	2	6 × 10 = 60	60 ÷ 2 = 30	30
25.	9, 15	▩	?	?	▩
26.	9, 12	▩	?	?	▩
27.	12, 18	▩	?	?	▩
28.	15, 20	▩	?	?	▩
29.	45, 30	▩	?	?	▩

Find the LCM of each pair of numbers.

30. $133{,}518 = 2 \cdot 3 \cdot 7 \cdot 11 \cdot 17^2$
$20{,}349 = 3^2 \cdot 7 \cdot 17 \cdot 19$

31. $8{,}993 = 17 \cdot 23^2$
$154{,}037 = 13 \cdot 17^2 \cdot 41$

THINK MATH

Mental Math

Use the table to find the products.

1. $2^3 \cdot 5^3 \cdot 139$
2. $2^4 \cdot 5^4 \cdot 1{,}103$
3. $2^2 \cdot 1{,}123 \cdot 5^2$
4. $2^5 \cdot 5^6 \cdot 7$
5. $2^6 \cdot 5^7 \cdot 11$
6. $2^6 \cdot 5^6 \cdot 18$

$2 \cdot 5 = 10$
$2^2 \cdot 5^2 = 100$
$2^3 \cdot 5^3 = 1{,}000$
$2^4 \cdot 5^4 = 10{,}000$
$2^5 \cdot 5^5 = 100{,}000$
$2^6 \cdot 5^6 = 1{,}000{,}000$

More Practice, page 426, Set D

Variables and Expressions

Jean is 4 years younger than her brother Dave. We can write an **expression** for Jean's age.

Let d = Dave's age.

$d - 4$ = Jean's age

The letter d is a **variable.** Any letter could be used as the variable.

To **evaluate an expression**, we substitute a number for the variable and perform the operations.

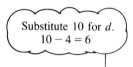

Substitute 10 for d.
$10 - 4 = 6$

Dave's Age	Jean's Age
10	6
13	9
20	16

If d is **10**, then $d - 4$ is **6**.

If d is **13**, then $d - 4$ is **9**.

If d is **20**, then $d - 4$ is **16**.

Study the examples in the table below.

Expression in words	**Variable**	**Expression with a variable**
5 times a number t	t	$5t$ or $5 \cdot t$
A number k divided by 3	k	$\frac{k}{3}$ or $k \div 3$
The sum of b and 7	b	$b + 7$
6 minus a number n	n	$6 - n$

Warm Up

Name the variable in each expression.

1. $n + 7$ **2.** $r - 6$ **3.** $9y$ **4.** $\frac{t}{7}$ **5.** $17z$

Evaluate each expression.

6. $5t$ if t is 7

7. $\frac{k}{3}$ if k is 24

8. $a + 7$ if a is 15

9. $14 - n$ if n is 9

10. $10z$ if z is 8

Evaluate each expression.

1. $n + 11$
   if n is 10

2. $8m$
   if m is 7

3. $100 - s$
   if s is 65

4. $\frac{h}{2}$
   if h is 40

5. $13b$
   if b is 4

Complete each sentence by evaluating the expression.

6. If t is 7, then $9t$ is __?__.
7. If c is 11, then $9c$ is __?__.
8. If x is 19, then $x - 12$ is __?__.
9. If f is 1, then $f + 99$ is __?__.
10. If m is 20, then $\frac{m}{5}$ is __?__.
11. If p is 87, then $\frac{p}{3}$ is __?__.
12. If y is 19, then $4y$ is __?__.
13. If j is 47, then $100 - j$ is __?__.
14. If x is 8, then $\frac{x}{8}$ is __?__.
15. If h is 9, then $16 - h$ is __?__.

Complete each table by evaluating the expressions.

	k	$k + 12$
16.	3	
17.	9	
18.	15	
19.	38	

	z	$4z$
20.	7	
21.	10	
22.	15	
23.	25	

	b	$17 - b$
24.	10	
25.	9	
26.	6	
27.	1	

Write an expression.

28. 4 times a number t
29. 8 less than a number x
30. a number z divided by 5
31. 20 decreased by a number q
32. 9 more than a number n
33. a number k divided by 7
34. the product of 8 and a number d
35. the sum of a number w and 11

THINK MATH

Number Clues

In what year did Neil Armstrong walk on the moon? Substitute the correct numbers for the variables in the expression, and you will find out.

$y + d + w + m$

y = year Columbus discovered America

d = days in a year that is not a leap year

w = weeks in a year

m = minutes in an hour

More Practice, page 427, Set A

Addition and Subtraction Equations

The same number of counters are on each side of the balance scale. How many counters are in the container?

To find the number of counters, we can write and solve an **equation** for the problem.

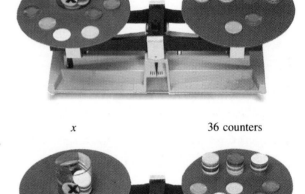

x + 6 counters   42 counters

Let x = the number of counters in the container.

Equation	$x + 6 = 42$
Subtract the same number from each side of the equation.	$x + 6 - 6 = 42 - 6$
Simplify.	$x = 36$
Substitute the number for the variable in the first equation.	$36 + 6 = 42$

x 36 counters

It checks. The solution to the equation is 36.

There are 36 counters in the container.

6   6

To solve some equations, we may need to *add* the same number to each side.

Equation	$y - 8 = 23$
Add 8 to each side of the equation.	$y - 8 + 8 = 23 + 8$
Simplify.	$y = 31$
Substitute 31 for y in the first equation.	$31 - 8 = 23$

Warm Up

Solve each equation.

1. $x + 8 = 23$
2. $t + 11 = 30$
3. $k + 9 = 27$
4. $n + 15 = 40$
5. $s - 7 = 16$
6. $q - 10 = 25$
7. $z - 6 = 14$
8. $m - 16 = 16$

Solve each equation.

1. $x + 9 = 16$
2. $n + 12 = 20$
3. $y + 6 = 21$
4. $a + 13 = 30$
5. $w + 7 = 15$
6. $m + 23 = 50$
7. $k + 14 = 26$
8. $g + 17 = 25$
9. $h + 8 = 33$
10. $q + 34 = 60$
11. $z + 58 = 74$
12. $s + 284 = 419$

Solve each equation.

13. $t - 4 = 11$
14. $r - 15 = 6$
15. $z - 8 = 14$
16. $j - 9 = 26$
17. $p - 21 = 15$
18. $f - 12 = 19$
19. $n - 10 = 13$
20. $x - 17 = 17$
21. $k - 6 = 0$
22. $s - 18 = 16$
23. $v - 15 = 38$
24. $z - 26 = 35$

Solve each equation. You may have to add or subtract.

25. $m - 7 = 18$
26. $w + 4 = 14$
27. $b - 12 = 8$
28. $x + 1 = 50$
29. $h - 9 = 27$
30. $j + 11 = 20$
31. $c - 40 = 60$
32. $g + 27 = 33$
33. $n - 41 = 39$
34. $x - 25 = 75$
35. $n - 33 = 167$
36. $g + 19 = 99$

Solve each equation.

37. $m + 19{,}384 = 21{,}928$
38. $k - 26{,}009 = 84{,}356$
39. $w + 6{,}587 = 12{,}063$

THINK MATH

Logical Reasoning

A quarter and a dime balance with three dimes. How many dimes would it take to balance four quarters?

More Practice, page 427, Set B

Multiplication and Division Equations

Four containers each have the same number of counters. The containers balance with 48 counters. How many counters are in each container?

$4p$ counters   48 counters

To find the number of counters, we can write and solve an equation.
Let p = the number of counters in one container.

Equation	$4p = 48$
Divide each side of the equation by the same number.	$\frac{4}{4}p = \frac{48}{4}$
Simplify.	$p = 12$
Substitute the number for the variable in the first equation.	$4 \cdot 12 = 48$

It checks. The solution to the equation is 12.

p 12 counters

There are 12 counters in each container.

To solve some equations, we may need to *multiply* each side of the equation by the same number.

$\frac{n}{3}$ means n divided by 3.	$\frac{n}{3} = 8$
Multiply each side of the equation by 3.	$\frac{n}{3} \cdot 3 = 8 \cdot 3$
Simplify.	$n = 24$
Substitute 24 for n in the first equation. It checks.	$\frac{24}{3} = 8$

Warm Up

Solve each equation.

1. $7p = 56$
2. $8y = 24$
3. $10k = 70$
4. $9z = 54$
5. $\frac{p}{5} = 30$
6. $\frac{n}{7} = 42$
7. $\frac{t}{9} = 9$
8. $\frac{b}{12} = 6$

Solve each equation.

1. $3c = 27$
2. $9n = 72$
3. $5k = 50$
4. $6m = 48$
5. $12b = 12$
6. $15t = 45$
7. $7d = 0$
8. $4x = 100$
9. $19w = 19$
10. $14k = 42$
11. $17z = 170$
12. $37h = 148$

Solve each equation.

13. $\frac{h}{4} = 5$
14. $\frac{j}{5} = 7$
15. $\frac{z}{10} = 4$
16. $\frac{x}{11} = 3$
17. $\frac{p}{9} = 6$
18. $\frac{q}{7} = 7$
19. $\frac{s}{6} = 0$
20. $\frac{n}{2} = 13$
21. $\frac{t}{5} = 12$
22. $\frac{m}{9} = 16$
23. $\frac{b}{7} = 23$
24. $\frac{g}{18} = 3$

Solve each equation. You may have to multiply or divide.

25. $4r = 32$
26. $\frac{h}{6} = 5$
27. $\frac{g}{9} = 2$
28. $\frac{n}{15} = 1$
29. $7p = 84$
30. $\frac{z}{4} = 25$
31. $12m = 48$
32. $3r = 51$
33. $65x = 3{,}640$
34. $18n = 36$
35. $\frac{x}{7} = 14$
36. $9r = 45$

Solve each equation.

37. $27{,}500\,x = 110{,}000$
38. $\frac{m}{19{,}392} = 671$
39. $55{,}655\,n = 4{,}841{,}985$

SKILLKEEPER

List all the factors of each number.

1. 10
2. 35
3. 17
4. 60
5. 23

Write P (prime) or C (composite) for each number.

6. 43
7. 59
8. 27
9. 19
10. 51

Write the prime factorization of each number.

11. 54
12. 16
13. 48
14. 30
15. 21

More Practice, page 427, Set C

PROBLEM SOLVING: Choosing the Right Equation

QUESTION
DATA
PLAN
ANSWER
CHECK

A case packer puts 24 cans of vegetables in a box and seals the box. How many boxes will be needed for 1,800 cans of vegetables?

An equation can be used to solve the problem. You need to choose an equation that uses the necessary data and the right operations. Which equation below can be used?

Let b = the number of boxes.

$b + 24 = 1,800$ $\qquad$ $1,800\,b = 24$

$24\,b = 1,800$ $\qquad$ $\frac{b}{24} = 1,800$

Since there are 24 cans in each box, 24 times the number of boxes needed must equal 1,800.

The correct equation is $24\,b = 1,800$.

$24\,b = 1,800$

$\dfrac{24\,b}{24} = \dfrac{1,800}{24}$ $\qquad$ Divide each side by 24.

$b = 75$ $\qquad$ Simplify.

$24 \times 75 = 1,800$ $\qquad$ It checks.

The case packer will need 75 boxes.

Choose the equation that will solve the problem. Use the equation to find the answer.

1. The sum of a certain number and 15 is 63. What is the number?
   Let n = the number.

   A $n + 15 = 63$ $\qquad$ **B** $15\,n = 63$

   C $n + 63 = 15$ $\qquad$ **D** $n - 15 = 63$

2. When a certain number is multiplied by 21, the product is 315. What is the number?
   Let t = the number.

   A $315\,t = 21$ $\qquad$ **B** $21\,t = 315$

   C $t + 21 = 315$ $\qquad$ **D** $315 - t = 21$

3. When a certain number is divided by 16, the quotient is 34. What is the number?
   Let d = the number.
   A $16d = 34$
   B $16 + d = 34$
   C $\frac{d}{16} = 34$
   D $d - 16 = 34$

4. If 35 is subtracted from a certain number, the difference is 77. What is the number?
   Let b = the number.
   A $77 + b = 35$
   B $35 - b = 77$
   C $35b = 77$
   D $b - 35 = 77$

5. Curtis bought a 69¢ ballpoint pen and some paper. He spent $1.48 altogether. What was the cost of the paper?
   Let p = the cost of the paper in cents.
   A $p + 69 = 148$
   B $69p = 148$
   C $p - 69 = 148$
   D $\frac{p}{69} = 148$

6. Lisa earns $8 an hour. If Lisa earns $312, how many hours must she work?
   Let h = the number of hours worked.
   A $312 - h = 8$
   B $8h = 312$
   C $h = 8 \times 312$
   D $h + 8 = 312$

7. Leonard has collected 792 stamps. This is 6 times as many stamps as Diane has in her collection. How many stamps does Diane have in her collection?
   Let s = the number of stamps Diane has.
   A $792 - s = 6$
   B $s = 6 \times 792$
   C $6s = 792$
   D $s + 6 = 792$

8. Mayeta buys a car that costs $6,779. She pays $2,150 down on the car. How much more does she owe?
   Let n = the amount owed.
   A $n = 6,779 + 2,150$
   B $n + 2,150 = 6,779$
   C $2,150n = 6,779$
   D $n - 2,150 = 6,779$

9. A certain number of eggs are packed in cartons of 12 each. All together there are 52 cartons of eggs. How many eggs are packed?
   Let e = the number of eggs.
   A $12e = 52$
   B $52 - e = 12$
   C $e + 12 = 52$
   D $\frac{e}{12} = 52$

10. Joan buys 3 cans of fruit for $2.07. What is the cost of each can?
    Let c = the cost of one can of fruit in cents.
    A $c + 3 = 207$
    B $3c = 207$
    C $207 - c = 3$
    D $c = 3 \times 207$

11. A company paid $12 (1200¢) in postage to mail letters. Each letter cost 20¢ to mail. How many letters were mailed?
    Let m = the number of letters.
    A $1,200m = 20$
    B $\frac{m}{20} = 1,200$
    C $1,200 + m = 20$
    D $20m = 1,200$

12. **Try This** Mountain View Swimming Pool uses chlorine that costs 40.9¢ per liter. Fairfield City Pool uses chlorine costing 34.9¢ per liter. The cities bought the same amount of chlorine, but Mountain View paid $120 more than Fairfield. How many liters did each city buy? Hint: Solve a simpler problem first.

PROBLEM SOLVING: Writing and Solving an Equation

| QUESTION |
| DATA |
| PLAN |
| ANSWER |
| CHECK |

James Fulton is a ranger in Mt. McKinley National Park. He knows that Mt. McKinley has a height of about 6,195 m. The height of Mt. McKinley is about 0.7 of the height of Mt. Everest. What is the height of Mt. Everest?

To solve this problem, you can write and solve an equation.

Choose a variable to represent the number you must find.	Let h = the height of Mt. Everest.	
Write an equation.	$0.7h = 6,195$	0.7 of the height of Mt. Everest equals the height of Mt. McKinley.
Solve the equation.	$\dfrac{0.7}{0.7}h = \dfrac{6,195}{0.7}$ $h = 8,850$ $0.7 \times 8,850 = 6,195.0$	Divide both sides by 0.7 and simplify. It checks.

Mt. Everest has a height of about 8,850 m.

Write and solve an equation for each problem.

1. Atlanta, Georgia is 336 m above sea level. This elevation is 0.21 of the elevation of Denver, Colorado. What is the elevation of Denver?
   Let d = the elevation of Denver.

2. If 339 m is subtracted from the elevation of Albuquerque, New Mexico, the result is the elevation of Salt Lake City. The elevation of Salt Lake City is 1,286 m. What is the elevation of Albuquerque?
   Let h = the elevation of Albuquerque.

3. Mt. McKinley is 6,195 m in height. Mt. Logan in Yukon Territory is the highest mountain in Canada. If 144 m is added to the height of Mt. Logan, the sum is the height of Mt. McKinley. What is the height of Mt. Logan?

4. Gannett Peak, with a height of 4,207 m, is the highest point in Wyoming. This height is 7 times the height of Mt. Curwood, the highest point in Michigan. What is the height of Mt. Curwood?

5. The highest point in Mississippi is Mt. Woodall with a height of 245 m. The highest point in Pennsylvania is Mt. Davis. If the height of Mt. Davis is divided by 4, the quotient is the height of Mt. Woodall. What is the height of Mt. Davis?

6. When 586 m is subtracted from the height of Wheeler Peak in New Mexico, the result is the height of Mt. Hood in Oregon. Mt. Hood is 3,425 m in height. What is the height of Wheeler Peak?

7. The highest point in California is Mt. Whitney with a height of about 4,410 m. This height is 42 times the height of the highest point in Florida. What is the height of the highest point in Florida?

8. The highest point in Colorado is Mt. Ebert and the lowest point is 1,021 m. This is 3,379 m less than the height of Mt. Ebert. What is the height of Mt. Ebert?

9. **DATA HUNT** What is the highest point in your state or province and a nearby state or province? Make up a problem about the data. Write and solve an equation for your problem.

10. **Try This** The highest point in South Dakota is Harney Peak. If 69 m is subtracted from its height and the difference is divided by 2, the quotient is 1,069 m. This is the height of White Butte, the highest point in North Dakota. What is the height of Harney Peak? Hint: Work backward.

PROBLEM SOLVING: Make a Table

QUESTION
DATA
PLAN
ANSWER
CHECK

Try This

In 1982 Eileen and Karen started new jobs. Eileen got $10,000 for the first year with a $3,000 raise each year after that. Karen got a starting salary of $16,000 a year with a $2,000 raise each year after that. In what year will Eileen and Karen be earning the same salary?

To solve some problems, you may need to use the same operation many times until you get the answer. To organize your work, it will help if you use the strategy **Make a Table**.

First make a table that shows the data in the problem.

Year	1982	1983	1984
Eileen	$10,000	$13,000	
Karen	$16,000	$18,000	

Now fill in the table for other years until you find the year when the salaries are the same.

Year	1982	1983	1984	1985	1986	1987	1988
Eileen	$10,000	$13,000	$16,000	$19,000	$22,000	$25,000	$28,000
Karen	$16,000	$18,000	$20,000	$22,000	$24,000	$26,000	$28,000

The salaries are the same in 1988.

Solve.

1. In 1980 Stella and Bob each started a job that paid $10,000 a year. Each year Stella got a raise of $5,000 and Bob got a raise of $2,000. In what year will Stella be earning twice as much as Bob?

2. In 1975 Johnny started working at a job that paid $8,500 the first year with a $1,000 raise each year. In the same year, Pearl started a job that paid $6,000 a year with a raise of $1,800 each year. How much more was Pearl earning in 1985 than Johnny?

CHAPTER REVIEW/TEST

List all the factors of each number.

1. 10
2. 23
3. 15
4. 24

5. Which number is divisible by 9?

   114   2,718   7,024

6. Which number is divisible by 3?

   233   87   173

7. Which number is a prime number?

   15, 27, 43, 57, or 81

8. Which number is a composite number?

   7, 17, 27, 37, or 47

Find the prime factorization of each number.

9. 20
10. 75
11. 42

Find the greatest common factor (GCF) of each pair of numbers.

12. 20, 45
13. 12, 36
14. 18, 45

Find the least common multiple (LCM) of each pair of numbers.

15. 6, 8
16. 8, 12
17. 10, 15

Complete each table by evaluating the expressions.

	n	$4n$
18.	3	
19.	5	
20.	10	

	x	$\frac{x}{5}$
21.	20	
22.	45	
23.	60	

	y	$y+9$
24.	17	
25.	4	
26.	21	

Solve each equation.

27. $x - 4 = 16$
28. $k + 12 = 23$
29. $7n = 70$
30. $\frac{b}{7} = 3$

31. Write the equation for the problem. Solve the equation.

    When 9 is subtracted from a certain number, the difference is 12. What is the number? Let $x =$ the number.

32. Write and solve an equation for the problem.

    Twelve times a certain number is 132. What is the number? Let $k =$ the number.

ANOTHER LOOK

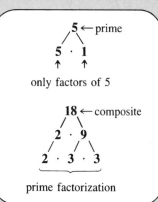

only factors of 5

prime factorization

$12 = 2 \cdot (2 \cdot 3)$
$30 = (2 \cdot 3) \cdot 5$
GCF = $2 \cdot 3 = 6$

$12 = 2 \cdot 2 \cdot 3$
$30 = 2 \cdot 3 \cdot 5$
LCM = $2 \cdot 2 \cdot 3 \cdot 5 = 60$

List all the factors of each number.

1. 12
2. 40
3. 35

Write (P) prime or (C) composite for each number.

4. 21
5. 23
6. 99

Find the prime factorization of each number.

7. 30
8. 28
9. 54

Find the greatest common factor (GCF) of each pair of numbers.

10. 8, 12
11. 9, 27
12. 30, 45

Find the least common multiple (LCM) of each pair of numbers.

13. 5, 6
14. 8, 24
15. 20, 30

variable 3 times y

y	$3y$
2	6
5	15
12	36

Complete each table by evaluating the expression.

	n	$8n$
16.	4	
17.	7	
18.	10	

	x	$x - 3$
19.	30	
20.	45	
21.	120	

$k - 7 = 8$
$k - 7 + 7 = 8 + 7$ Add 7 to both sides.
$k = 15$

$4y = 24$
$\dfrac{4y}{4} = \dfrac{24}{4}$ Divide each side by 4.
$y = 6$

Solve each equation.

22. $x + 19 = 35$
23. $7n = 56$
24. $y - 13 = 9$
25. $\dfrac{z}{6} = 13$
26. $p + 75 = 125$
27. $r - 17 = 29$
28. $5b = 250$
29. $\dfrac{s}{8} = 32$

ENRICHMENT

Number Relationships

Finding the greatest common factor of two large numbers can be difficult. The direction boxes below give a method that will make the problem easier. You may want to use a calculator.

What is the greatest common factor of 11,039 and 3,458?

> Divide the larger number by the smaller number. ➤ Divide the first divisor by the first remainder. ➤ Divide the second divisor by the second remainder. ➤ Continue the process until the remainder is 0. The last divisor is the GCF.

$$3{,}458 \overline{)11{,}039} \quad \begin{array}{r} 3 \\ \underline{10374} \\ 665 \end{array}$$

$$665 \overline{)3{,}458} \quad \begin{array}{r} 5 \\ \underline{3325} \\ 133 \end{array}$$

$$133 \overline{)665} \quad \begin{array}{r} 5 \\ \underline{665} \\ 0 \end{array}$$

The GCF of 11,039 and 3,458 is 133.

Check:

11,039 = 133 × 83
3,458 = 133 × 26

The GCF of 83 and 26 is 1, therefore 133 is the GCF of 11,039 and 3,458.

Find the GCF of each pair of numbers.

1. 1,001 and 4,466
2. 2,813 and 5,771
3. 642 and 25,787
4. 5,585 and 65,903
5. 1,541 and 67,603
6. 531,389 and 13,606,507
7. Choose two large numbers and find their greatest common factor.

TECHNOLOGY

Using INPUT in a Computer Program

A computer program can be written so that the computer can ask questions and receive data. This data is called INPUT.

The INPUT may be numbers that are used to solve problems.

A pollster collects data that will be used with a computer to prepare a public opinion survey.

Program

```
10 REM DIFFERENCE OF AGES
20 PRINT "GIVE YOUR MOTHER'S
   AGE."

30 INPUT A
40 PRINT "WHAT IS YOUR AGE?"

50 INPUT B

60 C = A - B
70 PRINT "YOUR MOTHER IS ";C
80 PRINT "YEARS OLDER THAN YOU."
90 END
```

Comments

REM (Remark) statements are ignored by the computer. Here, REM names the program.

INPUT is the number you choose.

Another INPUT number

C is the difference using INPUT numbers.

Here is an example RUN for the program.

```
RUN
GIVE YOUR MOTHER'S AGE.
?41
WHAT IS YOUR AGE?
?13
YOUR MOTHER IS 28
YEARS OLDER THAN YOU.
```

What would a RUN of this program be using your mother's age and your age?

Write a RUN for each program. Use the given INPUT numbers.

1. ```
   10 REM MULTIPLYING NUMBERS
   20 PRINT "CHOOSE A NUMBER."
   30 INPUT A
   40 PRINT "CHOOSE ANOTHER
      NUMBER."
   50 INPUT B
   60 C = A * B
   70 PRINT "THE PRODUCT IS ";C
   80 END
   ```
   INPUT: A = 27, B = 35

2. ```
   10 REM AVERAGE
   20 PRINT "CHOOSE 3 NUMBERS."
   30 INPUT A
   40 INPUT B
   50 INPUT C
   60 D = (A + B + C)/3
   70 PRINT "AVERAGE = ";D
   80 END
   ```
   INPUT: A = 3, B = 6, C = 21

3. ```
   10 REM SQUARES AND CUBES
   20 PRINT "CHOOSE A NUMBER."
   30 INPUT A
   40 B = A * A
   50 C = A * A * A
   60 PRINT A;" SQUARED = ";B
   70 PRINT A;" CUBED = ";C
   80 END
   ```
   INPUT: A = 5

4. ```
   10 REM YOUR AGE
   20 PRINT "WHAT IS THIS YEAR?"
   30 INPUT Y
   40 PRINT "WHAT IS YOUR
      BIRTH YEAR?"
   50 INPUT B
   60 A = Y - B
   70 PRINT "YOU ARE ";A
   80 END
   ```
   INPUT: Y = This year
   B = Your birth year

5. ```
   10 REM ESTIMATING
   20 PRINT "HOW MANY SECONDS
      DOES"
   30 PRINT "IT TAKE TO WALK
      100 METERS?"
   40 INPUT S
   50 T = 10 * S/60
   60 PRINT "YOU CAN WALK 1 KM"
   70 PRINT "IN ";T;" MINUTES."
   80 END
   ```
   INPUT: S = 54

6. Write a RUN for the program in exercise 5 using your own INPUT number.

7. Write a computer program that will find the sum of three INPUT numbers. Write a RUN of the program using your own INPUT.
   ```

CUMULATIVE REVIEW

Find the products.

1. 41.37 × 12
   - A 496.44
   - B 49.674
   - C 49.644
   - D not given

2. 742 × 0.18
   - A 135
   - B 134.37
   - C 133.56
   - D not given

3. 0.004 × 0.6
   - A 0.00024
   - B 2.40
   - C 0.024
   - D not given

4. Choose the best estimate.
   6.82 × 2.05
   - A 14
   - B 12
   - C 18
   - D not given

5. Multiply.
   0.7194 × 1,000
   - A 71.94
   - B 7,194
   - C 719.4
   - D not given

6. What is the standard numeral for 3.9×10^5?
   - A 3,900,000
   - B 39,000
   - C 390,000
   - D not given

7. What kind of angle is ∠ABC?

   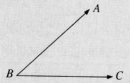

   - A acute
   - B right
   - C straight
   - D not given

8. What is the measure of the third angle?

   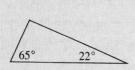

   - A 90°
   - B 87°
   - C 93°
   - D not given

9. What is the name of polygon ABCDE?

   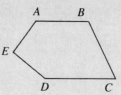

   - A hexagon
   - B quadrilateral
   - C pentagon
   - D not given

10. What is the name of the chord which passes through the center point of a circle?
    - A diameter
    - B arc
    - C radius
    - D not given

11. What is the symbol for ray AB?

    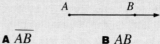

    - A $\overline{AB}$
    - B AB
    - C $\overrightarrow{AB}$
    - D not given

12. What is the measure of any right angle?
    - A 360°
    - B 90°
    - C 180°
    - D not given

13. How many sides does a hexagon have?
    - A 9
    - B 11
    - C 8
    - D not given

14. On a map, 1 cm represents 45.7 km. What distance would 7.9 cm represent?
    - A 360 km
    - B 361.03 km
    - C 36.103 km
    - D not given

8
Addition and Subtraction of Fractions

Speeding a patient to the emergency room of a hospital is no longer the only job of emergency medical service. A city of 50,000 people may have about 2,600 emergency medical calls in one year. About $\frac{1}{4}$ of the emergency calls will need a paramedic. The paramedic, in radio contact with a doctor or nurse, identifies the problem and starts treatment before the patient is taken to a hospital. This might mean giving heart massage or oxygen, splinting bones, or delivering a baby.

The training to become a paramedic has three parts. The first $\frac{1}{3}$ is classroom learning, followed by several months of training in a hospital. The last part is the field internship during which the student works with a team member in a paramedic van. This means working 24-hour shifts and always being ready for an emergency.

Fractions

We use fractions to describe a part of a region or a part of a set.

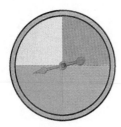 

1 of **4** equal parts of the spinner is blue.

One fourth of the spinner is blue.

$\dfrac{1}{4}$ ← numerator
← denominator

4 of the **7** markers are yellow.

Four sevenths of the markers are yellow.

$\dfrac{4}{7}$ ← numerator
← denominator

A fraction can also be related to division.

1 whole unit is divided into **3** equal parts.

Each part is $\dfrac{1}{3}$ of **1** whole unit.

2 whole units can be divided into **3** equal parts.

Each part is $\dfrac{2}{3}$ of **1** whole unit.

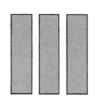
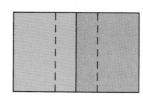
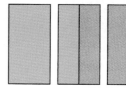

$1 \div 3$ $\quad 1 \div 3 = \dfrac{1}{3}$ $\quad 2 \div 3$ $\quad 2 \div 3 = \dfrac{2}{3}$

Warm Up

Name the fraction for the part of the region or set that is shaded.

1.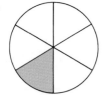
2.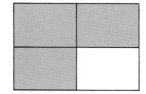
3.

Give a fraction for each division exercise.

4. $1 \div 4$ **5.** $2 \div 9$ **6.** $1 \div 10$

Write a fraction for the part of the region or set that is shaded.

1.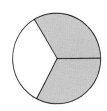
2.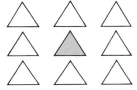
3.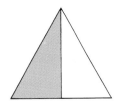
4.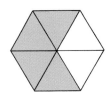
5.
6.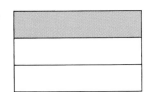
7.
8.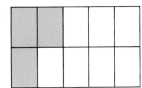
9.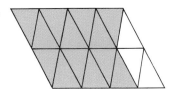

Write a fraction for each division exercise.

10. $2 \div 5$
11. 5 divided by 6
12. $9 \div 7$
13. 7 divided by 100
14. $8 \div 25$
15. $10 \div 16$

Write a division exercise for each fraction.

16. $\frac{5}{9}$
17. $\frac{11}{12}$
18. $\frac{5}{4}$
19. $\frac{23}{100}$
20. $\frac{7}{10}$
21. $\frac{3}{2}$

22. A game spinner is divided into 6 equal sections. What fractional part of the spinner is each section?

23. There are 5 players. What fractional part of 8 game markers are left if each player picks 1 marker?

24. Dana and Phil played 8 games and made a graph of the games they won. What fractional part of the games did Dana win?

Games Won

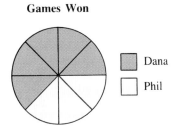

195

Equivalent Fractions

Gordon is trying to make one face of the puzzle cube all orange. What part of the face is orange?

6 out of 9 squares are orange.

$\frac{6}{9}$ of the top face is orange.

2 out of 3 rows of squares are orange.

$\frac{2}{3}$ of the top face is orange.

Two fractions which name the same part of a region or the same part of a set are **equivalent fractions**.

$\frac{2}{3}$ and $\frac{6}{9}$ are equivalent fractions. $\frac{2}{3} = \frac{6}{9}$

Equivalent fractions can be formed by multiplying the numerator and denominator by the same number that is not zero.

$\frac{2}{3}$ $\frac{4}{6}$ $\frac{6}{9}$ $\frac{8}{12}$

$\left(\frac{2 \times 2}{3 \times 2}\right)$ $\left(\frac{2 \times 3}{3 \times 3}\right)$ $\left(\frac{2 \times 4}{3 \times 4}\right)$

Warm Up

Give the missing numerator or denominator.

1. $\frac{5 \times 3}{8 \times 3} = \frac{\text{\rule{0.5cm}{0.15mm}}}{24}$

2. $\frac{2 \times 4}{3 \times 4} = \frac{8}{\text{\rule{0.5cm}{0.15mm}}}$

3. $\frac{3 \times 10}{5 \times 10} = \frac{\text{\rule{0.5cm}{0.15mm}}}{50}$

Give the next two equivalent fractions.

4. $\frac{1}{2}, \frac{2}{4}, \frac{\text{\rule{0.3cm}{0.15mm}}}{\text{\rule{0.3cm}{0.15mm}}}, \frac{\text{\rule{0.3cm}{0.15mm}}}{\text{\rule{0.3cm}{0.15mm}}}$

5. $\frac{3}{4}, \frac{6}{8}, \frac{\text{\rule{0.3cm}{0.15mm}}}{\text{\rule{0.3cm}{0.15mm}}}, \frac{\text{\rule{0.3cm}{0.15mm}}}{\text{\rule{0.3cm}{0.15mm}}}$

6. $\frac{1}{3}, \frac{2}{6}, \frac{\text{\rule{0.3cm}{0.15mm}}}{\text{\rule{0.3cm}{0.15mm}}}, \frac{\text{\rule{0.3cm}{0.15mm}}}{\text{\rule{0.3cm}{0.15mm}}}$

Write the missing numerator or denominator.

1. $\dfrac{2\times 4}{3\times 4} = \dfrac{\text{\textemdash}}{12}$
2. $\dfrac{5\times 5}{6\times 5} = \dfrac{25}{\text{\textemdash}}$
3. $\dfrac{3\times 6}{8\times 6} = \dfrac{\text{\textemdash}}{48}$
4. $\dfrac{1\times 9}{2\times 9} = \dfrac{9}{\text{\textemdash}}$
5. $\dfrac{4\times 6}{7\times 6} = \dfrac{\text{\textemdash}}{42}$

6. $\dfrac{3}{4} = \dfrac{\text{\textemdash}}{16}$
7. $\dfrac{3}{5} = \dfrac{21}{\text{\textemdash}}$
8. $\dfrac{3}{10} = \dfrac{\text{\textemdash}}{30}$
9. $\dfrac{1}{4} = \dfrac{\text{\textemdash}}{16}$
10. $\dfrac{2}{9} = \dfrac{\text{\textemdash}}{45}$

11. $\dfrac{5}{12} = \dfrac{\text{\textemdash}}{60}$
12. $\dfrac{8}{9} = \dfrac{16}{\text{\textemdash}}$
13. $\dfrac{4}{15} = \dfrac{\text{\textemdash}}{45}$
14. $\dfrac{2}{7} = \dfrac{\text{\textemdash}}{70}$
15. $\dfrac{7}{12} = \dfrac{21}{\text{\textemdash}}$

16. $\dfrac{3}{4} = \dfrac{\text{\textemdash}}{32}$
17. $\dfrac{1}{3} = \dfrac{\text{\textemdash}}{48}$
18. $\dfrac{3}{16} = \dfrac{27}{\text{\textemdash}}$
19. $\dfrac{5}{8} = \dfrac{\text{\textemdash}}{64}$
20. $\dfrac{19}{20} = \dfrac{\text{\textemdash}}{100}$

Write the next three equivalent fractions.

21. $\dfrac{3}{8}, \dfrac{6}{16}, \dfrac{9}{24}, \dfrac{\text{\textemdash}}{\text{\textemdash}}, \dfrac{\text{\textemdash}}{\text{\textemdash}}, \dfrac{\text{\textemdash}}{\text{\textemdash}}$
22. $\dfrac{1}{10}, \dfrac{2}{20}, \dfrac{3}{30}, \dfrac{\text{\textemdash}}{\text{\textemdash}}, \dfrac{\text{\textemdash}}{\text{\textemdash}}, \dfrac{\text{\textemdash}}{\text{\textemdash}}$
23. $\dfrac{5}{6}, \dfrac{10}{12}, \dfrac{15}{18}, \dfrac{\text{\textemdash}}{\text{\textemdash}}, \dfrac{\text{\textemdash}}{\text{\textemdash}}, \dfrac{\text{\textemdash}}{\text{\textemdash}}$

Write an equivalent fraction with a denominator of 24 for each fraction.

24. $\dfrac{1}{2}$
25. $\dfrac{3}{8}$
26. $\dfrac{3}{4}$
27. $\dfrac{5}{6}$
28. $\dfrac{11}{12}$

Write an equivalent fraction with a denominator of 100 for each fraction.

29. $\dfrac{3}{5}$
30. $\dfrac{9}{10}$
31. $\dfrac{7}{20}$
32. $\dfrac{3}{25}$
33. $\dfrac{3}{2}$

34. Keiko made 3 of the 9 small squares on one face of the cube yellow. How many ninths of the face is yellow?

35. One sixth of the small squares on the puzzle cube are orange. There are 54 small squares in all. How many 54ths of the small squares are orange?

THINK MATH

Fraction Concepts

What part of the large square is shaded?

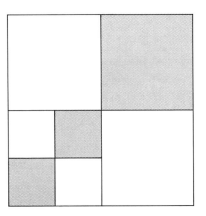

More Practice, page 428, Set A

Lowest-Terms Fractions

Renee skied a cross-country course in 24 min, or $\frac{24}{60}$ of an hour. What is the **lowest-terms fraction** for $\frac{24}{60}$?

A fraction is in lowest terms if the greatest common factor (GCF) of the numerator and denominator is 1.

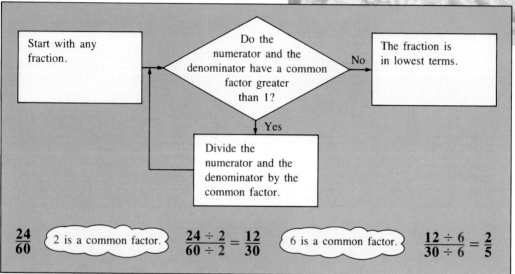

The lowest-terms fraction for $\frac{24}{60}$ is $\frac{2}{5}$.

Dividing by the GCF can save steps.

$$\frac{24}{60} = \frac{24 \div 12}{60 \div 12} = \frac{2}{5}$$

Other Examples

$\frac{5}{15} = \frac{5 \div 5}{15 \div 5} = \frac{1}{3}$ $\qquad$ $\frac{18}{27} = \frac{18 \div 9}{27 \div 9} = \frac{2}{3}$ $\qquad$ $\frac{11}{12}$ $\quad$ GCF = 1 $\quad$ $\frac{11}{12}$ is in lowest terms.

Warm Up

Give each fraction in lowest terms.

1. $\frac{6}{8}$
2. $\frac{12}{18}$
3. $\frac{35}{50}$
4. $\frac{24}{28}$
5. $\frac{20}{100}$
6. $\frac{8}{9}$
7. $\frac{12}{16}$
8. $\frac{56}{64}$

Write each fraction in lowest terms.

1. $\frac{5}{10}$
2. $\frac{8}{12}$
3. $\frac{3}{9}$
4. $\frac{3}{12}$
5. $\frac{10}{24}$

6. $\frac{8}{14}$
7. $\frac{6}{15}$
8. $\frac{25}{60}$
9. $\frac{6}{18}$
10. $\frac{18}{36}$

11. $\frac{6}{24}$
12. $\frac{16}{30}$
13. $\frac{32}{36}$
14. $\frac{40}{60}$
15. $\frac{75}{100}$

16. $\frac{15}{30}$
17. $\frac{250}{1,000}$
18. $\frac{2}{24}$
19. $\frac{30}{100}$
20. $\frac{8}{10}$

21. $\frac{750}{1,000}$
22. $\frac{48}{100}$
23. $\frac{20}{24}$
24. $\frac{560}{1,000}$
25. $\frac{15}{48}$

26. $\frac{27}{81}$
27. $\frac{42}{48}$
28. $\frac{120}{150}$
29. $\frac{55}{75}$
30. $\frac{48}{96}$

31. Marj skied 200 yards (yd) of a 400-yard course. What fraction of the course did she ski?

32. Brent finished a race in 56 s. What fraction of one minute is this?

33. A ski area was open for skiing 13 weeks one year. About what fractional part of the year was this?

34. **DATA BANK** What was the time for the winner of the 500-meter speed skating competition in the 1964 Winter Olympics? What was the time as a fraction of a minute? (See page 415.)

SKILLKEEPER

Find the GCF of each pair of numbers.

1. 35, 18
2. 24, 32
3. 20, 30
4. 30, 45
5. 12, 18
6. 80, 200
7. 51, 39
8. 56, 38
9. 48, 80
10. 8, 27

Find the LCM of each pair of numbers.

11. 7, 10
12. 16, 24
13. 25, 10
14. 36, 48
15. 8, 12
16. 30, 45
17. 40, 60
18. 49, 21
19. 8, 6
20. 12, 15

More Practice, page 428, Set B

Improper Fractions and Mixed Numbers

Judy keeps a record of the food she feeds each animal in a pet store. She uses 5 cans of dog food to feed 3 puppies.

Since the amount is 5 ÷ 3, she can write a fraction. $\frac{5}{3}$ is an **improper fraction**.

We can use the idea that $\frac{5}{3}$ means 5 ÷ 3 to write $\frac{5}{3}$ as a **mixed number**.

| Divide the numerator by the denominator. | Write the quotient as the whole number part. | Write the remainder over the divisor as the fraction part. |

$\frac{5}{3} \rightarrow 3\overline{)5}$ ← Whole number
$\phantom{\frac{5}{3} \rightarrow 3)}\underline{3}$
$\phantom{\frac{5}{3} \rightarrow 3)}2$ ← Number of thirds

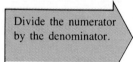 $1\frac{2}{3}$

Mixed numbers like $3\frac{1}{4}$ can be written as improper fractions.

| Multiply the whole number by the denominator. | Add the numerator to the product. | Write the sum over the denominator. |

$3\frac{1}{4}$ $4 \times 3 = 12$ $12 + 1 = 13$ $\frac{13}{4}$

Other Examples

$\frac{24}{8} = 3$ $\frac{12}{12} = 1$ $12\frac{1}{2} = \frac{25}{2}$ (reduce) $\frac{34}{6} = 5\frac{4}{6} = 5\frac{2}{3}$

Warm Up

Write each improper fraction as a mixed number or a whole number.

1. $\frac{11}{3}$ 2. $\frac{16}{4}$ 3. $\frac{23}{5}$ 4. $\frac{22}{8}$ 5. $\frac{40}{12}$

Write each mixed number as an improper fraction.

6. $3\frac{5}{6}$ 7. $2\frac{1}{4}$ 8. $5\frac{2}{3}$ 9. $7\frac{5}{8}$ 10. $2\frac{3}{16}$

Write each improper fraction as a mixed number or a whole number.

1. $\frac{7}{4}$
2. $\frac{9}{8}$
3. $\frac{13}{5}$
4. $\frac{14}{6}$
5. $\frac{6}{2}$

6. $\frac{15}{8}$
7. $\frac{21}{3}$
8. $\frac{18}{9}$
9. $\frac{24}{5}$
10. $\frac{33}{12}$

11. $\frac{60}{15}$
12. $\frac{17}{4}$
13. $\frac{100}{100}$
14. $\frac{56}{8}$
15. $\frac{27}{3}$

16. $\frac{80}{10}$
17. $\frac{64}{10}$
18. $\frac{15}{4}$
19. $\frac{16}{12}$
20. $\frac{30}{15}$

Write each mixed number as an improper fraction.

21. $2\frac{3}{4}$
22. $7\frac{1}{3}$
23. $9\frac{5}{8}$
24. $10\frac{1}{8}$
25. $6\frac{4}{5}$

26. $7\frac{3}{8}$
27. $6\frac{5}{24}$
28. $5\frac{7}{12}$
29. $4\frac{3}{10}$
30. $1\frac{1}{8}$

31. $12\frac{2}{3}$
32. $3\frac{7}{100}$
33. $2\frac{27}{100}$
34. $3\frac{15}{16}$
35. $6\frac{7}{12}$

36. $15\frac{23}{60}$
37. $15\frac{3}{10}$
38. $10\frac{2}{3}$
39. $7\frac{7}{24}$
40. $13\frac{2}{3}$

41. Judy used $3\frac{1}{2}$ bags of dog food. Write an improper fraction for this mixed number.

42. How many cans of food were fed to each hamster? Write the answer as a mixed number.

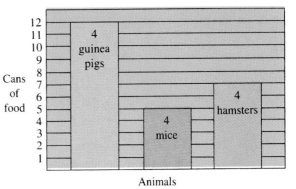

Animals

Write each mixed number as an improper fraction.

43. $27\frac{5}{8}$
44. $176\frac{7}{16}$
45. $92\frac{37}{64}$
46. $101\frac{100}{101}$
47. $2{,}974\frac{77}{127}$

Write each improper fraction as a mixed number or as a whole number.

48. $\frac{1{,}825}{175}$
49. $\frac{516}{387}$
50. $\frac{6{,}555}{228}$
51. $\frac{7{,}921}{801}$
52. $\frac{76{,}323}{5{,}871}$

More Practice, page 428, Set C

Comparing Fractions

In the United Nations, a resolution will pass when at least $\frac{2}{3}$ of the nations present vote in favor of the resolution. If $\frac{5}{8}$ of the nations present vote in favor of a resolution, will the resolution pass?

The resolution will pass if $\frac{5}{8}$ is equal to or greater than $\frac{2}{3}$. We need to compare fractions.

Look at the denominators.	Write equivalent fractions with a common denominator.	Compare the numerators.	The fractions compare the same way the numerators compare.

$\frac{5}{8}$
$\frac{2}{3}$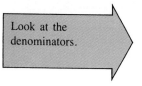

$\frac{5}{8} = \frac{15}{24}$
$\frac{2}{3} = \frac{16}{24}$

$15 < 16$

$\frac{15}{24} < \frac{16}{24}$
so $\frac{5}{8} < \frac{2}{3}$

Since $\frac{5}{8}$ is less than $\frac{2}{3}$, the resolution will not pass.

Other Examples

$\frac{7}{12} = \frac{14}{24}$
$\frac{5}{8} = \frac{15}{24}$ $\frac{5}{8} > \frac{7}{12}$

$\frac{1}{4} = \frac{4}{16}$
$\frac{5}{16} = \frac{5}{16}$ $\frac{1}{4} < \frac{5}{16}$

$\frac{11}{16} < \frac{13}{16}$

Same denominators

To compare mixed numbers, compare the whole numbers, then compare the fractions if necessary.

$3\frac{1}{2} > 2\frac{3}{4}$
$3 > 2$

$4\frac{2}{3} = 4\frac{8}{12}$
$4\frac{3}{4} = 4\frac{9}{12}$ $4\frac{2}{3} < 4\frac{3}{4}$

Warm Up

Compare the numbers. Write > or < for each ●.

1. $\frac{5}{8}$ ● $\frac{3}{4}$
2. $\frac{3}{4}$ ● $\frac{9}{16}$
3. $\frac{3}{5}$ ● $\frac{7}{12}$
4. $\frac{7}{10}$ ● $\frac{11}{15}$
5. $2\frac{1}{3}$ ● $2\frac{1}{4}$
6. $4\frac{1}{10}$ ● $3\frac{9}{10}$
7. $1\frac{5}{6}$ ● $1\frac{4}{7}$
8. $\frac{15}{16}$ ● 1

Compare the fractions. Write > or < for each ▦.

1. $\frac{7}{8}$ ▦ $\frac{13}{16}$
2. $\frac{3}{7}$ ▦ $\frac{1}{2}$
3. $\frac{7}{9}$ ▦ $\frac{5}{6}$
4. $\frac{7}{12}$ ▦ $\frac{3}{4}$

5. $\frac{2}{15}$ ▦ $\frac{1}{5}$
6. $\frac{6}{7}$ ▦ $\frac{23}{28}$
7. $\frac{5}{12}$ ▦ $\frac{1}{2}$
8. $\frac{8}{15}$ ▦ $\frac{2}{5}$

9. $\frac{21}{24}$ ▦ $\frac{3}{4}$
10. $\frac{7}{20}$ ▦ $\frac{2}{5}$
11. $\frac{5}{16}$ ▦ $\frac{3}{8}$
12. $\frac{5}{8}$ ▦ $\frac{25}{50}$

13. $\frac{7}{34}$ ▦ $\frac{1}{6}$
14. $\frac{7}{10}$ ▦ $\frac{33}{38}$
15. $\frac{17}{100}$ ▦ $\frac{7}{36}$
16. $\frac{39}{100}$ ▦ $\frac{11}{36}$

17. $\frac{13}{16}$ ▦ $\frac{7}{9}$
18. $\frac{4}{7}$ ▦ $\frac{13}{20}$
19. $\frac{13}{47}$ ▦ $\frac{5}{12}$
20. $\frac{23}{32}$ ▦ $\frac{17}{21}$

21. $\frac{17}{49}$ ▦ $\frac{9}{28}$
22. $\frac{33}{98}$ ▦ $\frac{7}{24}$
23. $\frac{12}{83}$ ▦ $\frac{7}{30}$
24. $\frac{6}{51}$ ▦ $\frac{4}{39}$

Compare the mixed numbers. Write > or < for each ▦.

25. $1\frac{4}{5}$ ▦ $1\frac{1}{2}$
26. $7\frac{1}{2}$ ▦ $5\frac{2}{3}$
27. $3\frac{1}{3}$ ▦ $3\frac{1}{4}$
28. $6\frac{9}{10}$ ▦ $6\frac{3}{4}$

29. 2 ▦ $2\frac{1}{10}$
30. $6\frac{7}{8}$ ▦ $6\frac{11}{12}$
31. $5\frac{3}{10}$ ▦ $5\frac{27}{100}$
32. $4\frac{4}{9}$ ▦ $4\frac{7}{15}$

33. If $\frac{7}{12}$ of the nations present voted to send aid to the victims of a flood, would the vote pass by a $\frac{2}{3}$ majority?

34.

Nations Present	Yes Vote	No Vote
19	12	5

Would the vote pass by a $\frac{2}{3}$ majority?

THINK MATH

Comparing Fractions

You can compare two fractions using a "cross-products" method.

first 4   5 → 15 × 5 = 75
second 15 18 → 4 × 18 = 72   72 < 75

The product for the first step is less than the product for the second step,

so $\frac{4}{15} < \frac{5}{18}$.

Use the cross-products method and a calculator to compare $\frac{267}{489}$ with each fraction below.

1. $\frac{578}{997}$
2. $\frac{754}{1,443}$
3. $\frac{2,103}{5,274}$

More Practice, page 429, Set A

Using Fractions in Estimation

Make an estimate for each picture.

1. About what part of the glass is filled?

2. About what part of the tank is full?

3. About what part of the notebook is green?

4. About what part of the mosaic is finished?

5. About what part of the pie is left?

6. About how full is the bowl?

7. The length of the short pencil is about what part of the length of the new pencil?

8. Jonathan's score is about what part of Nancy's score?

9. About what part of a minute is shown on the stop watch?

10. About what part of an hour has passed since 8:00?

PROBLEM SOLVING: Using Data from a Table

QUESTION
DATA
PLAN
ANSWER
CHECK

Martha Hines owns a T-shirt store. She keeps records of the monthly sales. Use the table, if necessary, to solve the problems below.

	V-neck	Crew Neck	Total
Jan	75	450	525
Feb	200	475	675
Mar	275	600	875
April	270	650	920

1. What part of the January sales were V-neck shirts? Give the answer as a fraction in lowest terms.

2. What part of the January sales were crew neck shirts? Give the answer as a fraction in lowest terms.

3. What part of the March sales were V-neck shirts?

4. What part of the March sales were crew neck shirts?

5. What was the total number of T-shirts sold during the four months?

6. What was the average number of T-shirts sold per month during the four-month period?

7. Were the January sales above or below the average?

8. What will be the total sales for May? Use the January through April sales to make a prediction.

9. In February the crew neck shirts were on sale for $9.00 each. The V-neck shirts sold for $10.99. What was the total number of dollars of sales for February?

10. How many more shirts were sold in April than in March?

11. The T-shirts in the store are either blue, yellow, or red. In February, 240 red shirts and 137 yellow shirts were sold. How many blue shirts were sold?

12. **Try This** In January a store sold 500 T-shirts and 150 tank tops. Each month after that the number of T-shirts sold decreased by 20 and the number of tank tops sold increased by 30. In what month did the number of tank tops sold equal the number of T-shirts sold? Hint: Make a table.

Month	Jan	Feb	March
T-shirts	500	480	
Tank tops	150	180	

Adding Fractions

A graphic artist is arranging advertisements on a magazine page. The advertisers have purchased $\frac{1}{4}$ of a page and $\frac{1}{8}$ of a page. What part of the page do the two advertisements cover?

To find the total part of the page, we add $\frac{1}{4}$ and $\frac{1}{8}$.

Look at the denominators. → **Find the least common denominator (LCD).** → **Write equivalent fractions with this denominator.** → **Add the numerators. Write the sum over the common denominator.**

$\frac{1}{4}$
$+\frac{1}{8}$
(Unlike denominators)

The LCD is the least common multiple of 4 and 8. The LCD is 8.

$\frac{1}{4} = \frac{2}{8}$
$+\frac{1}{8} = \frac{1}{8}$

$\frac{1}{4} = \frac{2}{8}$
$+\frac{1}{8} = \frac{1}{8}$
$\frac{3}{8}$

The two advertisements cover $\frac{3}{8}$ of a page.

Other Examples

$\frac{3}{10} + \frac{1}{10} = \frac{4}{10} = \frac{2}{5}$ (reduce)

$\frac{6}{8} = \frac{18}{24}$
$+\frac{1}{3} = \frac{8}{24}$
$\frac{26}{24} = 1\frac{1}{12}$ (rename)

$\frac{4}{5} + \frac{7}{10} = \frac{8}{10} + \frac{7}{10} = \frac{15}{10} = 1\frac{1}{2}$

$\frac{3}{4} = \frac{9}{12}$
$\frac{1}{2} = \frac{6}{12}$
$+\frac{2}{3} = \frac{8}{12}$
$\frac{23}{12} = 1\frac{11}{12}$

Warm Up

Add.

1. $\frac{4}{9} + \frac{2}{9}$
2. $\frac{3}{8} + \frac{2}{8}$
3. $\frac{1}{6} + \frac{2}{3}$
4. $\frac{5}{12} + \frac{1}{4}$

5. $\frac{1}{3}$
  $+\frac{3}{4}$

6. $\frac{2}{5}$
  $+\frac{3}{10}$

7. $\frac{5}{8}$
  $+\frac{1}{3}$

8. $\frac{1}{3}$
  $\frac{5}{6}$
  $+\frac{1}{2}$

Add.

1. $\frac{3}{16} + \frac{5}{16}$
2. $\frac{1}{5} + \frac{3}{5}$
3. $\frac{5}{8} + \frac{3}{8}$
4. $\frac{7}{12} + \frac{11}{12}$
5. $\frac{5}{9} + \frac{2}{9}$

6. $\frac{1}{10} + \frac{3}{10}$
7. $\frac{5}{24} + \frac{7}{24}$
8. $\frac{1}{4} + \frac{1}{4}$
9. $\frac{3}{16} + \frac{15}{16}$
10. $\frac{3}{4} + \frac{2}{3}$

11. $\frac{1}{2} + \frac{2}{3}$
12. $\frac{7}{8} + \frac{1}{4}$
13. $\frac{7}{10} + \frac{1}{5}$
14. $\frac{3}{4} + \frac{1}{3}$
15. $\frac{5}{6} + \frac{1}{4}$

16. $\frac{3}{8} + \frac{1}{6}$
17. $\frac{5}{12} + \frac{1}{3}$
18. $\frac{1}{2} + \frac{5}{16}$
19. $\frac{7}{8} + \frac{1}{10}$
20. $\frac{7}{100} + \frac{7}{10}$

21. $\frac{2}{3} + \frac{1}{4} + \frac{1}{12}$
22. $\frac{1}{2} + \frac{5}{6} + \frac{7}{12}$
23. $\frac{5}{8} + \frac{1}{2} + \frac{1}{4}$
24. $\frac{3}{100} + \frac{9}{10} + \frac{1}{2}$
25. $\frac{7}{8} + \frac{3}{4} + \frac{5}{16}$

26. A shoe company bought a $\frac{1}{2}$ page advertisement and a $\frac{1}{6}$ page advertisement. What total part of a page did the shoe company buy?

27. An advertisement for shoes took $\frac{1}{2}$ of a page, an advertisement for tires took $\frac{1}{3}$ of a page, and an advertisement for books took $\frac{1}{6}$ of a page. What part of the page was filled with advertisements?

28. Write and solve an addition problem for the drawings.

29. **DATA HUNT** Find a page in a magazine with two or more advertisements. What part of the page does each advertisement cover? Make an estimate. What part of the page do the combined advertisements cover? Make an estimate.

More Practice, page 429, Set B

Subtracting Fractions

Seth works in a plant store. He needs $\frac{3}{8}$ cup (c) of plant food to feed all the plants in the store. He has $\frac{1}{2}$ c of plant food. What fraction of a cup will be left?

To find the amount left, we subtract. Since the fractions have unlike denominators, a common denominator must be found.

| Look at the denominators. | Find the least common denominator (LCD). | Write equivalent fractions with this denominator. | Subtract the numerators. Write the difference over the common denominator. |

$\begin{array}{r}\frac{1}{2}\\-\frac{3}{8}\\\hline\end{array}$ 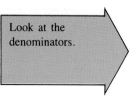 Unlike denominators

The LCD is 8.

$\begin{array}{r}\frac{1}{2}=\frac{4}{8}\\-\frac{3}{8}=\frac{3}{8}\\\hline\end{array}$

$\begin{array}{r}\frac{1}{2}=\frac{4}{8}\\-\frac{3}{8}=\frac{3}{8}\\\hline\frac{1}{8}\end{array}$

There will be $\frac{1}{8}$ c of plant food left.

Other Examples

$\frac{5}{9}-\frac{2}{9}=\frac{3}{9}=\frac{1}{3}$ (reduce)

$\frac{5}{12}-\frac{1}{3}=\frac{5}{12}-\frac{4}{12}=\frac{1}{12}$

$\begin{array}{r}\frac{3}{4}=\frac{15}{20}\\-\frac{2}{5}=\frac{8}{20}\\\hline\frac{7}{20}\end{array}$

Warm Up

Subtract.

1. $\frac{7}{8}-\frac{3}{8}$
2. $\frac{5}{16}-\frac{3}{16}$
3. $\frac{3}{4}-\frac{1}{4}$
4. $\frac{4}{5}-\frac{2}{6}$

5. $\frac{1}{2}-\frac{1}{4}$
6. $\frac{5}{8}-\frac{3}{16}$
7. $\frac{2}{3}-\frac{1}{4}$
8. $\frac{6}{12}-\frac{1}{3}$

Subtract.

1. $\dfrac{5}{6} - \dfrac{1}{6}$ 2. $\dfrac{3}{5} - \dfrac{1}{5}$ 3. $\dfrac{7}{10} - \dfrac{3}{5}$ 4. $\dfrac{5}{8} - \dfrac{3}{8}$ 5. $\dfrac{2}{3} - \dfrac{5}{9}$

6. $\dfrac{3}{4} - \dfrac{1}{4}$ 7. $\dfrac{5}{16} - \dfrac{3}{16}$ 8. $\dfrac{11}{12} - \dfrac{5}{6}$ 9. $\dfrac{2}{3} - \dfrac{1}{2}$ 10. $\dfrac{3}{4} - \dfrac{1}{6}$

11. $\dfrac{2}{5} - \dfrac{1}{10}$ 12. $\dfrac{9}{10} - \dfrac{3}{100}$ 13. $\dfrac{1}{4} - \dfrac{1}{5}$ 14. $\dfrac{2}{3} - \dfrac{1}{8}$ 15. $\dfrac{3}{8} - \dfrac{1}{3}$

16. $\dfrac{7}{10} - \dfrac{1}{6}$ 17. $\dfrac{5}{12} - \dfrac{1}{4}$ 18. $\dfrac{7}{16} - \dfrac{3}{8}$ 19. $\dfrac{7}{8} - \dfrac{3}{16}$ 20. $\dfrac{1}{3} - \dfrac{3}{36}$

21. $\dfrac{8}{9} - \dfrac{2}{3}$ 22. $\dfrac{33}{100} - \dfrac{1}{10}$ 23. $\dfrac{11}{12} - \dfrac{7}{8}$ 24. $\dfrac{4}{5} - \dfrac{1}{8}$ 25. $\dfrac{7}{16} - \dfrac{1}{4}$

26. Sue has $\dfrac{5}{8}$ c of potting mix. She uses $\dfrac{1}{2}$ c. What fraction of a cup of potting mix is left?

27. Aaron is going to use a mixture of bone meal and potting soil. He needs $\dfrac{3}{4}$ c of mixture altogether. He uses $\dfrac{1}{8}$ c of bone meal. How much potting soil must he use?

SKILLKEEPER

Write each fraction in lowest terms.

1. $\dfrac{12}{20}$ 2. $\dfrac{8}{20}$ 3. $\dfrac{24}{64}$ 4. $\dfrac{15}{24}$ 5. $\dfrac{28}{32}$

Write each improper fraction as a mixed number.

6. $\dfrac{22}{5}$ 7. $\dfrac{31}{3}$ 8. $\dfrac{23}{3}$ 9. $\dfrac{58}{7}$ 10. $\dfrac{43}{6}$

Write each mixed number as an improper fraction.

11. $7\dfrac{1}{8}$ 12. $6\dfrac{1}{9}$ 13. $2\dfrac{5}{8}$ 14. $5\dfrac{1}{7}$ 15. $9\dfrac{1}{4}$

More Practice, page 429, Set C

Adding Mixed Numbers

A plumber used two pieces of pipe. One piece was $15\frac{1}{2}$ in. long. The other piece was $8\frac{5}{8}$ in. long. What was the total length of pipe used?

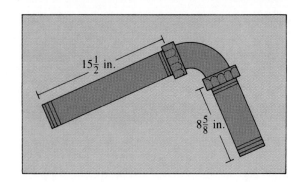

To find the total length, we add.

Find equivalent fractions with a common denominator.

$$15\frac{1}{2} = 15\frac{4}{8}$$
$$+\ 8\frac{5}{8} = 8\frac{5}{8}$$

Add the fractions.

$$15\frac{1}{2} = 15\frac{4}{8}$$
$$+\ 8\frac{5}{8} = 8\frac{5}{8}$$
$$\frac{9}{8}$$

Add the whole numbers.

$$15\frac{1}{2} = 15\frac{4}{8}$$
$$+\ 8\frac{5}{8} = 8\frac{5}{8}$$
$$23\frac{9}{8} = 24\frac{1}{8}$$

rename

The total length of pipe used was $24\frac{1}{8}$ in.

Other Examples

$$4\frac{1}{3} = 4\frac{4}{12}$$
$$+\ 2\frac{1}{4} = 2\frac{3}{12}$$
$$6\frac{7}{12}$$

$$4$$
$$+\ 6\frac{7}{8}$$
$$10\frac{7}{8}$$

$$3\frac{1}{2} = 3\frac{3}{6}$$
$$+\ \frac{2}{3} = \frac{4}{6}$$
$$3\frac{7}{6} = 4\frac{1}{6}$$

Warm Up

Find the sums.

1. $7\frac{1}{8}$
   $+\ 1\frac{3}{8}$

2. 4
   $+\ 5\frac{4}{5}$

3. $2\frac{5}{8}$
   $+\ 6\frac{3}{4}$

4. $5\frac{2}{3}$
   $+\ 2\frac{1}{6}$

5. $1\frac{7}{8}$
   $+\ 3\frac{1}{3}$

6. 2
   $+\ 4\frac{7}{10}$

7. $3\frac{2}{3}$
   $+\ \frac{1}{9}$

8. $12\frac{5}{16}$
   $+\ 23\frac{3}{4}$

Add.

1. $6\frac{1}{2}$
   $+ 3\frac{1}{4}$

2. $6\frac{3}{10}$
   $+ 10$

3. $8\frac{3}{4}$
   $+ 3\frac{1}{4}$

4. 15
   $+ 9\frac{3}{16}$

5. $7\frac{1}{2}$
   $+ \frac{3}{4}$

6. $2\frac{8}{12}$
   $+ 1\frac{1}{4}$

7. $12\frac{1}{2}$
   $+ 17\frac{1}{2}$

8. $9\frac{2}{3}$
   $+ 8\frac{5}{6}$

9. $\frac{1}{6}$
   $+ 7\frac{3}{4}$

10. $20\frac{1}{2}$
    $+ 19\frac{3}{6}$

11. $7\frac{1}{2}$
    $+ 4\frac{5}{6}$

12. $12\frac{1}{3}$
    $+ \frac{3}{5}$

13. $2\frac{3}{7}$
    $+ 5\frac{1}{3}$

14. $11\frac{3}{8}$
    $+ 13\frac{5}{16}$

15. $14\frac{7}{10}$
    $+ 5\frac{3}{100}$

16. $47\frac{4}{5} + 16$

17. $13\frac{5}{16} + 20\frac{5}{8}$

18. $16\frac{2}{3} + 5\frac{7}{9}$

19. $36\frac{2}{5} + 17\frac{7}{10}$

20. $\frac{7}{8} + 28\frac{3}{4}$

21. $4\frac{1}{2}$
    $3\frac{1}{4}$
    $+ 6\frac{1}{2}$

22. $17\frac{1}{8}$
    $11\frac{1}{4}$
    $+ 13\frac{1}{2}$

23. $6\frac{5}{12}$
    $3\frac{1}{4}$
    $+ \frac{1}{3}$

24. $7\frac{5}{6}$
    $2\frac{1}{8}$
    $+ 1\frac{1}{2}$

25. $57\frac{2}{3}$
    $15\frac{5}{6}$
    $+ 29\frac{1}{2}$

26. A plumber used pipes that were $29\frac{3}{4}$ in. long and $16\frac{7}{8}$ in. long. What was the total length of the pipe used?

27. A plumber used two pieces of copper tubing. One piece was $26\frac{1}{2}$ in. long. The other piece was $3\frac{1}{2}$ in. longer than that. What was the total length of the two pieces?

THINK MATH

Math History

Unit fractions, fractions with a numerator of one, were important in ancient Egyptian mathematics. All other fractions were written as the sum of unit fractions with different denominators.

$\frac{3}{5}$ became $\frac{1}{3} + \frac{1}{5} + \frac{1}{15}$ or $\frac{1}{2} + \frac{1}{10}$.

Write each fraction as the sum of unit fractions.

1. $\frac{2}{5}$ 2. $\frac{2}{3}$ 3. $\frac{2}{7}$ 4. $\frac{3}{7}$

More Practice, page 430, Set A

Subtracting Mixed Numbers

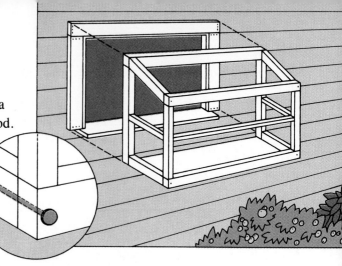

The plans for a greenhouse window show a $2\frac{1}{2}$-inch nail driven through $1\frac{7}{16}$ in. of wood. How far will the nail extend into the second piece of wood?

To find out how much longer the nail is than the first piece of wood, we subtract.

Find equivalent fractions with a common denominator.	Subtract the fractions.	Subtract the whole numbers.

$$\begin{array}{r} 2\frac{1}{2} = 2\frac{8}{16} \\ -1\frac{7}{16} = 1\frac{7}{16} \\ \hline \end{array} \qquad \begin{array}{r} 2\frac{1}{2} = 2\frac{8}{16} \\ -1\frac{7}{16} = 1\frac{7}{16} \\ \hline \frac{1}{16} \end{array} \qquad \begin{array}{r} 2\frac{1}{2} = 2\frac{8}{16} \\ -1\frac{7}{16} = 1\frac{7}{16} \\ \hline 1\frac{1}{16} \end{array}$$

The nail will extend $1\frac{1}{16}$ inches into the second piece of wood.

Other Examples

$$\begin{array}{r} 12\frac{1}{2} = 12\frac{5}{10} \\ -3\frac{2}{5} = 3\frac{4}{10} \\ \hline 9\frac{1}{10} \end{array} \qquad \begin{array}{r} 8\frac{3}{4} \\ -3 \\ \hline 5\frac{3}{4} \end{array} \qquad \begin{array}{r} 10\frac{5}{6} = 10\frac{5}{6} \\ -2\frac{1}{2} = 2\frac{3}{6} \\ \hline 8\frac{2}{6} = 8\frac{1}{3} \end{array}$$

Warm Up

Find the differences.

1. $5\frac{1}{2} - 1\frac{1}{4}$
2. $8\frac{2}{3} - 3\frac{1}{3}$
3. $12\frac{9}{10} - 6$
4. $20\frac{3}{4} - 9\frac{5}{8}$

5. $87\frac{1}{2} - 29\frac{1}{2}$
6. $99\frac{99}{100} - 83\frac{3}{4}$
7. $66\frac{2}{3} - 33\frac{1}{3}$
8. $5\frac{9}{10} - 5\frac{3}{4}$

Subtract.

1. $7\frac{1}{2} - 3\frac{1}{8}$
2. $16\frac{3}{5} - 9$
3. $12\frac{3}{4} - 5\frac{5}{8}$
4. $27\frac{2}{3} - 18\frac{1}{2}$
5. $30\frac{7}{10} - 15\frac{1}{5}$

6. $18\frac{5}{8} - 9\frac{5}{16}$
7. $42\frac{1}{5} - 29\frac{13}{100}$
8. $11\frac{15}{16} - 7\frac{1}{2}$
9. $2\frac{4}{5} - 2\frac{1}{6}$
10. $21\frac{7}{16} - 13\frac{1}{4}$

11. $33\frac{5}{8} - 18$
12. $7\frac{11}{12} - 4\frac{2}{3}$
13. $8\frac{5}{9} - 1\frac{1}{6}$
14. $60\frac{1}{2} - 37\frac{1}{6}$
15. $41\frac{3}{8} - 27\frac{3}{16}$

16. $79\frac{3}{4} - 38\frac{3}{10}$
17. $7\frac{1}{2} - 3\frac{1}{7}$
18. $96\frac{5}{6} - 21\frac{1}{5}$
19. $96\frac{1}{4} - 8\frac{1}{8}$
20. $98\frac{1}{2} - 66\frac{1}{3}$

21. $32\frac{3}{4} - 27\frac{1}{3}$
22. $75\frac{2}{5} - 29\frac{1}{4}$
23. $66\frac{2}{3} - 39$
24. $98\frac{9}{16} - 75\frac{1}{4}$

25. A nail is $2\frac{1}{4}$ in. long. A piece of wood is $1\frac{2}{16}$ in. thick. How much longer is the nail than the thickness of the wood?

26. Mitch drove a nail $2\frac{1}{2}$ in. long into a piece of wood $2\frac{11}{16}$ in. thick. How close was the point of the nail to the other side of the wood?

27. Cesar had a piece of lumber that was $42\frac{15}{16}$ in. long. He sawed off $14\frac{1}{4}$ in. He then had to saw off $1\frac{1}{2}$ in. more. How long was the piece that was left?

28. **DATA BANK** How many inches longer is a 16-penny nail than a 7-penny nail? (See page 413.)

THINK MATH

Shape Perception

The 16 nails form 5 squares. Move only 3 of the nails to new positions so that just 4 squares are formed.

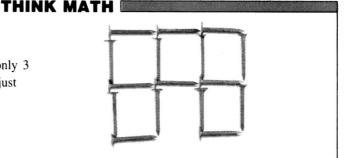

More Practice, page 430, Set B

Subtracting Mixed Numbers with Renaming

A computer science student used $5\frac{1}{2}$ hours (h) of computer time to complete two assignments. The first assignment took $3\frac{3}{4}$ h. How long did the second assignment take?

To find the amount of time for the second assignment, we can subtract.

| Find equivalent fractions with a common denominator. | Rename the fractions if necessary. Subtract the fractions. | Subtract the whole numbers. |

$$5\frac{1}{2} = 5\frac{2}{4}$$
$$-3\frac{3}{4} = 3\frac{3}{4}$$

$$5\frac{2}{4} = 4\frac{6}{4}$$
$$-3\frac{3}{4} = 3\frac{3}{4}$$
$$\phantom{-3\frac{3}{4} = }\frac{3}{4}$$

$$5\frac{2}{4} = 4\frac{6}{4}$$
$$-3\frac{3}{4} = 3\frac{3}{4}$$
$$\phantom{-3\frac{3}{4} = }1\frac{3}{4}$$

$$5\frac{2}{4} = 4 + \frac{4}{4} + \frac{2}{4} = 4\frac{6}{4}$$

The second assignment took $1\frac{3}{4}$ h.

Other Examples

$$3\frac{1}{10} = 2\frac{11}{10}$$
$$-1\frac{7}{10} = 1\frac{7}{10}$$
$$\phantom{-1\frac{7}{10} = }1\frac{4}{10} = 1\frac{2}{5}$$

$$6 = 5\frac{3}{3}$$
$$-2\frac{1}{3} = 2\frac{1}{3}$$
$$\phantom{-2\frac{1}{3} = }3\frac{2}{3}$$

$$10\frac{1}{4} = 10\frac{2}{8} = 9\frac{10}{8}$$
$$-\phantom{10\frac{1}{4} = 10}\frac{5}{8} = \frac{5}{8} = \frac{5}{8}$$
$$\phantom{-10\frac{1}{4} = 10\frac{2}{8} = }9\frac{5}{8}$$

Warm Up

1. $3\frac{1}{4}$
   $-1\frac{3}{4}$

2. 4
   $-\frac{5}{6}$

3. 6
   $-3\frac{7}{10}$

4. $10\frac{1}{5}$
   $-2\frac{3}{5}$

5. $15\frac{1}{2}$
   $-3\frac{5}{6}$

6. $9\frac{3}{8}$
   $-8\frac{3}{4}$

7. $5\frac{3}{16}$
   $-2\frac{3}{4}$

8. $7\frac{1}{3}$
   $-2\frac{1}{2}$

Subtract.

1. $4\frac{1}{5}$
   $-2\frac{1}{3}$

2. $7\frac{1}{4}$
   $-3\frac{3}{4}$

3. 13
   $-5\frac{2}{3}$

4. $7\frac{5}{16}$
   $-\frac{9}{16}$

5. 4
   $-\frac{1}{2}$

6. $14\frac{1}{2}$
   $-9\frac{9}{10}$

7. $9\frac{2}{5}$
   $-1\frac{3}{4}$

8. 99
   $-33\frac{1}{3}$

9. 13
   $-6\frac{7}{8}$

10. $9\frac{1}{2}$
    $-3\frac{2}{3}$

11. $6\frac{1}{4}$
    $-1\frac{3}{4}$

12. 18
    $-10\frac{9}{10}$

13. $2\frac{17}{100}$
    $-\frac{1}{4}$

14. $10\frac{1}{5}$
    $-3\frac{4}{5}$

15. $14\frac{3}{16}$
    $-12\frac{7}{16}$

16. 8
    $-2\frac{2}{3}$

17. $16\frac{1}{4}$
    $-4\frac{1}{2}$

18. $12\frac{1}{4}$
    $-10\frac{5}{8}$

19. 25
    $-19\frac{11}{16}$

20. $17\frac{1}{3}$
    $-11\frac{5}{8}$

21. $9\frac{2}{5} - 1\frac{7}{10}$

22. $17\frac{3}{4} - 6\frac{7}{8}$

23. $15 - 3\frac{3}{8}$

24. $13\frac{2}{3} - 8\frac{7}{8}$

25. Ellen used the computer $4\frac{5}{6}$ h on Monday and $6\frac{1}{3}$ h on Tuesday. How much longer did she use the computer on Tuesday?

26. Brad used the computer a total of 9 h. He used the computer $2\frac{1}{4}$ h on Monday and $2\frac{1}{2}$ h on Tuesday. How many more hours did he use the computer?

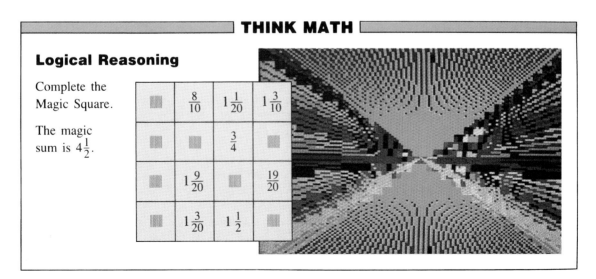

THINK MATH

Logical Reasoning

Complete the Magic Square.

The magic sum is $4\frac{1}{2}$.

	$\frac{8}{10}$	$1\frac{1}{20}$	$1\frac{3}{10}$
		$\frac{3}{4}$	
	$1\frac{9}{20}$		$\frac{19}{20}$
	$1\frac{3}{20}$	$1\frac{1}{2}$	

Skills Practice

Add.

1. $\frac{6}{8} + \frac{1}{4}$
2. $\frac{4}{5} + \frac{11}{25}$
3. $\frac{5}{21} + \frac{1}{7}$
4. $\frac{1}{3} + \frac{4}{9}$

5. $\frac{4}{9} + \frac{3}{4}$
6. $\frac{5}{6} + \frac{8}{9}$
7. $\frac{2}{8} + \frac{2}{3}$
8. $\frac{1}{3} + \frac{3}{4}$
9. $7\frac{1}{10} + 6\frac{9}{10}$

10. $13\frac{5}{8} + 9\frac{2}{5}$
11. $45\frac{6}{8} + 10\frac{6}{7}$
12. $9\frac{4}{5} + 17\frac{1}{3}$
13. $34\frac{3}{9} + 27\frac{2}{4}$
14. $19\frac{3}{6} + 11\frac{2}{8}$

Subtract.

15. $\frac{4}{8} - \frac{1}{4}$
16. $\frac{1}{2} - \frac{1}{3}$
17. $\frac{3}{4} - \frac{3}{10}$
18. $\frac{8}{9} - \frac{3}{4}$

19. $7\frac{3}{9} - 6\frac{2}{6}$
20. $15 - 11\frac{1}{2}$
21. $8\frac{3}{8} - 3\frac{1}{3}$
22. $16\frac{4}{5} - 4\frac{2}{3}$
23. $9\frac{2}{10} - 4\frac{1}{3}$

24. $6\frac{1}{6} - 5\frac{3}{4}$
25. $17\frac{2}{5} - 8\frac{1}{2}$
26. $33\frac{1}{5} - 13\frac{3}{8}$
27. $14\frac{4}{16} - 13\frac{1}{4}$
28. $28\frac{3}{7} - 26\frac{1}{2}$

Add or subtract.

29. $\frac{3}{4} - \frac{1}{5}$
30. $\frac{1}{9} + \frac{4}{3}$
31. $\frac{5}{7} + \frac{1}{3}$
32. $\frac{1}{2} - \frac{1}{13}$

33. $5\frac{7}{8} + 10\frac{1}{6}$
34. $9\frac{3}{16} - 8\frac{3}{4}$
35. $22\frac{3}{10} - 17\frac{4}{5}$
36. $61\frac{2}{5} - 59\frac{3}{4}$
37. $7 - 5\frac{8}{56}$

38. $26\frac{1}{2} + 8\frac{2}{8} + 14\frac{3}{4}$
39. $10\frac{2}{3} + 19\frac{4}{6} + 7\frac{1}{4}$
40. $55\frac{1}{2} + 37\frac{5}{9} + 81\frac{2}{6}$
41. $22\frac{2}{3} + 25\frac{3}{4} + 23\frac{1}{2}$
42. $10\frac{4}{5} + 19\frac{5}{6} + 1\frac{2}{3}$

PROBLEM SOLVING: Practice

QUESTION
DATA
PLAN
ANSWER
CHECK

Solve.

1. The school record for the standing broad jump was $20\frac{1}{2}$ ft. Bonnie made a jump of $18\frac{5}{6}$ ft. How much shorter than the school record was her jump?

2. Franklin ran in a 2-mile race. He ran the first mile in $5\frac{1}{2}$ min. He ran the second mile in $6\frac{1}{10}$ min. What was his total time for the race?

3. In the high jump, Keith jumped $4\frac{3}{4}$ ft. The world record for the high jump was $6\frac{7}{12}$ ft. How much below that world record was Keith's jump?

4. A baseball game lasted for 9 innings. Lefty McGuire pitched the first $6\frac{1}{3}$ innings. Roberto Macias pitched the next $1\frac{1}{3}$ innings. Bob Blazer finished the game as pitcher. How many innings did Bob pitch?

5. Delia had times of 11.2 s, 10.8 s, 11.0 s, and 10.6 s for the 100-yard dash. Find the average time to the nearest tenth of a second.

6. Halfback Rick Freeman had a total of 168 yards gained in carrying the ball 27 times. How many yards per carry did Rick average in the game? Round your answer to the nearest tenth.

7. The mile relay team ran its fastest race in 4 min 6 s. The four runners ran the next race in 59.6 s, 66.4 s, 60.2 s, and 59.5 s. What was their total time for the race? Did the team break their record? If so, by how much?

8. **Try This** In the 880-yard run, Willie's time was 0.2 s faster than Carlton's. Mike's time for the race was 2 min 21.4 s. Ben was 0.9 s slower than Mike but 0.1 s faster than Willie. Jerome was 0.3 s behind Willie. What was the order of finish in the race and what was each person's time for the race?

PROBLEM SOLVING: Choose the Operations

| QUESTION |
| DATA |
| **PLAN** |
| ANSWER |
| CHECK |

Try This

Jennifer is a clerk in a produce market. She had a box with 214 peaches that she wanted to put into bags, 8 peaches to a bag. She had to throw away 17 peaches that were damaged. How many peaches were left after she had filled as many bags as possible?

When you try to solve a problem like the one above, you must decide which operations are needed and the order in which to use them. Sometimes you may find different ways to solve the same problem. We call this strategy **Choose the Operations**.

Addition +	**Subtraction −**	**Multiplication ×**	**Division ÷**
• Total • Combining • How many in all? • Sum	• How many more or less? • Compare • Take away • Difference	• Total of same size groups • Repeated addends • Product	• How many same size groups? • How many in each group? • How many times? • Quotient

Use this strategy to solve the problem above.

To find the number of peaches to be put in bags, you will need to subtract.

To find how many bags, you need to divide by the number of peaches per bag.

The remainder tells how many peaches are left over. This is the answer.

There were 5 peaches left over.

```
Total peaches    Throw away    Peaches to be
                               put into bags
    214       −     17      =      197
```

```
      24
   8)197
     16
     ---
      37
      32
      ---
       5 ← remainder
```

Solve.

1. Roland picked 85 cherries. He ate 9 cherries and shared the rest evenly with Maggie. She ate 12 cherries and shared the rest evenly with Amy. How many cherries did Amy get?

2. Niagara Falls moved upstream about 1.4 m each year between 1700 and 1900. The falls moved about half as far each year since 1900. How far did the falls move between 1700 and 1984?

218

CHAPTER REVIEW/TEST

Write a fraction for the part of the region or set that is shaded.

1.
2.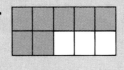
3.
4.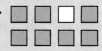

Write the missing numerator or denominator.

5. $\frac{2}{5} = \frac{\square}{20}$
6. $\frac{8}{12} = \frac{\square}{3}$
7. $\frac{5}{6} = \frac{\square}{60}$
8. $\frac{3}{4} = \frac{27}{\square}$

Write each fraction in lowest terms.

9. $\frac{4}{12}$
10. $\frac{20}{24}$
11. $\frac{15}{18}$
12. $\frac{24}{100}$

Write each improper fraction as a mixed number or a whole number.

13. $\frac{8}{3}$
14. $\frac{22}{8}$
15. $\frac{28}{7}$
16. $\frac{38}{16}$

Write each mixed number as an improper fraction.

17. $5\frac{2}{3}$
18. $3\frac{5}{8}$
19. $10\frac{2}{5}$
20. $2\frac{9}{10}$

Compare the fractions. Write > or < for each ●.

21. $\frac{5}{9}$ ● $\frac{1}{2}$
22. $\frac{7}{12}$ ● $\frac{3}{4}$
23. $\frac{3}{8}$ ● $\frac{3}{10}$
24. $\frac{2}{3}$ ● $\frac{11}{16}$

Add or subtract.

25. $\frac{1}{8} + \frac{5}{8}$
26. $\frac{3}{4} + \frac{7}{12}$
27. $\frac{11}{12} - \frac{5}{12}$
28. $\frac{4}{5} - \frac{3}{10}$

29. $4\frac{2}{3} + 2\frac{3}{5}$
30. $6 - 3\frac{2}{3}$
31. $23\frac{9}{10} + 18\frac{1}{2}$
32. $10\frac{1}{3} - 3\frac{5}{6}$

33. Megan ran a mile in $7\frac{3}{4}$ min. Then she took $8\frac{1}{2}$ min to run another mile. What was the total time for the two miles?

34. A recipe calls for $3\frac{1}{2}$ c of flour. Penny has only $2\frac{2}{3}$ c of flour. How much more flour does she need for the recipe?

ANOTHER LOOK

$\dfrac{2 \times 3}{3 \times 3} = \dfrac{6}{9}$ Multiply to find equivalent fraction.

$\dfrac{6 \div 3}{9 \div 3} = \dfrac{2}{3}$ Divide to rewrite in lowest terms.

Write the missing numerator or denominator.

1. $\dfrac{4}{6} = \dfrac{\square}{12}$
2. $\dfrac{2}{5} = \dfrac{\square}{25}$
3. $\dfrac{1}{3} = \dfrac{7}{\square}$

Write each fraction in lowest terms.

4. $\dfrac{2}{6}$
5. $\dfrac{8}{24}$
6. $\dfrac{16}{100}$

$1\dfrac{1}{3} = \dfrac{4}{3}$
↑ ↑
mixed   improper
number fraction

Write each improper fraction as a whole number or a mixed number.

7. $\dfrac{11}{3}$
8. $\dfrac{21}{5}$
9. $\dfrac{35}{8}$
10. $\dfrac{23}{4}$
11. $\dfrac{15}{12}$
12. $\dfrac{36}{12}$

Write each mixed number as an improper fraction.

13. $5\dfrac{5}{6}$
14. $3\dfrac{1}{8}$
15. $7\dfrac{2}{3}$
16. $1\dfrac{7}{12}$
17. $10\dfrac{2}{3}$
18. $6\dfrac{7}{8}$

Compare $\dfrac{2}{3}$ and $\dfrac{3}{5}$.
↓ ↓
$\dfrac{10}{15} > \dfrac{9}{15}$
10 > 9
so
$\dfrac{2}{3} > \dfrac{3}{5}$

Compare the numbers. Write > or < for each ●.

19. $\dfrac{9}{11}$ ● $\dfrac{4}{5}$
20. $\dfrac{5}{8}$ ● $\dfrac{7}{16}$
21. $\dfrac{3}{4}$ ● $\dfrac{20}{24}$
22. $3\dfrac{5}{8}$ ● $3\dfrac{7}{8}$
23. $1\dfrac{5}{6}$ ● $1\dfrac{1}{2}$
24. $6\dfrac{8}{15}$ ● $7\dfrac{5}{8}$

$5\dfrac{3}{4} = 5\dfrac{3}{4}$
$+ 2\dfrac{1}{2} = 2\dfrac{2}{4}$
$\overline{}$
$7\dfrac{5}{4} = 8\dfrac{1}{4}$

$2\dfrac{1}{3} = 1\dfrac{4}{3}$
$- 1\dfrac{2}{3} = 1\dfrac{2}{3}$
$\overline{}$
$\dfrac{2}{3}$

Add or subtract.

25. $\dfrac{5}{6}$
    $+ \dfrac{1}{3}$

26. $\dfrac{11}{12}$
    $- \dfrac{2}{3}$

27. $\dfrac{7}{10}$
    $+ \dfrac{2}{3}$

28. $2\dfrac{1}{2}$
    $+ 2\dfrac{7}{8}$

29. 6
    $- 4\dfrac{5}{8}$

30. $3\dfrac{3}{4}$
    $- 1\dfrac{5}{8}$

31. $8\dfrac{7}{8} + 5\dfrac{5}{6}$
32. $4\dfrac{3}{8} - 2\dfrac{3}{4}$
33. $6\dfrac{2}{5} - 3\dfrac{7}{10}$

ENRICHMENT

Fractions

"Transjovian Pipeline"
Computer-generated
artwork by David Em
Copyright © 1979

Use the clues to name the mystery fractions.

1. My numerator is 1 less than my denominator. Both my numerator and denominator are prime. Which fraction am I?

2. I am greater than 1 but less than 2. The sum of my numerator and denominator is 9. My denominator subtracted from my numerator is 1. Which fraction am I?

3. I am greater than $\frac{1}{3}$ but less than $\frac{1}{2}$. My denominator is 12. Which fraction am I?

4. I am greater than $\frac{1}{2}$ but less than $\frac{3}{4}$. My denominator is 8. Which fraction am I?

5. My denominator is a prime number less than 10. My numerator is a perfect square. I am equal to the whole number 5. Which fraction am I?

6. I am greater than 5 but less than 6. My numerator is a perfect square. The sum of my numerator and denominator is 19. Which fraction am I?

7. I am greater than $\frac{1}{2}$ but less than $\frac{2}{3}$. The difference between my numerator and denominator is 5. The product of my numerator and denominator is 84. Which fraction am I?

8. I am greater than 1 but less than $1\frac{1}{2}$. My numerator and denominator are prime numbers less than 20. The sum of my numerator and denominator is a multiple of 5. Which fraction am I?

9. We are two fractions. Our sum is 1. Our difference is $\frac{1}{5}$. Who are we?

10. We are two numbers. Our sum is 2. Our difference is $\frac{1}{2}$. Who are we?

CUMULATIVE REVIEW

Find the quotients.

1. $16\overline{)6.96}$

   A 0.475   B 0.435
   C 47.5   D not given

2. $226\overline{)33.9}$

   A 0.15   B 1.5
   C 0.20   D not given

Find each quotient to the nearest hundredth.

3. $29\overline{)24.7}$

   A 0.86   B 0.851
   C 0.85   D not given

4. $41\overline{)94.87}$

   A 2.41   B 2.313
   C 0.231 D not given

5. Divide.

   $7{,}297 \div 1{,}000$

   A 729.7   B 72.97
   C 7.297   D not given

6. Estimate the quotient.

   $0.247\overline{)7.57}$

   A 3.5   B 40
   C 4   D not given

7. What is the prime factorization of 88?

   A $2 \cdot 2 \cdot 2 \cdot 11$ B $2 \cdot 4 \cdot 11$
   C $2 \cdot 2 \cdot 11$ D not given

8. What is the GCF of 24 and 180?

   A 24   B 12
   C 18   D not given

Solve each equation.

9. $x + 12 = 39$

   A $x = 22$ B $x = 12$
   C $x = 32$ D not given

10. $5t = 125$

    A $t = 25$ B $t = 55$
    C $t = 625$ D not given

11. $\frac{y}{6} = 60$

    A $y = 6$ B $y = 360$
    C $y = 10$ D not given

12. $z - 2 = 63$

    A $z = 65$ B $z = 61$
    C $z = 62$ D not given

13. Natalie ran a 10 km race in 45.2 min. How long did it take her to run 1 km?

    A 4.52 min   B 4 min
    C 45 s   D not given

14. The same number of nickels are in each of 7 stacks. The stacks balance with 84 nickels. Which equation shows this situation?

    A $7x = 84$ B $7 + x = 84$
    C $\frac{7}{x} = 84$ D not given

Multiplication and Division of Fractions

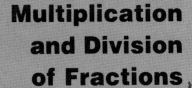

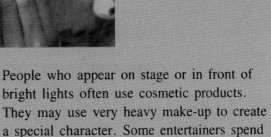

People who appear on stage or in front of bright lights often use cosmetic products. They may use very heavy make-up to create a special character. Some entertainers spend as much as $\frac{1}{8}$ of the day wearing make-up.

Cosmetic products are also used to protect the skin, but tests have shown that when a substance is applied to the skin, it is absorbed into the body. Researchers study make-up and cosmetics to make sure they are not dangerous. This is very important for people who wear make-up for many hours every day.

Cosmetic manufacturers list the contents of their products on each package. Consumers can study the ingredients before they buy anything to apply to their skin.

Finding a Fraction of a Whole Number

Jack tumble-polished 12 gemstones in a machine. He saw that $\frac{1}{3}$ of the gemstones were turquoise. How many gemstones were turquoise?

To find $\frac{1}{3}$ of a number, divide the number by 3.

To find $\frac{2}{3}$ of a number, find $\frac{1}{3}$ of the number, then multiply by 2.

We think: $\frac{1}{3}$ of **12** is **4**.
We write: $\frac{1}{3} \times 12 = 4$.

$12 \div 3 = 4$

We think: $\frac{2}{3}$ of **12** is **8**.
We write: $\frac{2}{3} \times 12 = 8$.

$\frac{1}{3}$ of 12 is 4.
$2 \times 4 = 8$

There were 4 turquoise gemstones.

There were 8 gemstones that were not turquoise.

Other Examples

$\frac{1}{4}$ of **16** is **4**. $\frac{1}{4} \times 16 = 4$ $\frac{1}{8}$ of **24** is **3**. $\frac{3}{8} \times 24 = 9$

$\frac{3}{4} \times 16 = 12$ $\frac{5}{8} \times 24 = 15$

Warm Up

Find the fraction of the number.

1. $\frac{1}{3}$ of 9   **2.** $\frac{1}{4}$ of 8   **3.** $\frac{2}{3}$ of 6

4. $\frac{1}{3} \times 9$ **5.** $\frac{1}{7} \times 14$ **6.** $\frac{9}{10} \times 10$

Find the fraction of the number.

1. $\frac{2}{3}$ of 9
2. $\frac{4}{5}$ of 10

3. $\frac{1}{6}$ of 12
4. $\frac{5}{6}$ of 12

5. $\frac{1}{2} \times 6$
6. $\frac{2}{5} \times 20$
7. $\frac{3}{8} \times 8$
8. $\frac{5}{6} \times 18$

9. $\frac{5}{8} \times 32$
10. $\frac{1}{7} \times 28$
11. $\frac{5}{9} \times 27$
12. $\frac{1}{4} \times 40$

13. $\frac{9}{10} \times 30$
14. $\frac{1}{3} \times 27$
15. $\frac{1}{8} \times 56$
16. $\frac{2}{3} \times 18$

17. $\frac{1}{4} \times 100$
18. $\frac{3}{4} \times 80$
19. $\frac{1}{16} \times 32$
20. $\frac{5}{12} \times 24$

21. $\frac{6}{7} \times 21$
22. $\frac{3}{4} \times 60$
23. $\frac{5}{8} \times 48$
24. $\frac{7}{10} \times 50$

25. $\frac{1}{100} \times 500$
26. $\frac{1}{16} \times 64$
27. $\frac{2}{3} \times 300$
28. $\frac{9}{10} \times 10$

29. $\frac{5}{8} \times 80$
30. $\frac{1}{6} \times 126$
31. $\frac{3}{20} \times 80$
32. $\frac{5}{6} \times 84$

33. Ernie bought a sack of 30 round flint rocks called geodes. He found that $\frac{2}{15}$ of them contained fossils. How many of the geodes contained fossils?

34. Simona bought $\frac{5}{6}$ of a dozen geodes that were to have purple quartz inside and $\frac{1}{2}$ of a dozen geodes that were to have a brown flint inside. How many geodes did she buy?

Multiplying Fractions

Gary is restoring a fresco for a museum of ancient art. He began work on $\frac{1}{2}$ of the fresco. During the past month he has restored $\frac{3}{4}$ of the $\frac{1}{2}$ section. What fractional part of the whole fresco has he restored?

Since we want to find $\frac{3}{4}$ of $\frac{1}{2}$, we can multiply.

Multiply the numerators. → Multiply the denominators.

$\frac{3}{4} \times \frac{1}{2} = \frac{3}{}$ $\qquad$ $\frac{3}{4} \times \frac{1}{2} = \frac{3}{8}$

Gary has restored $\frac{3}{8}$ of the whole fresco.

Other Examples

$\frac{2}{5} \times \frac{3}{8} = \frac{6}{40} = \frac{3}{20}$ $\qquad$ $\frac{3}{4} \times 12 = \frac{3}{4} \times \frac{12}{1} = \frac{36}{4} = 9$ $\qquad$ $7 \times \frac{1}{2} = \frac{7}{1} \times \frac{1}{2} = \frac{7}{2} = 3\frac{1}{2}$

Multiply.

1. $\frac{1}{5} \times \frac{2}{3}$
2. $\frac{3}{5} \times \frac{3}{4}$
3. $\frac{1}{2} \times \frac{1}{2}$
4. $\frac{3}{4} \times \frac{1}{6}$

5. $6 \times \frac{1}{4}$
6. $\frac{9}{10} \times \frac{1}{3}$
7. $8 \times \frac{2}{3}$
8. $\frac{1}{3} \times 10$

9. $\frac{5}{8} \times \frac{3}{5}$
10. $\frac{1}{6} \times \frac{4}{5}$
11. $\frac{7}{10} \times \frac{5}{6}$
12. $\frac{3}{8} \times \frac{2}{9}$

13. $\frac{3}{2} \times \frac{2}{3}$
14. $\frac{3}{4} \times 5$
15. $\frac{1}{2} \times \frac{5}{8}$
16. $\frac{3}{10} \times 16$

17. $\frac{8}{10} \times \frac{5}{6}$
18. $\frac{2}{3} \times \frac{1}{6}$
19. $\frac{1}{8} \times \frac{8}{3}$
20. $\frac{5}{9} \times \frac{3}{10}$

21. Gary had $\frac{3}{4}$ of a can of paint. He used $\frac{2}{3}$ of it. What fractional part of a can of paint did he use?

22. Gary had a box of 24 brushes. He used $\frac{5}{6}$ of them. How many brushes did he use?

Using a Multiplication Shortcut

Jean and Al were asked to find $\frac{3}{4}$ of $\frac{5}{12}$, and to give the answer in lowest terms.

Jean's Method

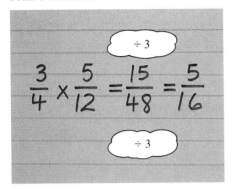

Al's Shortcut

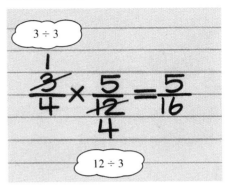

Jean divided the numerator and denominator of $\frac{15}{48}$ by 3 **after multiplying**.

Al divided a numerator by 3 and a denominator by 3 **before multiplying**.

The product, in lowest terms, is the same by either method.

Other Examples

$\frac{\overset{1}{\cancel{5}}}{\underset{2}{\cancel{8}}} \times \frac{\overset{1}{\cancel{4}}}{\underset{5}{\cancel{25}}} = \frac{1}{10}$
$\qquad \frac{\overset{1}{\cancel{3}}}{\underset{2}{\cancel{10}}} \times \frac{\overset{1}{\cancel{2}}}{\underset{3}{\cancel{9}}} \times \frac{\overset{1}{\cancel{5}}}{\underset{4}{\cancel{8}}} = \frac{1}{24}$

Find the products. Use the shortcut when possible.

1. $\frac{2}{7} \times \frac{5}{8}$
2. $\frac{1}{8} \times \frac{4}{9}$
3. $\frac{3}{4} \times \frac{1}{6}$
4. $\frac{3}{8} \times \frac{2}{15}$

5. $\frac{5}{12} \times \frac{4}{5}$
6. $\frac{5}{16} \times \frac{2}{10}$
7. $\frac{2}{3} \times \frac{5}{6}$
8. $\frac{3}{10} \times \frac{5}{6}$

9. $\frac{3}{2} \times \frac{8}{9}$
10. $\frac{3}{10} \times \frac{4}{7}$
11. $\frac{3}{4} \times \frac{8}{15}$
12. $\frac{2}{3} \times \frac{2}{3}$

13. $\frac{5}{8} \times \frac{4}{5} \times \frac{1}{2}$
14. $\frac{7}{16} \times \frac{4}{7} \times \frac{2}{3}$
15. $\frac{1}{2} \times \frac{2}{3} \times \frac{3}{4}$
16. $\frac{5}{12} \times \frac{9}{20} \times \frac{4}{15}$

17. $\frac{9}{10} \times \frac{5}{3} \times \frac{3}{8}$
18. $\frac{1}{2} \times \frac{1}{3} \times \frac{2}{5}$
19. $\frac{7}{8} \times \frac{2}{7} \times \frac{4}{9}$
20. $\frac{15}{16} \times \frac{4}{5} \times \frac{1}{3}$

21. $\frac{21}{30} \times \frac{3}{4} \times \frac{5}{7}$
22. $\frac{9}{20} \times \frac{5}{12} \times \frac{4}{15}$
23. $\frac{3}{16} \times \frac{4}{27} \times \frac{9}{12}$
24. $\frac{3}{2} \times \frac{2}{3} \times \frac{5}{5}$

More Practice, page 431, Set A

Multiplying with Mixed Numbers

The reduced picture of a model airplane is $2\frac{3}{4}$ in. long. The actual model is $3\frac{1}{2}$ times as long as the picture. How long is the model?

$\longmapsto 2\frac{3}{4}$ in. $\longrightarrow$

To find the actual length of the model, we multiply $3\frac{1}{2}$ times $2\frac{3}{4}$.

> Write the mixed numbers as improper fractions. ⟹ Multiply the fractions.

$3\frac{1}{2} \times 2\frac{3}{4} = \frac{7}{2} \times \frac{11}{4}$ $\frac{7}{2} \times \frac{11}{4} = \frac{77}{8} = 9\frac{5}{8}$

The model is $9\frac{5}{8}$ in. long.

Other Examples

Dividing by common factors before multiplying may make the work easier.

$6 \times 3\frac{3}{8} = \frac{\overset{3}{\cancel{6}}}{1} \times \frac{27}{\underset{4}{\cancel{8}}} = \frac{81}{4} = 20\frac{1}{4}$ $2\frac{2}{3} \times 15 = \frac{8}{\underset{1}{\cancel{3}}} \times \frac{\overset{5}{\cancel{15}}}{1} = \frac{40}{1} = 40$

Warm Up

Find the products.

1. $2\frac{2}{3} \times \frac{1}{6}$
2. $3\frac{2}{5} \times 1\frac{3}{10}$
3. $6\frac{1}{4} \times 8$
4. $3\frac{5}{8} \times \frac{4}{5}$
5. $7\frac{1}{2} \times 3\frac{2}{3}$
6. $8 \times 2\frac{1}{6}$
7. $1\frac{7}{12} \times \frac{2}{3}$
8. $2\frac{1}{12} \times 5\frac{2}{6}$

Find the products.

1. $1\frac{2}{3} \times 2\frac{3}{4}$
2. $6 \times 4\frac{1}{3}$
3. $6\frac{1}{3} \times \frac{3}{8}$
4. $\frac{4}{5} \times 2\frac{3}{4}$
5. $1\frac{1}{3} \times 2\frac{1}{4}$
6. $\frac{2}{3} \times 1\frac{1}{2}$
7. $5 \times 7\frac{1}{2}$
8. $4\frac{1}{8} \times 2\frac{2}{3}$
9. $\frac{2}{5} \times 2\frac{1}{2}$
10. $6\frac{5}{12} \times 1\frac{6}{9}$
11. $5\frac{2}{3} \times 2\frac{1}{10}$
12. $3\frac{2}{10} \times 1\frac{1}{10}$
13. $1\frac{2}{3} \times 1\frac{4}{5}$
14. $1\frac{1}{3} \times \frac{3}{5}$
15. $8 \times \frac{11}{8}$
16. $3\frac{3}{4} \times 2\frac{2}{5}$
17. $3\frac{1}{2} \times 2\frac{1}{4}$
18. $\frac{2}{3} \times 4\frac{1}{5}$
19. $2\frac{3}{8} \times 4$
20. $5 \times 3\frac{1}{3}$
21. $\frac{3}{5} \times 3\frac{1}{3}$
22. $5\frac{3}{5} \times 2\frac{1}{6}$
23. $6\frac{2}{3} \times 1\frac{1}{9}$
24. $\frac{6}{7} \times 2\frac{2}{6}$
25. $10\frac{1}{2} \times 1\frac{5}{7}$
26. $2\frac{3}{4} \times 2\frac{3}{4}$
27. $6 \times 2\frac{1}{12}$
28. $4\frac{2}{3} \times \frac{1}{14}$

29. A model ship is $6\frac{1}{2}$ times as long as the picture. How long is the actual model?

30. The length of a model spaceship is $\frac{1}{4}$ of an inch less than 3 times the length shown in the picture. What is the actual length of the model?

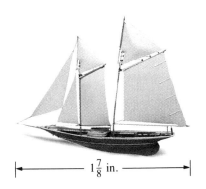

$1\frac{7}{8}$ in.

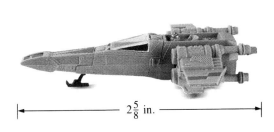

$2\frac{5}{8}$ in.

SKILLKEEPER

1. $\frac{5}{6} + \frac{1}{6}$
2. $\frac{3}{4} + \frac{5}{6}$
3. $\frac{4}{5} + \frac{9}{10}$
4. $\frac{2}{3} + \frac{5}{12}$
5. $\frac{1}{4} + \frac{5}{6}$

6. $5\frac{1}{3}$ $+ 6\frac{1}{3}$
7. $4\frac{1}{2}$ $+ 7\frac{1}{2}$
8. $4\frac{1}{2}$ $+ 2\frac{1}{10}$
9. $3\frac{1}{5}$ $+ 7\frac{4}{5}$
10. $5\frac{3}{4}$ $+ 7\frac{1}{2}$

11. $\frac{3}{4} - \frac{2}{3}$
12. $\frac{4}{5} - \frac{1}{4}$
13. $\frac{2}{3} - \frac{5}{8}$
14. $\frac{8}{9} - \frac{1}{2}$
15. $\frac{1}{3} - \frac{1}{4}$

16. 8 $- 5\frac{3}{5}$
17. $6\frac{1}{2}$ $- 5\frac{2}{3}$
18. $4\frac{1}{4}$ $- 2\frac{1}{2}$
19. $18\frac{1}{2}$ $- 9\frac{7}{12}$
20. $15\frac{1}{2}$ $- 8\frac{1}{8}$

More Practice, page 431, Set B

Skills Practice

Add.

1. $\frac{3}{4} + \frac{1}{2}$
2. $\frac{1}{2} + \frac{1}{5}$
3. $\frac{3}{8} + \frac{3}{4}$
4. $\frac{9}{16} + \frac{1}{4}$

5. $\frac{1}{4} + \frac{2}{5}$
6. $\frac{2}{3} + \frac{1}{2}$
7. $\frac{4}{6} + \frac{2}{4}$
8. $\frac{6}{8} + \frac{1}{3}$

9. $3\frac{1}{3} + 7\frac{2}{3}$
10. $12\frac{3}{4} + 27\frac{1}{2}$
11. $12\frac{3}{10} + 15\frac{1}{2} + 9\frac{4}{5}$
12. $27\frac{1}{2} + 9\frac{3}{4} + 56\frac{5}{8}$

Subtract.

13. $\frac{2}{3} - \frac{1}{2}$
14. $\frac{9}{10} - \frac{1}{2}$
15. $\frac{7}{8} - \frac{3}{4}$
16. $\frac{9}{10} - \frac{3}{4}$

17. $12 - 3\frac{1}{2}$
18. $9\frac{3}{4} - 2\frac{1}{3}$
19. $21\frac{1}{2} - 16\frac{3}{4}$
20. $9\frac{3}{10} - 3\frac{4}{5}$

21. $20\frac{1}{4} - 17$
22. $17\frac{2}{5} - 13\frac{1}{3}$
23. $7\frac{5}{16} - 3\frac{3}{4}$
24. $28\frac{3}{10} - 26\frac{1}{3}$

Multiply.

25. $\frac{7}{10} \times \frac{2}{3}$
26. $\frac{3}{8} \times \frac{4}{5}$
27. $10 \times \frac{1}{5}$
28. $\frac{3}{4} \times 8$

29. $4 \times 2\frac{1}{2}$
30. $\frac{1}{6} \times 30$
31. $2\frac{1}{8} \times 1\frac{1}{2}$
32. $\frac{8}{3} \times \frac{3}{16}$

33. $4\frac{1}{4} \times \frac{4}{5}$
34. $2\frac{2}{3} \times 4\frac{1}{2}$
35. $10 \times 1\frac{1}{10}$
36. $2\frac{1}{4} \times \frac{8}{9}$

Reciprocals

There are 24 students on a gymnastics team. They are divided into 6 groups to work out on the gymnastics equipment. How can we find the number of students in each group?

To find the number of students in each group, divide 24 by 6 or multiply 24 times $\frac{1}{6}$.

Dividing a number by 6 gives the same answer as multiplying the number by $\frac{1}{6}$.

$$24 \times \frac{1}{6} = \frac{24}{6} = 4 \qquad 24 \div 6 = 4$$

Two numbers are **reciprocals** if their product is 1.
$6 \times \frac{1}{6} = 1$ 6 and $\frac{1}{6}$ are reciprocals.

Other Examples

$\frac{3}{4} \times \frac{4}{3} = 1$ $2\frac{1}{2} \times \frac{2}{5} = \frac{5}{2} \times \frac{2}{5} = 1$ $0 \times ? = 1$

$\frac{3}{4}$ and $\frac{4}{3}$ are reciprocals. $2\frac{1}{2}$ and $\frac{2}{5}$ are reciprocals.   Zero has no reciprocal.

Give the reciprocal of each number.

1. $\frac{1}{2}$
2. $\frac{2}{3}$
3. $\frac{5}{8}$
4. $\frac{7}{2}$
5. $\frac{1}{10}$
6. 9
7. $1\frac{1}{2}$
8. 1
9. $2\frac{1}{4}$
10. $\frac{9}{10}$

Give the number for n in each equation.

11. $\frac{3}{5} \times n = 1$
12. $n \times \frac{7}{8} = 1$
13. $1\frac{1}{3} \times n = 1$
14. $4 \times n = 1$
15. $2\frac{1}{6} \times \frac{6}{13} = n$
16. $\frac{3}{8} \times n = 1$
17. $n \times 3\frac{1}{4} = 1$
18. $1\frac{7}{9} \times \frac{9}{16} = n$

19. There are 36 gymnasts training in 4 groups. How many gymnasts are in each group? Solve the problem using both multiplication and division.

20. There are 35 gymnasts on a team. If 5 are absent, how many groups of 5 can be formed?

More Practice, page 431, Set C

Dividing Fractions

The dolphin tank at a city aquarium was emptied and is to be refilled until it is $\frac{9}{10}$ full. It takes 1 h to fill $\frac{1}{5}$ of the pool. How long will it take to fill the pool $\frac{9}{10}$ full?

We need to find the number of fifths in $\frac{9}{10}$. To do this, we divide $\frac{9}{10}$ by $\frac{1}{5}$.

Dividing by a number gives the same answer as multiplying by the reciprocal of the number.

Look at the divisor.	Find the reciprocal of the divisor.	Multiply the dividend by the reciprocal of the divisor.

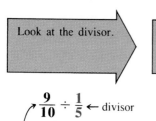

$$\frac{9}{10} \div \frac{1}{5} \leftarrow \text{divisor}\qquad \frac{5}{1} \qquad \frac{9}{\overset{}{\underset{2}{10}}} \times \frac{\overset{1}{\cancel{5}}}{1} = \frac{9}{2} = 4\frac{1}{2} \leftarrow \text{quotient}$$

dividend

To check the problem, multiply the quotient by the divisor.

$4\frac{1}{2} \times \frac{1}{5} = \frac{9}{2} \times \frac{1}{5} = \frac{9}{10}$ It checks.

It will take $4\frac{1}{2}$ h to fill the tank $\frac{9}{10}$ full.

Other Examples

$\frac{1}{2} \div \frac{3}{4} = \frac{1}{\underset{1}{\cancel{2}}} \times \frac{\overset{2}{\cancel{4}}}{3} = \frac{2}{3} \qquad \frac{3}{5} \div 6 = \frac{\overset{1}{\cancel{3}}}{5} \times \frac{1}{\underset{2}{\cancel{6}}} = \frac{1}{10} \qquad 4 \div \frac{3}{8} = \frac{4}{1} \times \frac{8}{3} = \frac{32}{3} = 10\frac{2}{3}$

Warm Up

Find the quotients.

1. $\frac{3}{5} \div \frac{2}{3}$
2. $\frac{5}{8} \div \frac{1}{6}$
3. $\frac{3}{16} \div \frac{1}{4}$
4. $\frac{7}{10} \div \frac{5}{16}$
5. $3 \div \frac{1}{2}$
6. $\frac{7}{8} \div \frac{3}{8}$
7. $\frac{3}{4} \div \frac{1}{2}$
8. $\frac{2}{3} \div 4$

Find the quotients.

1. $\frac{3}{4} \div \frac{1}{5}$
2. $\frac{4}{5} \div \frac{1}{2}$
3. $\frac{5}{9} \div \frac{2}{3}$
4. $\frac{3}{8} \div \frac{2}{3}$
5. $\frac{3}{8} \div 3$
6. $\frac{3}{4} \div \frac{2}{5}$
7. $\frac{5}{12} \div \frac{3}{8}$
8. $\frac{7}{10} \div 5$
9. $\frac{7}{8} \div \frac{3}{16}$
10. $\frac{1}{2} \div 7$
11. $\frac{5}{8} \div \frac{1}{16}$
12. $\frac{2}{3} \div \frac{1}{6}$
13. $6 \div \frac{2}{3}$
14. $\frac{7}{12} \div \frac{3}{15}$
15. $\frac{3}{5} \div \frac{3}{8}$
16. $\frac{11}{12} \div \frac{3}{4}$
17. $\frac{1}{12} \div \frac{1}{4}$
18. $3 \div \frac{7}{10}$
19. $\frac{1}{4} \div 5$
20. $\frac{5}{16} \div \frac{2}{3}$
21. $16 \div \frac{4}{5}$
22. $\frac{3}{8} \div \frac{3}{16}$
23. $\frac{7}{12} \div \frac{5}{6}$
24. $\frac{9}{10} \div 10$
25. $\frac{3}{7} \div \frac{7}{12}$
26. $1 \div \frac{5}{8}$
27. $\frac{4}{5} \div \frac{2}{5}$

28. A class aquarium is to be filled $\frac{2}{3}$ full of water. It takes 1 min to fill $\frac{1}{4}$ of it. How long will it take to fill the aquarium $\frac{2}{3}$ full?

29. An aquarium tide pool is to be filled $\frac{7}{8}$ full of water. It takes 5 min to fill the tide pool $\frac{1}{4}$ full. How many minutes will it take to fill the tide pool $\frac{7}{8}$ full?

THINK MATH

Guess and Check

Fill in the six circles with the fractions $\frac{1}{8}$, $\frac{2}{8}$, $\frac{3}{8}$, $\frac{4}{8}$, $\frac{5}{8}$, and $\frac{6}{8}$ so that the sum of the three fractions on each side of the triangle is the same.

There is more than one way to do this. Find as many ways as you can.

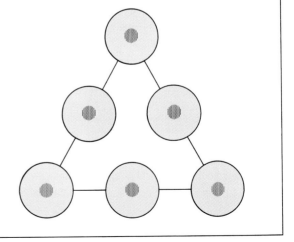

More Practice, page 431, Set D

Dividing with Mixed Numbers

It is 60 ft from home plate to first base on a softball diamond. Polly ran this distance taking strides of about $3\frac{1}{3}$ ft each. About how many strides did she take to run from home plate to first base?

To find how many strides of $3\frac{1}{3}$ ft are in 60 ft, we divide 60 by $3\frac{1}{3}$.

Write mixed numbers or whole numbers as improper fractions. → Divide the fractions.

$60 \div 3\frac{1}{3} = \frac{60}{1} \div \frac{10}{3} \quad \frac{\cancel{60}^{6}}{1} \times \frac{3}{\cancel{10}_{1}} = \frac{18}{1} = 18$

Polly took about 18 strides to run from home plate to first base.

Other Examples

$6\frac{2}{3} \div 5\frac{1}{3} = \frac{20}{3} \div \frac{16}{3} = \frac{\cancel{20}^{5}}{\cancel{3}_{1}} \times \frac{\cancel{3}^{1}}{\cancel{16}_{4}} = \frac{5}{4} = 1\frac{1}{4}$

$3\frac{1}{4} \div 2 = \frac{13}{4} \div \frac{2}{1} = \frac{13}{4} \times \frac{1}{2} = \frac{13}{8} = 1\frac{5}{8}$

$2\frac{2}{5} \div \frac{5}{6} = \frac{12}{5} \div \frac{5}{6} = \frac{12}{5} \times \frac{6}{5} = \frac{72}{25} = 2\frac{22}{25}$

Warm Up

Find the quotients.

1. $5\frac{2}{3} \div 1\frac{1}{4}$
2. $1\frac{3}{4} \div 4\frac{1}{2}$
3. $5\frac{1}{2} \div \frac{4}{5}$
4. $4 \div 1\frac{1}{9}$
5. $8\frac{1}{6} \div \frac{6}{7}$
6. $2\frac{3}{4} \div 6$
7. $4\frac{1}{3} \div 2\frac{1}{2}$
8. $5 \div 3\frac{5}{8}$

Find the quotients.

1. $2\frac{7}{8} \div 1\frac{1}{4}$
2. $2 \div 7\frac{1}{2}$
3. $6\frac{1}{2} \div 2\frac{2}{3}$
4. $1\frac{2}{5} \div \frac{2}{5}$
5. $9 \div 1\frac{1}{2}$
6. $\frac{2}{3} \div 1\frac{1}{5}$
7. $1\frac{3}{10} \div 2\frac{1}{2}$
8. $5\frac{1}{3} \div 2\frac{2}{3}$
9. $1\frac{7}{8} \div 5\frac{1}{4}$
10. $1\frac{3}{16} \div 1\frac{1}{3}$
11. $3\frac{3}{8} \div 4$
12. $3\frac{2}{9} \div \frac{2}{3}$
13. $2\frac{5}{12} \div \frac{3}{8}$
14. $6\frac{1}{2} \div 7$
15. $3\frac{7}{8} \div 7\frac{1}{2}$
16. $7\frac{2}{3} \div 3\frac{1}{4}$
17. $6 \div 1\frac{3}{16}$
18. $\frac{5}{8} \div 3\frac{1}{3}$
19. $3\frac{1}{12} \div 2\frac{2}{3}$
20. $1\frac{5}{9} \div 3$
21. $2\frac{3}{8} \div 2\frac{1}{4}$
22. $4\frac{1}{6} \div 1\frac{4}{9}$
23. $12 \div 5\frac{1}{3}$
24. $\frac{5}{6} \div 2\frac{3}{4}$
25. $\frac{2}{3} \div 2\frac{2}{5}$
26. $7\frac{1}{3} \div 3\frac{2}{3}$
27. $\frac{5}{16} \div 1\frac{5}{8}$
28. $1\frac{9}{12} \div 1\frac{4}{5}$

29. Mick ran 60 ft from first base to second base. Each stride was about $3\frac{3}{4}$ ft long. About how many strides did he take?

30. Chuy hit a home run in baseball. It is 90 ft between the bases. If each stride was about $4\frac{1}{6}$ ft long, how many strides was it around all the bases?

31. One mile is 5,280 ft. About how many strides would it take to walk a mile if each stride was about $2\frac{7}{10}$ ft long?

32. **DATA HUNT** Measure the length of one of your strides. About how many strides would you take to walk one mile?

SKILLKEEPER

Find the fraction of the number.

1. $\frac{1}{5}$ of 25
2. $\frac{4}{5}$ of 10
3. $\frac{2}{3}$ of 15
4. $\frac{3}{5}$ of 25
5. $\frac{1}{5}$ of 60
6. $\frac{2}{3} \times 24$
7. $\frac{5}{6} \times 30$
8. $\frac{7}{8} \times 48$
9. $\frac{3}{4} \times 32$
10. $\frac{3}{10} \times 50$

Multiply.

11. $\frac{1}{8} \times \frac{1}{2}$
12. $\frac{2}{3} \times \frac{1}{4}$
13. $\frac{3}{4} \times \frac{2}{3}$
14. $\frac{5}{12} \times \frac{1}{2}$
15. $\frac{7}{8} \times \frac{2}{7}$
16. $\frac{1}{10} \times \frac{5}{12}$
17. $\frac{2}{3} \times \frac{5}{8}$
18. $\frac{7}{8} \times \frac{4}{5}$
19. $\frac{8}{15} \times \frac{5}{9}$
20. $\frac{9}{10} \times \frac{5}{6}$

More Practice, page 432, Set A

PROBLEM SOLVING: Using Data from a Table

QUESTION
DATA
PLAN
ANSWER
CHECK

Electricity is used in homes for lighting, cooking, heating, and for electrical appliances. The amount of electricity we use is measured in kilowatt-hours (kWh). The table shows the amount of electricity certain appliances use in one hour.

Appliance	Electricity (kWh) used per hour
Broiler, portable	$1\frac{1}{2}$
Color TV	$\frac{3}{20}$
Clothes dryer	$2\frac{4}{5}$
Frying pan	$\frac{1}{2}$
Hair dryer	$1\frac{1}{4}$
Stove oven	$1\frac{1}{3}$
Stove burners	$1\frac{1}{4}$
Steam iron	$\frac{1}{3}$
Vacuum cleaner	$\frac{3}{4}$

A broiler was used for $\frac{3}{4}$ of an hour. How much electricity did it use?

$$\frac{3}{4} \times 1\frac{1}{2} = \frac{3}{4} \times \frac{3}{2} = \frac{9}{8} = 1\frac{1}{8}$$

The broiler used $1\frac{1}{8}$ kWh of electricity.

Use the table to solve these problems.

1. It took $2\frac{1}{2}$ h to roast a small turkey in an oven. How many kilowatt-hours of electricity were used?

2. Carey used the stove burners $\frac{1}{2}$ h at breakfast. How many kilowatt-hours of electricity did she use?

3. Sanjay used the vacuum cleaner, the oven, and the clothes dryer each for 1 h. How much electricity did he use in all?

4. An electric company estimated that an electric hot water heater will use 6,935 kWh of electricity per year to heat water for a household. About how many kilowatt-hours is that per day?

5. Debbie estimated that she uses about 30 kWh of electricity each month with her electric stove burners. About how many hours does she use the burners?

6. A color TV set was used for 5 h. How much did it cost for the electricity if each kilowatt-hour cost 8¢?

7. How many color TV sets, each running for 1 h, would it take to use the same amount of electricity it takes to use a clothes dryer for 1 h?

8. An oven and burners were each used for 3 h. How much more electricity was used by the oven than by the burners?

9. A family uses their clothes dryer 15 h each month. Their electricity costs 9¢ per kilowatt-hour. What is the cost per month for drying clothes?

10. It cost a family $564 a year to heat water by electricity. They changed to a natural gas water heater which cost $207 a year to heat their water. How much money did they save a year on water heating?

11. How much electricity would be used if the frying pan, the steam iron, and the vacuum cleaner were all used for 2 h each?

12. How many hours would a clothes dryer be in service to use 70 kWh of electricity? What is the cost of this amount of electricity at 9¢ per kilowatt-hour?

13. **DATA HUNT** What is the average cost per kilowatt-hour for electricity in your area? Find the amount of a home electric bill for one month. Divide by the number of kilowatt-hours used to find the average cost per kilowatt-hour.

14. **DATA BANK** What would be the total cost of using a broiler and a frying pan for 1 h each in San Juan, Puerto Rico? (See page 415.)

15. **Try This** The Hinson family uses their electric clothes dryer 4 h a week. They use their stove burners 2 h a day. They pay 8¢ per kilowatt-hour for electricity. What is the total cost (to the nearest cent) of electricity per year for the stove burners and clothes dryer?
Hint: Choose the operations.

Fractions and Decimals

There are 10 cars on a roller coaster. The seats are taken in 8 of the 10 cars. What part of the cars are filled?

8 **tenths** of the cars are filled.

We can write 8 tenths as a decimal or a fraction.

$$8 \text{ tenths} = \underset{\text{Decimal}}{0.8} = \underset{\text{Fraction}}{\frac{8}{10}} = \underset{\text{Lowest-terms fraction}}{\frac{4}{5}}$$

Other Examples

36 hundredths = $0.36 = \frac{36}{100} = \frac{9}{25}$ 125 thousandths = $0.125 = \frac{125}{1,000} = \frac{1}{8}$

2 and 3 tenths = $2.3 = 2\frac{3}{10}$ 4 and 75 hundredths = $4.75 = 4\frac{75}{100} = 4\frac{3}{4}$

We can write decimals for some fractions by finding equivalent fractions with denominators that are 10, 100, 1,000, and so on.

$\frac{1}{2} = \frac{1 \times 5}{2 \times 5} = \frac{5}{10} = 0.5$ $\frac{3}{4} = \frac{3 \times 25}{4 \times 25} = \frac{75}{100} = 0.75$

$\frac{11}{20} = \frac{11 \times 5}{20 \times 5} = \frac{55}{100} = 0.55$ $\frac{7}{25} = \frac{7 \times 4}{25 \times 4} = \frac{28}{100} = 0.28$

Warm Up

Write each number as a decimal and as a lowest-terms fraction or a mixed number.

1. 4 tenths
2. 35 hundredths
3. 1 and 6 tenths
4. 408 thousandths

Write each decimal as a lowest-terms fraction.

5. 0.28
6. 0.2
7. 0.060
8. 2.25

Write each fraction or mixed number as a decimal.

9. $\frac{4}{5}$
10. $\frac{37}{50}$
11. $2\frac{3}{4}$
12. $\frac{117}{250}$

Write each number as a decimal and as a lowest-terms fraction or a mixed number.

1. 6 tenths
2. 2 and 5 tenths
3. 24 hundredths
4. 50 hundredths
5. 250 thousandths
6. 3 and 36 hundredths
7. 75 thousandths
8. 10 and 8 tenths
9. 125 ten-thousandths

Write each decimal as a lowest-terms fraction or a mixed number.

10. 0.4
11. 0.14
12. 0.06
13. 0.48
14. 0.1
15. 0.04
16. 3.28
17. 5.7
18. 0.355
19. 0.80
20. 8.12
21. 12.5
22. 0.006
23. 7.64
24. 8.45
25. 24.2
26. 0.0008
27. 0.125
28. 5.55
29. 0.00001

Write each fraction as a decimal.

30. $\frac{1}{4}$
31. $\frac{8}{10}$
32. $\frac{3}{4}$
33. $3\frac{1}{4}$
34. $\frac{3}{5}$
35. $\frac{7}{20}$
36. $2\frac{3}{4}$
37. $\frac{13}{25}$
38. $\frac{37}{100}$
39. $\frac{33}{50}$
40. $\frac{9}{10}$
41. $5\frac{4}{5}$
42. $\frac{327}{1,000}$
43. $25\frac{7}{10}$
44. $15\frac{3}{4}$

45. A waiting line for an amusement park ride had 100 spaces. The line had only 0.64 of the spaces filled. What fraction of the spaces were filled?

46. A large parking lot was completely filled with 1,000 cars. Then 200 cars left the lot. What part of the parking spaces were now filled? Give the answer as a decimal and as a lowest-terms fraction.

THINK MATH

Fraction Patterns

Find the sums. Do you see a pattern for the sums?

What would the sum be if the last addend were $\frac{1}{99 \times 100}$?

1. $\frac{1}{1 \times 2} + \frac{1}{2 \times 3}$
2. $\frac{1}{1 \times 2} + \frac{1}{2 \times 3} + \frac{1}{3 \times 4}$
3. $\frac{1}{1 \times 2} + \frac{1}{2 \times 3} + \frac{1}{3 \times 4} + \frac{1}{4 \times 5}$
4. $\frac{1}{1 \times 2} + \frac{1}{2 \times 3} + \frac{1}{3 \times 4} + \frac{1}{4 \times 5} + \frac{1}{5 \times 6}$

More Practice, page 432, Set B

Skills Practice

Find the quotients.

1. $\frac{2}{5} \div \frac{1}{8}$
2. $\frac{5}{11} \div 5$
3. $\frac{4}{9} \div \frac{1}{3}$
4. $\frac{4}{6} \div \frac{2}{9}$

5. $10 \div \frac{15}{17}$
6. $2\frac{1}{3} \div 4\frac{1}{5}$
7. $4\frac{2}{4} \div 5\frac{1}{3}$
8. $8\frac{1}{4} \div 1\frac{3}{8}$

9. $5\frac{2}{6} \div \frac{1}{3}$
10. $1\frac{4}{12} \div \frac{8}{11}$
11. $5\frac{5}{7} \div 1\frac{1}{14}$
12. $\frac{1}{10} \div 2\frac{2}{8}$

Divide.

13. $9\frac{2}{6} \div \frac{2}{3}$
14. $11\frac{4}{7} \div 4\frac{1}{2}$
15. $12\frac{1}{4} \div 10\frac{1}{2}$
16. $14\frac{2}{5} \div \frac{9}{10}$

17. $5\frac{1}{3} \div \frac{4}{5}$
18. $10\frac{2}{4} \div 11\frac{2}{3}$
19. $12\frac{4}{5} \div 13\frac{1}{3}$
20. $2\frac{8}{14} \div 2\frac{4}{7}$

21. $5\frac{8}{11} \div 1\frac{4}{5}$
22. $1\frac{1}{8} \div 18$
23. $9\frac{1}{3} \div 1\frac{2}{5}$
24. $9\frac{3}{5} \div \frac{6}{10}$

Add, subtract, multiply, or divide.

25. $1\frac{3}{5} \times 2\frac{1}{16}$
26. $2\frac{1}{3} + 4\frac{6}{8}$
27. $4 - \frac{4}{10}$
28. $1\frac{3}{10} \times 2\frac{1}{2}$

29. $1\frac{7}{8} \div 1\frac{1}{4}$
30. $4\frac{6}{7} \times \frac{8}{17}$
31. $11\frac{6}{8} + 3\frac{1}{5}$
32. $2\frac{9}{12} \div \frac{11}{48}$

33. $36 - \frac{10}{52}$
34. $1\frac{1}{5} - \frac{4}{10}$
35. $6\frac{5}{6} + 7\frac{1}{5}$
36. $1\frac{1}{36} \div \frac{5}{18}$

37. $42\frac{8}{9} + 52\frac{5}{6}$
38. $1\frac{1}{11} \div 8\frac{4}{7}$
39. $72\frac{1}{9} - 69\frac{3}{4}$
40. $1\frac{2}{8} \times 3\frac{1}{5}$

Solve.

41. $22\frac{5}{8} - 21\frac{2}{3}$
42. $11\frac{4}{5} + 10\frac{2}{3}$
43. $\frac{8}{9} \div 14\frac{2}{5}$
44. $1\frac{1}{10} \times \frac{3}{22}$

45. $3\frac{3}{5} \div \frac{9}{13}$
46. $3\frac{9}{11} \times \frac{5}{6}$
47. $40\frac{1}{8} + 63\frac{2}{7}$
48. $29\frac{1}{8} - 13\frac{5}{6}$

49. $1\frac{13}{20} \times 3\frac{2}{11}$
50. $\frac{6}{25} \div \frac{7}{30}$
51. $5\frac{1}{3} - 4\frac{4}{18}$
52. $18\frac{2}{5} + 1\frac{11}{15}$

Terminating and Repeating Decimals

The Riverside Symphony has played $\frac{7}{8}$ of its concerts.

$$\frac{7}{8} \to 8\overline{)7.000} \begin{array}{l} 0.875 \\ 64 \\ 60 \\ 56 \\ 40 \\ 40 \\ 0 \end{array} \leftarrow \text{Terminating decimal}$$

← Zero remainder

The Lincoln Philharmonic has played $\frac{5}{6}$ of its concerts.

$$\frac{5}{6} \to 6\overline{)5.000} \begin{array}{l} 0.833\ldots \\ 48 \\ 20 \\ 18 \\ 20 \\ 18 \\ 2 \end{array} \leftarrow \text{Repeating decimal}$$

← Remainder is never zero.

Every fractional number can be written as a **terminating decimal** or a **repeating decimal**.

To show a repeating decimal, place a bar over the digits that repeat.

$\frac{5}{6} = 0.833\ldots = 0.8\overline{3}$

$\frac{4}{33} = 0.1212\ldots = 0.\overline{12}$

Write each repeating decimal using a bar.

1. 0.434343...
2. 1.077077...
3. 0.0575757...
4. 20.55555...
5. 2.345345...
6. 0.81188118...
7. 0.231231231...
8. 3.27276666...

Find the decimal for each fraction. Use a bar to show repeating decimals.

9. $\frac{5}{9}$
10. $\frac{1}{6}$
11. $\frac{23}{25}$
12. $\frac{7}{12}$
13. $\frac{2}{3}$
14. $\frac{8}{15}$
15. $\frac{9}{8}$
16. $\frac{21}{28}$
17. $\frac{4}{27}$
18. $\frac{29}{50}$
19. $\frac{5}{3}$
20. $\frac{7}{22}$
21. $\frac{11}{12}$
22. $\frac{9}{18}$
23. $\frac{13}{5}$
24. $\frac{55}{33}$
25. $\frac{19}{20}$
26. $\frac{17}{68}$
27. $\frac{13}{15}$
28. $\frac{13}{18}$

Compare the two fractions by comparing their decimals. Use > or <.

29. $\frac{11}{15}$ $\frac{7}{12}$
30. $\frac{11}{27}$ $\frac{26}{64}$
31. $\frac{11}{12}$ $\frac{89}{90}$
32. $\frac{37}{80}$ $\frac{127}{275}$

More Practice, page 432, Set C

PROBLEM SOLVING: Use Logical Reasoning

Try This

A health-class teacher asked her students to raise their hands if they liked carrots. 11 students raised their hands. There were 9 students who raised their hands as liking spinach. There were 7 students who did not like either of the vegetables. All 24 students in the class voted. How many students like both carrots and spinach?

With some problems it may be difficult to organize the data or to decide what to do. To solve such problems, it may be helpful to use a strategy called **Use Logical Reasoning**.

To help you reason logically, you can draw a special diagram called a **Venn diagram**.

Draw and label a diagram that shows how all the data in the problem are related.

Use the diagram to help you think logically about the problem.

There were **24 − 7 = 17** students who voted for either spinach or carrots.

Since **11 + 9 = 20**, there must have been **20 − 17 = 3** students that voted for both vegetables.

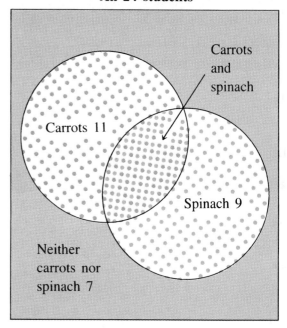

Solve.

1. In a class of 30 students, 18 students play soccer. There are 15 students that play softball. There are 9 students that play neither soccer nor softball. How many students play both soccer and softball?

2. Margarita made a list of 25 numbers. There were 10 numbers that were not divisible by 2 or by 3. There were 12 numbers that were divisible by 2 and among these 12 numbers, 8 were divisible by 2 and by 3. How many of the numbers were divisible by 3 but not by 2?

CHAPTER REVIEW/TEST

Find the fraction of the number.

1. $\frac{1}{4}$ of 20
2. $\frac{1}{5}$ of 35
3. $\frac{2}{3}$ of 12
4. $\frac{3}{4}$ of 16

Multiply.

5. $\frac{1}{2} \times \frac{2}{3}$
6. $\frac{5}{8} \times \frac{2}{3}$
7. $\frac{7}{16} \times \frac{4}{3}$
8. $\frac{9}{10} \times \frac{5}{12}$
9. $2\frac{2}{3} \times 2\frac{1}{2}$
10. $8 \times 4\frac{3}{4}$
11. $10 \times 2\frac{2}{3}$
12. $1\frac{7}{8} \times 3\frac{3}{4}$
13. $\frac{3}{4} \times 3\frac{1}{2}$
14. $\frac{4}{3} \times \frac{3}{4}$
15. $1\frac{1}{3} \times 18$
16. $6\frac{1}{4} \times \frac{1}{2}$

Give the reciprocal of each number.

17. $\frac{7}{8}$
18. $\frac{5}{2}$
19. $1\frac{3}{4}$
20. 6

Divide.

21. $\frac{2}{3} \div \frac{5}{6}$
22. $\frac{7}{12} \div \frac{3}{4}$
23. $3 \div \frac{1}{2}$
24. $\frac{3}{4} \div 2$
25. $2\frac{2}{3} \div \frac{8}{9}$
26. $5\frac{1}{4} \div \frac{7}{10}$
27. $16 \div 3\frac{1}{5}$
28. $5\frac{1}{2} \div 2\frac{1}{2}$
29. $1 \div \frac{1}{4}$
30. $\frac{5}{8} \div \frac{5}{8}$
31. $3\frac{1}{4} \div 2\frac{1}{2}$
32. $12\frac{1}{2} \div 5$

Write each decimal as a lowest-terms fraction.

33. 0.5
34. 0.24
35. 0.025
36. 0.375

Write each fraction as a repeating or terminating decimal.

37. $\frac{13}{15}$
38. $\frac{7}{9}$
39. $\frac{4}{15}$
40. $1\frac{1}{3}$

41. An oven uses $1\frac{1}{3}$ kWh of electricity each hour. How much electricity would it use for a roast that cooked $2\frac{1}{2}$ h?

42. Bill ran with a stride that measured $3\frac{3}{4}$ ft. How many strides would he take while running 120 ft?

ANOTHER LOOK

$\frac{2}{3}$ of 12 = 8

$\frac{2}{\cancel{3}} \times \frac{\cancel{12}^{4}}{1} = 8$

Find the fraction of the number.

1. $\frac{1}{5}$ of 10
2. $\frac{1}{4}$ of 24
3. $\frac{1}{8}$ of 40
4. $\frac{3}{4}$ of 12
5. $\frac{2}{3}$ of 9
6. $\frac{9}{10}$ of 20

$\frac{\cancel{5}^{1}}{\cancel{8}_{2}} \times \frac{\cancel{4}^{1}}{\cancel{15}_{3}} = \frac{1 \times 1}{2 \times 3} = \frac{1}{6}$

Look for a shortcut.
Multiply the numerators.
Multiply the denominators.

Multiply.

7. $\frac{1}{2} \times \frac{1}{5}$
8. $\frac{3}{4} \times \frac{8}{9}$
9. $\frac{5}{6} \times \frac{9}{10}$
10. $2\frac{1}{4} \times 4$
11. $1\frac{7}{8} \times \frac{7}{15}$
12. $1\frac{2}{3} \times 1\frac{2}{3}$
13. $20 \times 1\frac{3}{4}$
14. $\frac{1}{2} \times 3\frac{3}{4}$
15. $5 \times 3\frac{3}{5}$

$\frac{3}{4} \div \frac{5}{8}$

$\frac{3}{\cancel{4}_{1}} \times \frac{\cancel{8}^{2}}{5} = \frac{6}{5} = 1\frac{1}{5}$

reciprocal of $\frac{5}{8}$

To divide by a number, multiply by its reciprocal.

Divide.

16. $\frac{2}{3} \div \frac{2}{5}$
17. $\frac{15}{16} \div \frac{5}{4}$
18. $3 \div \frac{3}{4}$
19. $1\frac{1}{2} \div 2$
20. $2 \div 1\frac{1}{2}$
21. $6 \div \frac{2}{3}$
22. $8\frac{3}{4} \div 3\frac{3}{4}$
23. $10 \div \frac{1}{3}$
24. $15 \div 1\frac{1}{4}$

$\frac{1}{3}$ $\quad 3\overline{)1.00}$ $\frac{1}{3} = 0.\overline{3}$
$\underline{9}$ repeating decimal
10
$\underline{9}$
1

$\frac{3}{4}$ $\quad 4\overline{)3.00}$ $\frac{3}{4} = 0.75$
$\underline{2\,8}$ terminating decimal
20
$\underline{20}$
0 ←zero remainder

Write a repeating decimal or a terminating decimal for each fraction.

25. $\frac{2}{3}$
26. $\frac{11}{25}$
27. $\frac{1}{8}$
28. $\frac{5}{6}$
29. $\frac{8}{15}$
30. $\frac{7}{3}$
31. $1\frac{1}{2}$
32. $\frac{7}{8}$
33. $\frac{4}{6}$

ENRICHMENT

Converting Fractions to Decimals

Stock	Today's high	Today's low	Today's close	Change
AMRO	$34\frac{1}{2}$	$28\frac{1}{2}$	$29\frac{3}{4}$	$-\frac{3}{4}$
Betins	59	$51\frac{1}{2}$	$54\frac{1}{2}$	$-1\frac{3}{8}$
Computer	$13\frac{5}{8}$	$13\frac{1}{2}$	$13\frac{1}{2}$	$+\frac{1}{2}$
ITE	$26\frac{1}{2}$	$14\frac{1}{2}$	26	$+\frac{1}{4}$
New Gas	$47\frac{3}{4}$	$47\frac{1}{4}$	$47\frac{1}{2}$	$-1\frac{1}{4}$
PITT	$26\frac{7}{8}$	$26\frac{5}{8}$	$26\frac{3}{4}$	$+\frac{1}{8}$
Servco	31	$29\frac{1}{2}$	$29\frac{3}{4}$	$-1\frac{5}{8}$

A newspaper reports stock prices in fractions of a dollar. A price of $24\frac{3}{4}$ means $24.75.

Use a calculator to do these problems. Record the answers as decimals.

1. What is the difference between today's high and low for each of the following stocks?

   PITT   Betins   Computer   Servco

2. What is the cost of buying 100 shares of each of the following stocks at today's closing price?

   Computer   ITE   AMRO   New Gas

3. The "change" column indicates the difference between the closing price yesterday and today. A plus sign means a gain in price and a minus sign means a loss in price. What was the closing price yesterday for each of these stocks?

   PITT   Servco   AMRO   Betins

4. Suppose you sell 10 shares of ITE stock and 10 shares of AMRO stock at today's low. How many dollars do you receive?

5. Suppose you sell 200 shares of Computer stock and 100 shares of Betins stock at today's closing price. How much money do you receive?

6. Each share of New Gas stock is worth $1\frac{1}{4}$ dollars less at today's closing price than yesterday's closing price. How much less is 100 shares of New Gas stock worth at today's closing price than yesterday's closing price?

7. What is the total cost of buying 50 shares each of PITT, AMRO, and Betins at today's low?

CUMULATIVE REVIEW

1. What is the symbol for segment *AB*?

   A $\overrightarrow{AB}$
   B $\overleftrightarrow{AB}$
   C $\overline{AB}$
   D not given

2. What kind of angle is ∠*CED*?

   A acute
   B obtuse
   C right
   D not given

3. What is the measure of an angle that is complementary to ∠*XOY*?

   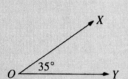

   A 55°
   B 50°
   C 145°
   D not given

4. What kind of quadrilateral is figure *QRST*?

   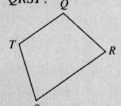

   A triangle
   B trapezoid
   C rectangle
   D not given

5. What is the sum of the measures of the angles in any triangle?

   A 360°
   B 180°
   C 90°
   D not given

6. How many lines of symmetry does a square have?

   A 4
   B 8
   C 2
   D not given

7. If the length of a radius of a circle is 2.4 cm, what is the length of the diameter?

   A 6.9
   B 1.2
   C 4.8
   D not given

8. What is $\frac{18}{20}$ in lowest terms?

   A $\frac{8}{10}$
   B $\frac{9}{10}$
   C $\frac{2}{5}$
   D not given

9. What is the mixed number for $\frac{28}{5}$?

   A $4\frac{5}{8}$
   B $5\frac{3}{5}$
   C $5\frac{2}{5}$
   D not given

10. Which is correct?

    A $\frac{5}{8} > \frac{2}{3}$
    B $\frac{3}{4} > \frac{4}{5}$
    C $\frac{2}{5} < \frac{5}{8}$
    D not given

Add or subtract.

11. $\frac{1}{8}$
    $+\frac{1}{2}$

    A $\frac{3}{4}$
    B $\frac{3}{16}$
    C $\frac{5}{8}$
    D not given

12. $3\frac{1}{2}$
    $+8\frac{3}{8}$

    A $11\frac{3}{4}$
    B $11\frac{7}{8}$
    C $12\frac{1}{8}$
    D not given

13. $7\frac{1}{2}$
    $-5\frac{3}{4}$

    A $2\frac{3}{4}$
    B $1\frac{1}{4}$
    C $2\frac{1}{2}$
    D not given

14. Elinor has $\frac{3}{4}$ c of flour. She uses $\frac{5}{8}$ c. What fraction of a cup of flour is left?

    A $\frac{1}{8}$ c
    B $\frac{1}{4}$ c
    C $\frac{1}{2}$ c
    D not given

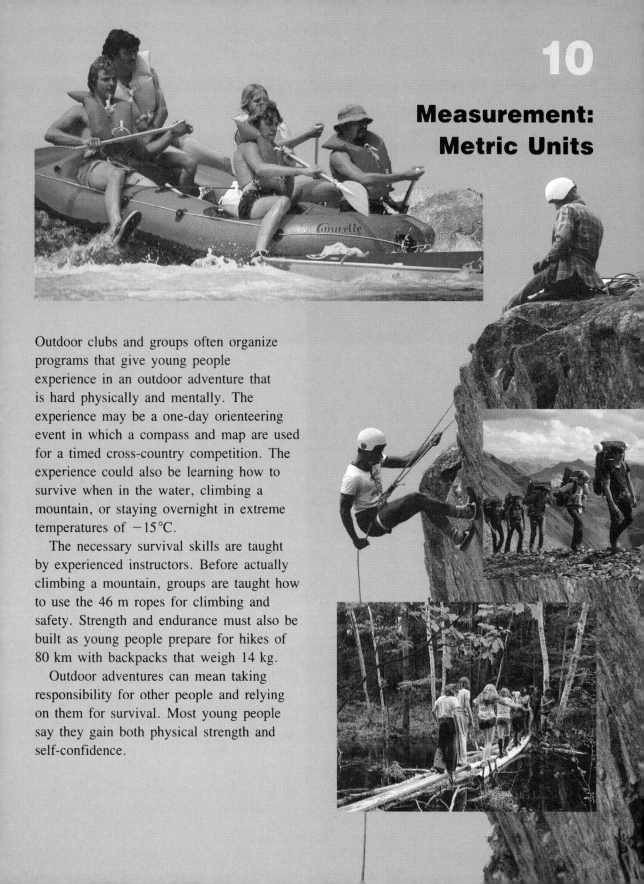

10

Measurement: Metric Units

Outdoor clubs and groups often organize programs that give young people experience in an outdoor adventure that is hard physically and mentally. The experience may be a one-day orienteering event in which a compass and map are used for a timed cross-country competition. The experience could also be learning how to survive when in the water, climbing a mountain, or staying overnight in extreme temperatures of −15°C.

The necessary survival skills are taught by experienced instructors. Before actually climbing a mountain, groups are taught how to use the 46 m ropes for climbing and safety. Strength and endurance must also be built as young people prepare for hikes of 80 km with backpacks that weigh 14 kg.

Outdoor adventures can mean taking responsibility for other people and relying on them for survival. Most young people say they gain both physical strength and self-confidence.

Units of Length

The kite is about 1 **meter (m)** long.

The basic unit of length in the metric system is the meter.

Other units are related to the meter.

1 kilometer (km) = 1,000 m
1 hectometer (hm) =   100 m
1 dekameter (dam) =   10 m

1 decimeter (dm) = 0.1 m
1 centimeter (cm) = 0.01 m
1 millimeter (mm) = 0.001 m

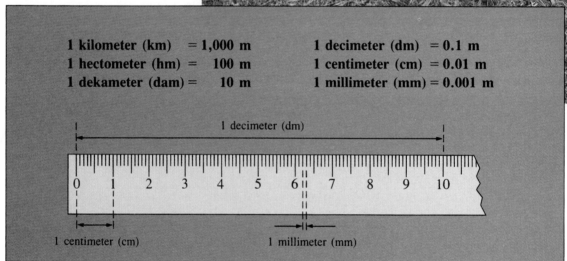

Warm Up

1. How many millimeters are in a centimeter?
2. How many centimeters are in a decimeter?
3. How many millimeters are in a meter?
4. How many decimeters are in a meter?
5. How many centimeters are in a meter?
6. A centimeter is what decimal part of a meter?
7. A meter is what decimal part of a kilometer?
8. A millimeter is what decimal part of a meter?

Estimate the length.

1. The height of a door
   - A 2 dm
   - B 2 m
   - C 2 cm
   - D 2 km

2. The height of a basketball goal
   - A 3 m
   - B 3 dm
   - C 3 km
   - D 3 hm

3. The flying altitude for a large jet airplane
   - A 10 m
   - B 10 km
   - C 10 cm
   - D 10 hm

4. The length of a baseball bat
   - A 100 m
   - B 100 km
   - C 100 mm
   - D 100 cm

5. The diameter of a penny
   - A 2 cm
   - B 2 dm
   - C 2 mm
   - D 2 m

6. The thickness of a dime
   - A 1 mm
   - B 10 mm
   - C 1,000 mm
   - D 1 dm

7. The height of the seat of a chair from the floor
   - A 45 m
   - B 45 mm
   - C 45 dm
   - D 45 cm

8. The distance from Detroit, Michigan to Memphis, Tennessee
   - A 1,000 dm
   - B 1,000 hm
   - C 1,000 dam
   - D 1,000 km

9. The length of a swimming pool
   - A 25 km
   - B 25 m
   - C 25 cm
   - D 25 dm

10. The length of a compact car
    - A 4,900 mm
    - B 4,900 hm
    - C 4,900 m
    - D 4,900 dm

Write the missing units.

11. 1 cm = 10 ▒
12. 100 ▒ = 1 m
13. 10 dm = 1 ▒
14. 1,000 m = 1 ▒
15. 0.1 m = 1 ▒
16. 0.01 m = 1 ▒
17. 100 m = 1 ▒
18. 0.001 km = 1 ▒
19. 0.001 m = 1 ▒

Write the missing numbers.

20. 1 m = ▒ cm
21. 1 cm = ▒ mm
22. 1 hm = ▒ m
23. 1 dam = ▒ m
24. 1 km = ▒ dam
25. 1 cm = ▒ m
26. ▒ mm = 1 cm
27. 1,000 mm = ▒ m
28. ▒ cm = 1 m
29. ▒ m = 1 km
30. 1 cm = ▒ dm
31. 10 dm = ▒ m

32. Which metric unit would you use to measure the distance between two cities?

33. How many meters more than 1 km is 1,011 m?

Changing Units of Length

A biology teacher asked students to measure the length of a ladybug. They found that it was 7 mm long.

7 mm = 0.7 cm
7 mm = 0.07 dm
7 mm = 0.007 m

To change from one unit to another in the metric system, we shift the position of the decimal point in the measure. This is because each unit in the table is 10 times as large as the unit on its right.

km	hm	dam	m	dm	cm	mm

Change kilometers to meters.

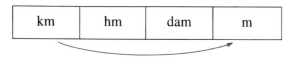

1.827 km = 1,827 m
shift 3 places right

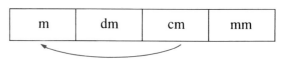

284.3 cm = 2.843 m
shift 2 places left

Change meters to millimeters.

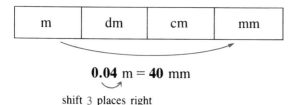

0.04 m = 40 mm
shift 3 places right

Change meters to kilometers.

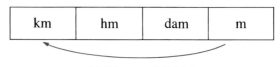

6,745 m = 6.745 km
shift 3 places left

Warm Up

1. Change 2,745 m to kilometers.
2. Change 3.587 m to centimeters.
3. Change 32.7 cm to meters.
4. Change 3.5 dm to centimeters.
5. Change 0.184 km to meters.
6. Change 2,354 mm to meters.
7. Change 54,000 cm to kilometers.
8. Change 838 cm to meters.

Change each length to meters.

1. 163 cm
2. 38.4 dm
3. 5 km
4. 6,388 mm
5. 93.5 cm
6. 2.85 dam
7. 18.42 hm
8. 0.754 km
9. 2 km
10. 2.5 km
11. 500 cm
12. 2,500 mm

Change each length to centimeters.

13. 2.53 m
14. 29 mm
15. 8.1 dm
16. 0.75 m
17. 2.73 km
18. 0.06 m
19. 0.2 dm
20. 484 mm
21. 2 m
22. 5 dm
23. 1 km
24. 65 mm

Change each length to kilometers.

25. 2,000 m
26. 2,135 m
27. 23,000 cm
28. 807 m
29. 1,000,000 mm
30. 9,800 dm
31. 54 hm
32. 200 dam
33. 750 m
34. 58.5 m
35. 20,000 m
36. 150,000 m

Change each length to millimeters.

37. 3 cm
38. 1.7 cm
39. 0.6 cm
40. 0.01 cm
41. 0.05 m
42. 25 cm
43. 2 dm
44. 3 m
45. 0.002 m
46. 0.95 cm
47. 6.2 cm
48. 0.06 dm

49. A measurement is 35 mm. What is it in centimeters, decimeters, and meters?

50. **DATA HUNT** What is your height in each of the six metric units in this lesson? Find your height to the nearest centimeter. Change the centimeter measurement to each of the other units.

THINK MATH

Using Metric Prefixes

Suppose that metric prefixes were used for money, with the dollar as the unit. A decidollar would be one tenth of a dollar or 10 cents. How much money would each of the following be?

1. 1 centidollar
2. 1 dekadollar
3. 1 hectodollar
4. 1 kilodollar
5. 3 decidollars
6. 56 centidollars
7. 2.5 decidollars
8. 2,000 millidollars

Perimeter

The base of the Parthenon in Greece is a rectangle. The length is 72 m and the width is 34 m. What is the distance around the base of the Parthenon?

The **perimeter** of a geometric figure is the distance around the figure.

We could find the perimeter by adding the lengths of the sides:

```
  34
  72
  34
+ 72
 ---
 212
```

$w = 34$ m
$l = 72$ m

For rectangles we can also use a perimeter formula.

Perimeter = 2 × (length + width)
$P = 2(l + w)$
$P = 2(72 + 34)$
$P = 2 \times 106$
$P = 212$

The perimeter of the Parthenon is 212 m.

What is the perimeter of the triangle in meters?

84 cm = 0.84 m
63 cm = 0.63 m
$P = 1.27 + 0.84 + 0.63$
$P = 2.74$

84 cm, 63 cm, 1.27 m

When adding or subtracting measures, all measures must be expressed in the same unit.

The perimeter of the triangle is 2.74 m.

Warm Up

Find the perimeter of each figure.

1. $w = 7.4$ cm, $l = 16.1$ cm

2. 1.55 m (top), 82 cm (left), 95 cm (right), 2.34 m (bottom)

252

Find the perimeter of each figure.

1.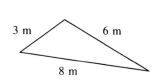

2.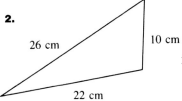

3.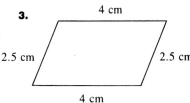

4.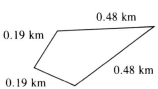

5.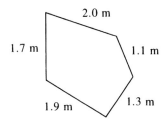

6.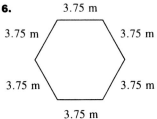

Find the perimeter of each rectangle.

7. $l = 120$ cm
$w = 36$ cm

8. $l = 4$ m
$w = 4$ m

9. $l = 273$ km
$w = 139$ km

10. $l = 64$ mm
$w = 35$ mm

11. $l = 100$ m
$w = 25$ m

12. $l = 83$ dm
$w = 75$ dm

13. What is the perimeter of a window frame that is 91 cm wide and 208 cm high?

14. What is the total length of weather stripping needed for 12 windows that each measure 90 cm by 120 cm?

★ **15.** How long is the piece of tape if it goes completely around the box? Write the answer in centimeters.

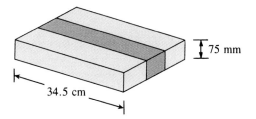

---SKILLKEEPER---

1. $\frac{1}{8} \times 24$ **2.** $\frac{2}{3} \times 6$ **3.** $\frac{1}{2} \times \frac{5}{6}$ **4.** $\frac{3}{4} \times \frac{1}{3}$ **5.** $\frac{2}{3} \times \frac{9}{10}$

6. $1\frac{6}{7} \times 3$ **7.** $6\frac{2}{3} \times 2$ **8.** $9 \times 1\frac{2}{3}$ **9.** $\frac{3}{5} \times 1\frac{3}{4}$ **10.** $3\frac{3}{8} \times 3\frac{1}{3}$

11. $\frac{4}{9} \div \frac{2}{3}$ **12.** $\frac{1}{10} \div \frac{2}{5}$ **13.** $1\frac{1}{3} \div \frac{3}{4}$ **14.** $3\frac{5}{6} \div 3$ **15.** $3\frac{1}{3} \div \frac{1}{10}$

More Practice, page 433, Set A

Area of Rectangles and Parallelograms

The floor in a tent is a rectangle 6.0 m long and 5.4 m wide. What is the **area** of the floor?

The **area** of a region is the number of unit squares it takes to cover the region.

Area of a rectangle = *length* × *width*

$A = lw$
$A = 6.0 \times 5.4$
$A = 32.40$

The area is 32.40 square meters (m²).

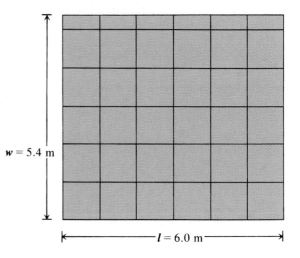

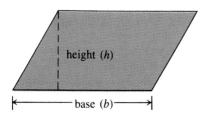

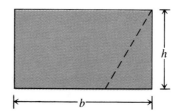

The area of the parallelogram = the area of the rectangle.

Area of a parallelogram = *base* × *height*
$A = bh$

Find the area of each rectangle.

1. $l = 12$ cm
   $w = 8$ cm

2. $l = 25$ m
   $w = 10$ m

3. $l = 3.8$ m
   $w = 2.0$ m

4. $l = 22$ mm
   $w = 9$ mm

5. $l = 6.5$ cm
   $w = 2.8$ cm

6. $l = 30$ cm
   $w = 30$ cm

7. $l = 7.4$ m
   $w = 0.5$ m

8. $l = 2.8$ km
   $w = 1.6$ km

Find the area of each parallelogram.

9. $b = 15$ cm
   $h = 8$ cm

10. $b = 24$ cm
    $h = 12$ cm

11. $b = 3.1$ m
    $h = 2.0$ m

12. $b = 32$ m
    $h = 9$ m

13. $b = 4.9$ cm
    $h = 3.0$ cm

14. $b = 16.8$ m
    $h = 5.2$ m

15. $b = 18$ cm
    $h = 12$ cm

16. $b = 32.5$ m
    $h = 11.7$ m

More Practice, page 433, Set B

PROBLEM SOLVING: Practice

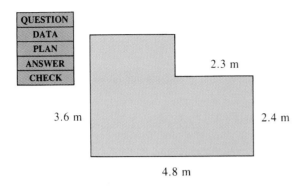

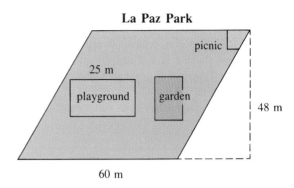

Use the figure above for problems 1–4.

1. A drawing was made of a part of a park lawn that needed new sod. What was the area of sod needed?

2. The city bought 20 m² of sod for the lawn. The cost of the sod was $3.98 per square meter. What was the total cost of the sod?

3. How many square meters of sod were left over?

4. About what was the cost of the leftover sod?

Use the figure above for problems 5–8.

5. The park is on a lot in the shape of a parallelogram with a base of 60 m and a height of 48 m. What is the area of the lot?

6. A playground in the park is rectangular and is 25 m long. If the area of the rectangle is 350 m², what is the width of the playground?

7. A garden in the park covers 160 m² and a picnic area covers 42 m². What is the total area covered by the garden, the picnic area, and the playground?

8. What is the area of the park that is not covered by the garden, the picnic area, and the playground?

9. **Try This** Dee Lewis bought 30 m² of carpeting at $22.50 per square meter. The carpet was used to cover a rectangular floor 5.3 m long and 4.1 m wide. Dee was able to return 4 m² of the carpeting for a full refund. What was the cost (to the nearest cent) of the remainder of the carpet that was not returned and was not used on the floor?

255

Area of Triangles and Trapezoids

The jib, a triangular sail on a sailboat, has a **base** of 6.6 m and a **height** of 2.1 m. What is the area of the jib?

We need to find the area of a triangle.

The area of $\triangle ABC$ is $\frac{1}{2}$ the area of parallelogram $ABCD$.

Area of parallelogram $ABCD = bh$
Area of $\triangle ABC = \frac{1}{2}bh$

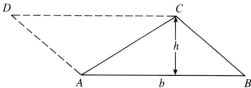

Area of a trapezoid $= \frac{1}{2}$ the area of a parallelogram

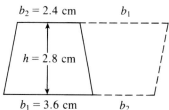

$A = \frac{1}{2}bh$
$A = \frac{1}{2} \times 6.6 \times 2.1$
$A = 6.93$

The area of the jib is 6.93 m².

$A = \frac{1}{2}(b_1 + b_2)h$
$A = \frac{1}{2}(3.6 + 2.4)2.8$
$A = \frac{1}{2} \times 6 \times 2.8$
$A = 8.4$ cm²

b_1 (b sub 1) and b_2 (b sub 2) are the upper and lower bases of the trapezoid.

The area of the trapezoid is 8.4 cm².

Warm Up

1. Find the area of the triangle.

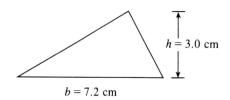

2. Find the area of the trapezoid.

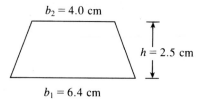

Find the area of each triangle.

1.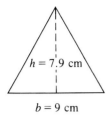
2.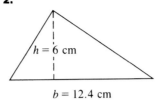
3.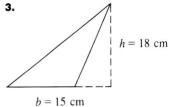

4. $b = 2.7$ cm, $h = 0.08$ cm
5. $b = 2$ m, $h = 1$ m
6. $b = 25$ m, $h = 10$ m

Find the area of each trapezoid.

7.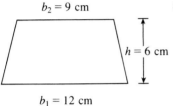
8.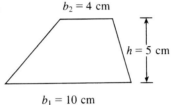
9.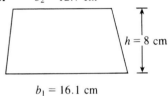

10. $b_1 = 20$ dm, $b_2 = 14$ dm, $h = 10$ dm
11. $b_1 = 5.7$ km, $b_2 = 1.3$ km, $h = 2.0$ km
12. $b_1 = 80$ m, $b_2 = 60$ m, $h = 10.5$ m

13. What is the area of a triangle with a base of 5 m and a height of 4 m?

14. Find the height of a triangle with a base of 12 cm and an area of 18 cm².

★ 15. Find the area of the path around the flowerbed.

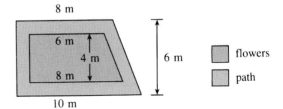

THINK MATH

Shape Perception

Each small square is 1 square unit.

How many square units of area does the shaded triangular region cover?

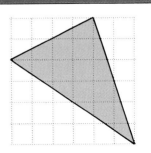

More Practice, page 433, Set C

Surface Area

A sewing box has the shape of a rectangular prism. What is the total surface area of the box?

The **surface area** of the box is the sum of the areas of all faces of the box. Since the opposite faces have the same area, we can find the area of the front, top, and one end and then multiply their sum by 2.

Area of front = lh
Area of top   = lw
Area of end   = wh

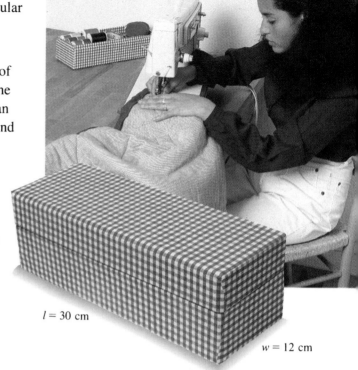

$h = 10$ cm
$l = 30$ cm
$w = 12$ cm

Surface area = $2(lh + lw + wh)$
$A = 2(30 \times 10 + 30 \times 12 + 12 \times 10)$
$A = 2(300 + 360 + 120)$
$A = 2(780) = 1{,}560$

The total surface area is 1,560 cm².

Find the total surface area of each box or rectangular prism.

1.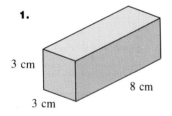
3 cm, 3 cm, 8 cm

2. 10 cm, 5 cm, 0.9 cm

3.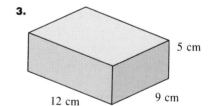
5 cm, 12 cm, 9 cm

4. $l = 9$ cm
   $w = 5.6$ cm
   $h = 4$ cm

5. $l = 2$ m
   $w = 1.5$ m
   $h = 0.8$ m

6. $l = 2.7$ cm
   $w = 2$ cm
   $h = 1.4$ cm

7. $l = 25$ cm
   $w = 25$ cm
   $h = 20$ cm

8. $l = 4.5$ m
   $w = 3.2$ m
   $h = 2.0$ m

9. $l = 0.8$ cm
   $w = 0.6$ cm
   $h = 0.5$ cm

PROBLEM SOLVING: Practice

QUESTION
DATA
PLAN
ANSWER
CHECK

Solve.

1. One face of a cube has an area of 12.25 cm². What is the total surface area of the cube?

2. The total surface area of a cube is 2,400 cm². What is the length of each edge of the cube?

3. What is the total surface area of a rectangular prism with a length of 2.5 m, a width of 1.5 m, and a height of 3 m?

4. A box in the shape of a rectangular prism has a length of 0.5 m, a width of 0.4 m, and a height of 0.3 m. Can all sides of the box be covered with 1 m² of contact paper? How much contact paper will be left over or how much more is needed?

5. Joseph Carillo estimates that a large building has 330 m² of area to be painted. A can of paint covers about 15 m² and costs $10.95. What will be the cost of the paint for the building?

6. A rectangular room has a length of 7 m and a width of 4 m. The height of the wall is 2.5 m. A door and two windows are about 8 m² of area. What is the remaining area of the four walls of the room?

★ 7. A carpenter cut out this shape from a piece of wood. What is the total surface area of the shape?

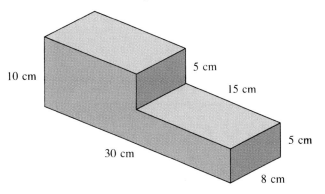

★ 8. A solid rectangular block of wood, 10 cm by 8 cm by 5 cm, has a square 2-cm groove cut in the top face as shown in the figure. How much greater is the total surface area of the grooved block than the original block?

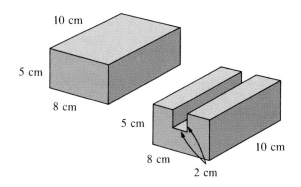

9. **Try This** The total number of dots on opposite faces of a die is always 7. Lonnie tossed 3 dice and a 3, 5, and 2 came up on top. What is the total number of dots on the 12 faces that were not on the top or bottom of Lonnie's dice? Hint: Use logical reasoning.

Volume

Volume is the measure of a region of space. A unit for volume is a cube which has 1 unit on each edge. To find volume, we count the number of **cubic units**.

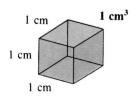

The box will hold 2 layers of 12 cubes or 24 cubic centimeters (cm^3). The volume is 24 cm^3.

The volume of any rectangular prism can be found by using a formula.

Volume = length × width × height

$V = lwh$
$V = 4 \times 3 \times 2 = 24$
$V = 24$ cm^3

Since $l \times w$ = area of the base, the volume formula is also

Volume = Area of Base × height

$V = Bh$
$V = 12 \times 2 = 24$
$V = 24$ cm^3

Other Examples

The formula $V = Bh$ can be used to find the volume of any prism.

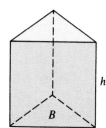

$B = 30$ cm^2
$h = 10$ cm
$V = Bh$
$V = 30 \times 10 = 300$
$V = 300$ cm^3

Triangular prism

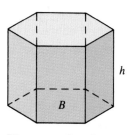

$B = 40.5$ cm^2
$h = 12$ cm
$V = Bh$
$V = 40.5 \times 12 = 486.0$
$V = 486.0$ cm^3

Hexagonal prism

Warm Up

1. Find the volume. Use $V = lwh$.

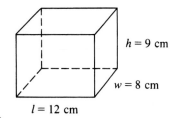

2. Find the volume. Use $V = Bh$.

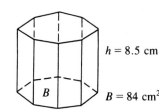

3. Find the volume. Use $V = Bh$.

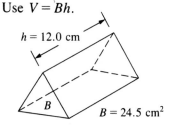

Find the volume of each rectangular prism.

1. $l = 11$ cm
   $w = 8$ cm
   $h = 6$ cm

2. $l = 20$ cm
   $w = 20$ cm
   $h = 12$ cm

3. $l = 2$ m
   $w = 1.5$ m
   $h = 0.8$ m

4. $l = 7$ mm
   $w = 6$ mm
   $h = 5$ mm

5. $l = 8.5$ cm
   $w = 5.4$ cm
   $h = 4.0$ cm

6. $l = 4$ dm
   $w = 3$ dm
   $h = 3$ dm

7. $l = 25$ km
   $w = 10$ km
   $h = 4$ km

8. $l = 0.8$ m
   $w = 0.8$ m
   $h = 0.4$ m

Find the volume of each prism.

9.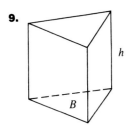
   $B = 18$ cm^2
   $h = 10$ cm

10.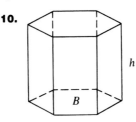
    $B = 56$ cm^2
    $h = 7$ cm

11.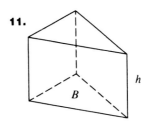
    $B = 20.0$ cm^2
    $h = 6.5$ cm

12.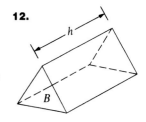
    $B = 4.5$ cm^2
    $h = 4.8$ cm

13. What is the volume of concrete needed for a patio in the shape of a rectangle 9.7 m long, 7.3 m, wide, and 0.15 m thick?

14. The cost to excavate the dirt for a pool that will be a rectangular prism 25 m by 15 m, and 4 m deep, is $4,125. What is the cost per cubic meter?

★ 15. Which has the greater volume, two boxes the same size or one box with double the dimensions of the two boxes?

16. **DATA HUNT** What is the volume of your classroom? Measure the length, the width, and the height of the room to the nearest tenth of a meter.

SKILLKEEPER

Write the missing numbers.

1. 92 cm = ▧ m
2. 6.25 mm = ▧ cm
3. 23 km = ▧ m
4. 1,956 m = ▧ km
5. 5 m = ▧ dm
6. 72,600 dm = ▧ km
7. 2,592 m = ▧ km
8. 912 m = ▧ mm
9. 0.05 hm = ▧ km
10. 257 m = ▧ km
11. 9,240 dm = ▧ km
12. 29.6 km = ▧ m

More Practice, page 433, Set D

Capacity

Mario has some mugs that each hold 200 **milliliters (mL)** of hot chocolate. Milliliters are units of liquid measure, or **capacity**.

1 kiloliter (kL) = 1,000 liters (L)
1 liter (L)   = 1,000 milliliters (mL)

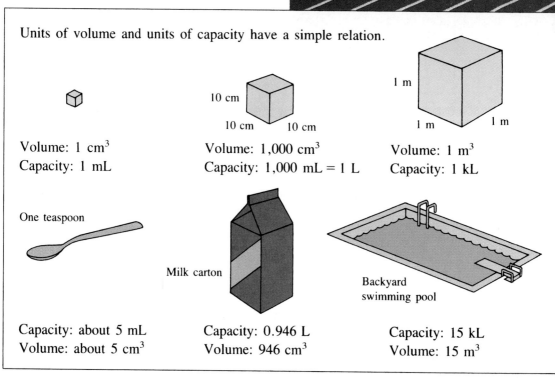

Give the missing numbers.

1. Juice jug: 1,500 cm³
   Capacity: ▓ mL

2. Can of paint: 873 cm³
   Capacity: ▓ mL

Which unit of capacity (mL, L, or kL) would you use to measure the capacity of each?

3. A juice glass

4. Home hot water heater

5. The amount of water in Lake Michigan

6. The amount of water you drink each day

7. The fuel tank of an automobile

8. The amount of liquid in an eyedropper

PROBLEM SOLVING: Practice

Solve.

1. A large milk carton holds 3.46 L of milk. How many milliliters does it hold?

2. The sun evaporated about 0.8 kL of water from a swimming pool each day. How many liters of water evaporated?

3. A jug of apple juice holds 1.75 L. What is the capacity of the jug in milliliters?

4. There are 2 L of fruit punch. If 20 servings of 72 mL each are served, about how many servings will be left?

5. A hummingbird feeder holds 475 mL of liquid. The hummingbirds drink about 75 mL each day. The feeder was last filled on Tuesday morning. When will it need to be filled again?

6. A small can of frozen orange juice has a capacity of 175 mL. There are 48 cans in a case of juice. How many liters of juice are there in a case?

7. A can of frozen lemonade has a capacity of 468 mL. To make lemonade, the directions say to add 3 full cans of water. How many milliliters of lemonade would this make? How many liters is that?

8. Leone bought 1.89 L of low-fat milk, 0.95 L of non-fat milk, and 0.47 L of whipping cream. What was the total amount of milk and cream?

9. A can of condensed soup has a capacity of 325 mL. To make the soup, a full can of milk must be added to the condensed soup. What is the total amount of soup that can be made from 3 cans of condensed soup?

★ 10. How high will the water level be in this aquarium if it is filled with 16 L of water?

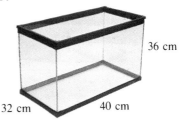

11. A 1.89-liter carton of milk costs $1.08. What is the cost of 1 liter of milk? Round your answer to the nearest cent.

12. **Try This** A glass holds 120 mL. A faucet starts dripping water into the glass. The first minute it drips 1 mL. The second minute it drips 2 mL, the third minute it drips 3 mL, and so on. How many minutes will it be until the glass is filled with water?

263

Units of Weight

Adrian Bishop likes to bowl. His bowling ball has a **weight** of a little more than 7 **kilograms (kg)**. A kilogram is a unit of weight.

1 metric ton (t) = 1,000 kilograms (kg)
1 kilogram (kg) = 1,000 grams (g)
1 gram (g)   = 1,000 milligrams (mg)

One drop of water	Nickel	Large book	School bus

About 50 mg	About 5 g	About 6 kg	About 5 t

There is a simple relation between the volume of pure water and the weight of the water.

 1 L of water has a weight of 1 kg.

 1 mL of water has a weight of 1 g.

Give the unit (mg, g, kg, t) that would be used to measure the weight of each object.

1. A penny　　**2.** A person　　**3.** A truck　　**4.** A drop of oil

Give the weight of each amount of water.

5. 3 L　　**6.** 500 mL　　**7.** 1.5 L　　**8.** 27 mL

Give the missing numbers.

9. 2,000 g = ▥ kg
10. 3,174 g = ▥ kg
11. 5 kg = ▥ g
12. 1.8 kg = ▥ g
13. 0.001 kg = ▥ g
14. 0.001 g = ▥ mg
15. 2 g = ▥ mg
16. 500 mg = ▥ g
17. 0.454 kg = ▥ g
18. 0.7 kg = ▥ g
19. 7,200 mg = ▥ g
20. 2.938 kg = ▥ g

PROBLEM SOLVING: Using Data from a Table

King-size Hamburger Ingredients			
Roll	28 g	Onions	14 g
Patty	113 g	Pickles	12 g
Tomatoes	22 g	Ketchup	8 g
Mayonnaise	3 g	Lettuce	20 g

Solve.

1. About how many hamburger patties does it take to equal 1 kg? Round your answer to the nearest whole number.

2. The rolls come in packages of 12. What is the weight of a package of rolls?

3. What is the total weight of all the ingredients in one king-size hamburger?

4. Sal ate two king-size hamburgers and a fruit drink. The fruit drink had a weight of 350 g. How much less than a kilogram did Sal eat?

5. Wakenda ordered a king-size hamburger without mayonnaise, onions, or tomatoes. What was the weight of her hamburger?

6. Robbie ordered a king-size hamburger, onion rings (weight 188 g), and a small fruit drink (weight 240 g). What was the total weight of Robbie's order?

7. The weight of ketchup in a full bottle is 336 g. How many king-size hamburgers could be served with one bottle of ketchup?

8. A head of lettuce has a weight of about 500 g. How many portions of lettuce for king-size hamburgers can be made from one head of lettuce?

9. On Saturday 250 king-size hamburgers were sold. How many kilograms of patties is this?

10. A large jar of mayonnaise has a weight of 1,560 g. About how many hamburgers could be served with this amount of mayonnaise?

11. **Try This** Doug ordered an avocado sandwich, a fruit drink, and some carrot sticks. The fruit drink cost 25¢ more than the carrot sticks. The sandwich cost 50¢ more than the fruit drink. Doug paid for his meal with a $5 bill and got $1.75 back in change. How much did each kind of food cost?

Celsius Temperature

The thermometer shows a room temperature of 20 **degrees Celsius** (20°C).

On a Celsius scale, water freezes at 0°C and boils at 100°C.

A temperature of ⁻8°C means 8 degrees "below zero." We can read ⁻8 as "minus 8" or "negative 8."

Common Celsius Temperatures			
Water freezes	0°C	Very hot day	40°C
Water boils	100°C	Very cold day	⁻30°C
Room temperature	20°C	Refrigerator	5°C
Body temperature	37°C	Freezer	⁻6°C
Moderate oven	175°C	Hot oven	250°C

Give the temperature on each Celsius scale.

1.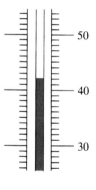

2.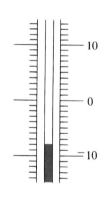

3.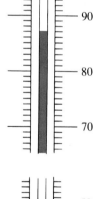

4.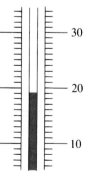

5.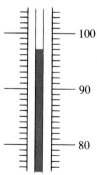

6.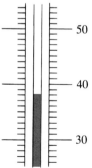

PROBLEM SOLVING: Practice

Solve.

1. The oven temperature for fruit cobbler should start at 215°C. After 15 min, the temperature should be lowered 35°C. What is the lower temperature?

2. Butter will melt at about 31°C. Paraffin will melt at about 55°C. How much higher is the melting temperature of paraffin?

3. Pure iron melts at 1,535°C. Pure gold will melt at a temperature that is 473°C lower. What is the melting temperature of gold?

4. The normal body temperature of a spiny anteater is 23.2°C. How much lower than normal human body temperature is this?

5. The normal body temperature of a songbird is about 13°C higher than the normal body temperature of a human. What is the normal body temperature of a songbird?

6. Lead will melt at about 327°C. The melting point of aluminum is 6°C higher than twice the melting temperature of lead. What is the melting temperature of aluminum?

7. Silver will boil at a temperature of 1,950°C. It will become solid at a temperature which is 15°C less than one half its boiling temperature. What is the temperature at which silver becomes solid?

★ 8. The coldest recorded weather temperature on earth was 88°C below zero, or ⁻88°C. The highest temperature ever recorded on earth was 58°C. How many degrees would the temperature have to rise from ⁻88°C to reach 58°C?

9. **DATA BANK** Find the lowest recorded temperatures in Bismarck, North Dakota and Honolulu, Hawaii. How many degrees difference is there between the two temperatures? (See page 416.)

10. **Try This** The temperature went up 3°C per hour from 6:00 a.m. to 1:00 p.m. At 1:00 p.m. the temperature was 7°C. What was the temperature at 6 a.m.?

267

PROBLEM SOLVING: Using the Strategies

Problem Solving Strategies

- Guess and Check
- Make an Organized List
- Find a Pattern
- Draw a Picture
- Work Backward
- Solve a Simpler Problem
- Make a Table
- Choose the Operations
- Use Logical Reasoning

The problems on this page can be solved with the help of one or more of the different strategies you have learned.

1. Amy, Vicki, Francine, and Magdalena have a double-Dutch rope jumping team. While two girls turn the rope, two other girls jump as partners. How many different combinations of partners could jump?

2. Paul is making a wooden window box to start seedlings. He cut a long board into two pieces of equal length. He cut 30 cm off one of the pieces and then cut the same piece into two more pieces of equal length of 52 cm. How long was the original board?

3. While riding the bus to a natural history museum, Ben and Joshua discover they have exactly $10.00. Ben has $2.70 more than Joshua. How much does each person have?

4. Water Tower Place has a height of 190 m less than the John Hancock Center. The Sears Tower is 204 m higher than the Standard Oil Building. The John Hancock Center is 96 m taller than the Standard Oil Building, which is 346 m tall. What are the heights of the other three buildings?

CHAPTER REVIEW/TEST

Write the missing numbers.

1. 1 cm = ▓ mm
2. 1 m = ▓ cm
3. 1,000 mm = ▓ m
4. 1 km = ▓ m
5. 2,000 m = ▓ km
6. 0.1 m = ▓ dm

Change each length to centimeters.

7. 2.75 m
8. 68 mm
9. 0.53 m

Change each length to meters.

10. 384 cm
11. 3.5 km
12. 27 dm

Find the perimeter and the area of each figure.

13.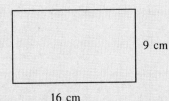

14.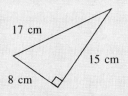

15.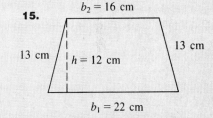

16. Find the total surface area.
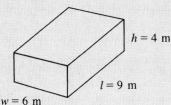

17. Find the volume.
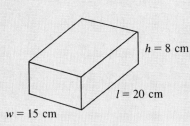

18. Find the volume.
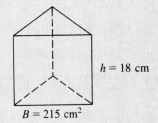

Write the missing numbers.

19. 42 L = ▓ mL
20. 220 cm^3 = ▓ mL
21. 6,252 cm^3 = ▓ L
22. 2.5 kg = ▓ g
23. 600 mg = ▓ g
24. 5,249 g = ▓ kg

25. A can of frozen orange juice concentrate has a capacity of 175 mL. If 3 cans of water are added to the frozen juice, how many milliliters of juice will be made?

26. A pie was baked for 10 min at 225°C. Then the temperature was lowered 38°C. What was the new baking temperature?

ANOTHER LOOK

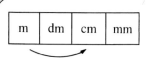

2.75 m = 275 cm
shift 2 places right

Write the missing numbers.

1. 185 cm = ▓ m
2. 67 mm = ▓ cm
3. 3,000 m = ▓ km
4. 0.9 cm = ▓ mm

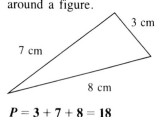

$P = 3 + 7 + 8 = 18$
$P = \mathbf{18\ cm}$

Perimeter is the distance around a figure.

Find the perimeter of each figure.

5.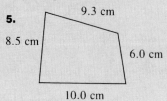
(9.3 cm, 8.5 cm, 6.0 cm, 10.0 cm)

6. (27 cm, 14 cm rectangle)

7. A triangle with sides 2.3 mm, 6.9 mm, and 1.6 mm

8. A parallelogram with sides 2.7 cm, 6.8 cm, 2.7 cm, and 6.8 cm

Area formulas:
Rectangle: $A = lw$
Parallelogram: $A = bh$
Triangle: $A = \frac{1}{2}bh$
Trapezoid: $A = \frac{1}{2}h(b_1 + b_2)$

Find the area of each figure.

9.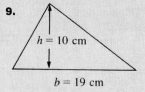
$h = 10$ cm, $b = 19$ cm

10. $b_2 = 8$ cm, $h = 6$ cm, $b_1 = 12$ cm

11. A rectangle with length 8.5 cm and width 2.4 cm

12. A parallelogram with base 16 m and height 7 m

Volume of a rectangular prism: $V = lwh$
Volume of a prism: $V = Bh$

Find the volume of each prism.

13.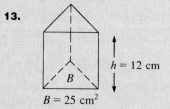
$h = 12$ cm, $B = 25$ cm^2

14.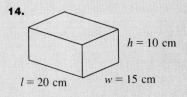
$h = 10$ cm, $l = 20$ cm, $w = 15$ cm

1,000 mL = 1 L

1 cm^3 has a capacity of 1 mL.

Write the capacity in liters.

15. 1,750 mL
16. 3,200 cm^3

ENRICHMENT

Measurement

Archimedes (287–212 B.C.) discovered the principle that an object floating on water displaces a weight of water equal to the weight of the floating object.

Here is a way you can repeat the experiment for the discovery made by Archimedes.

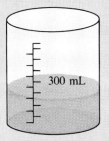

Put water in a metric container level to some mark.

Put in an object that floats in the water.

Find the amount of water displaced.

```
  500 mL
- 300 mL
  200 mL
```

200 mL of water has a weight of 200 g.
The weight of the floating object is 200 g.

What is the weight of each floating object?
The amount of water in each container at the start was 300 mL.

1.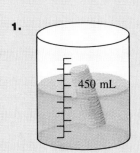
2.
3.
4.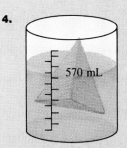

TECHNOLOGY

Computer Decisions

A computer can be given a program so that it can make a **decision** about INPUT data.

This program instructs the computer to make a decision about the voting age of a person.

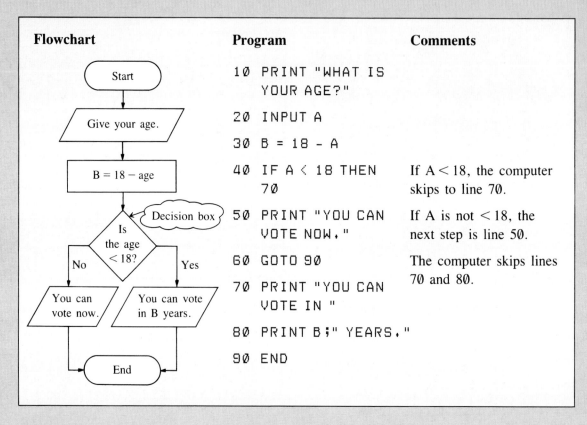

Compare each RUN of the program below.

RUN
WHAT IS YOUR AGE?
?11   INPUT: Age = 11
YOU CAN VOTE IN
7 YEARS.

RUN
WHAT IS YOUR AGE?
?23   INPUT: Age = 23
YOU CAN VOTE NOW.

What would a RUN of this program be if you used your own age as INPUT A?

Write the RUN for each program. Use the given INPUT numbers.

1.
```
10   PRINT "DOLLARS IN YOUR
     ACCOUNT?"
20   INPUT A
30   PRINT "AMOUNT TO
     WITHDRAW?"
40   INPUT B
50   C = A - B
60   IF A < B THEN 90
70   PRINT "YOUR BALANCE IS
     $";C
80   GOTO 100
90   PRINT "YOU DON'T HAVE
     ENOUGH MONEY."
100  END
```

A INPUT A = $549.12
   INPUT B = $277.95
B INPUT A = $549.12
   INPUT B = $549.15

2.
```
10   PRINT "GUESS A NUMBER
     < 20."
20   INPUT A
30   N = 12
40   IF A = N THEN 100
50   IF A < N THEN 80
60   PRINT "TOO LARGE!"
70   GOTO 10
80   PRINT "TOO SMALL!"
90   GOTO 10
100  PRINT "THAT'S IT!"
110  END
```

A INPUT A = 7
B INPUT A = 15
C INPUT A = 12

3.
```
10   PRINT "WHAT NUMBER
     TIMES"
20   PRINT "8 EQUALS 296?"
30   INPUT A
40   B = 296/8
50   IF A = B THEN 80
60   PRINT "TRY AGAIN."
70   GOTO 10
80   PRINT "THAT'S RIGHT!"
90   END
```

A INPUT A = 32
B INPUT A = 37

4.
```
10   PRINT "SOLVE THE
     EQUATION."
20   PRINT "7N = 112."
30   PRINT "N = ?"
40   INPUT A
50   N = 112/7
60   IF A = N THEN 90
70   PRINT "TRY AGAIN."
80   GOTO 20
90   PRINT "RIGHT! N = ";A
100  END
```

A INPUT A = 13
B INPUT A = 16

5. Write a program similar to the one in exercise 4. Choose your own equation. Give a RUN of your program.

6. Write a computer program which uses an IF . . . THEN statement and GOTO commands. Give a RUN of your program.

CUMULATIVE REVIEW

1. What is the prime factorization of 144?

   A $2^4 \cdot 3^2$ **B** $2^2 \cdot 6^2$
   C $2^3 \cdot 3^2$ **D** not given

2. What is the GCF of 25 and 150?

   A 5   **B** 10
   C 25   **D** not given

3. What is the LCM of 11 and 15?

   A 11   **B** 165
   C 55   **D** not given

Solve each equation.

4. $5a = 75$

   A $a = 20$ **B** $a = 80$
   C $a = 375$ **D** not given

5. $\frac{h}{9} = 54$

   A $h = 6$ **B** $h = 63$
   C $h = 486$ **D** not given

6. $z - 17 = 62$

   A $z = 36$ **B** $z = 79$
   C $z = 45$ **D** not given

Multiply.

7. $\frac{3}{8} \times \frac{4}{5}$

   A $\frac{3}{10}$ **B** $1\frac{1}{2}$
   C $\frac{12}{35}$ **D** not given

8. $6\frac{1}{8} \times 10$

   A $\frac{3}{40}$ **B** $61\frac{1}{4}$
   C 60   **D** not given

9. Solve.

   $n \times 2\frac{2}{3} = 1$

   A $n = \frac{3}{8}$ **B** $n = \frac{7}{2}$
   C $n = \frac{2}{7}$ **D** not given

10. Find the quotient.

    $\frac{5}{6} \div \frac{3}{8}$

    A 3   **B** $2\frac{2}{9}$
    C $2\frac{1}{9}$ **D** not given

11. What is the lowest-terms fraction for 0.56?

    A $\frac{14}{25}$ **B** $\frac{28}{50}$
    C $\frac{7}{5}$ **D** not given

12. What is the decimal for $\frac{17}{20}$?

    A 0.85   **B** 1.176
    C 11.76   **D** not given

13. The sum of a certain number and 27 is 95. What is the number?

    A 122   **B** 70
    C 40   **D** not given

14. A small movie theater can seat 200 people. Only 0.35 of the theater was filled. What fraction of the theater was filled?

    A $\frac{7}{20}$ **B** $\frac{12}{25}$
    C $\frac{1}{2}$ **D** not given

11
Ratio and Proportion

Benjamin Banneker was born in 1731 in Maryland. He attended school until he was 15 years old. Once he was out of school, he had to work hard on a farm. He continued to study, however, and mathematics became his greatest interest.

Through a variety of accomplishments, Benjamin Banneker became well-known for his brilliant mind. By studying the series of connected gear wheels in a tiny watch, he learned the ratio of the number of teeth on each gear wheel to the circumference of the gear wheels. He used the ratios to make larger gear wheels and built the first clock made entirely of American parts. Banneker observed nature and published almanacs which became popular and useful throughout America. Thomas Jefferson appointed Banneker to a commission to plan the city of Washington, D.C.

Ratio

On a nature hike, species of birds were identified. There were 2 sparrow hawks, 1 barn owl, and 6 gray partridges. The students observed 9 birds in all.

We can use a **ratio** to compare the numbers.

2 out of 9 birds were sparrow hawks.

2 to 9, **2:9**, or $\frac{2}{9}$ Read: "2 to 9"

A **rate** is a ratio which compares different kinds of units.

A field guidebook of American birds costs $8.

Ratio: $\frac{\text{Dollars} \rightarrow}{\text{Books} \rightarrow} \frac{\$8}{1}$ or $8 per guide

It took 3 hours to walk 6 km.

Ratio: $\frac{\text{Km} \rightarrow}{\text{Hours} \rightarrow} \frac{6}{3}$ or 2 km/h

Sparrow hawk

Barn owl

Gray partridge

Warm Up

Write each ratio as a fraction.

1. 3 to 8
2. 1:7
3. 4 out of 5
4. 7 to 10
5. 40 m to 5 s
6. $24 in 3 h
7. 100 L in 4 min
8. $6 for 2 L

Write each ratio as a fraction.

1. 8 to 9
2. 5:11
3. 3 out of 4
4. 9 to 7
5. 5 to 12
6. 4 out of 12
7. 16:6
8. 2 out of 6
9. 4 out of 7
10. 1.2 to 4.3
11. 0.76 to 1.34
12. 1 out of 8
13. 10 to 1
14. 2.5 to 7
15. 9:100
16. 1:1,000

Write each rate as a fraction.

17. 24 m in 6 s
18. $8.00 for 50 L
19. 8 cups in 2 L
20. 66 words per min
21. 85 km in 1 h
22. 3 tickets for $9.75
23. $225 for 4 tires
24. 3 c flour for 2 eggs
25. 1,400 km in 2 h
26. $105 for 7 days

27. What is the ratio of swamp sparrows to king rails?

28. What is the ratio of king rails to total birds identified?

29. Tommy identified 4 American kestrels, 1 barn owl, and 12 gray partridges. What was the ratio of American kestrels to gray partridges?

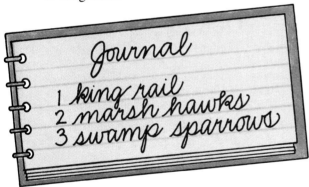

Journal
1 king rail
2 marsh hawks
3 swamp sparrows

THINK MATH

Estimation

John won a contest which awarded him a prize of $1,000 a week for life. Estimate how many years it will take John to collect $1,000,000. Then calculate the total number of years and weeks. Use 52 weeks for 1 year.

Equal Ratios

Ken saw 2 cartons of white eggs and 3 cartons of brown eggs on the grocery shelf. Each carton had 12 eggs. What is the ratio of white eggs to brown eggs?

White eggs → $\dfrac{24}{36} = \dfrac{24 \div 12}{36 \div 12} = \dfrac{2}{3}$
Brown eggs →

The ratio is $\dfrac{24}{36}$. In **lowest terms** it is $\dfrac{2}{3}$.

Both ratios give the same comparison of white eggs to brown eggs. They are **equal ratios**. $\dfrac{24}{36} = \dfrac{2}{3}$

A statement that two ratios are equal is called a **proportion**. $\dfrac{24}{36} = \dfrac{2}{3}$ is a proportion.

If two ratios are equal, then their "cross products" are equal. If the cross products are equal, then the ratios are equal.

$\dfrac{24}{36} \stackrel{\times}{=} \dfrac{2}{3}$ → $36 \times 2 = 72$
→ $24 \times 3 = 72$

Other Examples

$\dfrac{3}{4} = \dfrac{9}{12}$ because $\begin{array}{l} 4 \times 9 = 36 \\ 3 \times 12 = 36 \end{array}$

The cross products are equal.

The ratio 8 to 10 as a fraction in lowest terms is $\dfrac{4}{5}$.

$\dfrac{8}{10} = \dfrac{8 \div 2}{10 \div 2} = \dfrac{4}{5}$

means "is not equal to"

$\dfrac{1}{3} \neq \dfrac{3}{10}$ because $\begin{array}{l} 3 \times 3 = 9 \\ 1 \times 10 = 10 \end{array}$

The cross products are not equal.

The ratio 30:10 as a fraction in lowest terms is $\dfrac{3}{1}$.

$\dfrac{30}{10} = \dfrac{3}{1}$

Warm Up

Write = or ≠ for each ●. Use cross products to decide.

1. $\dfrac{8}{3}$ ● $\dfrac{16}{6}$
2. $\dfrac{3}{8}$ ● $\dfrac{4}{10}$
3. $\dfrac{14}{5}$ ● $\dfrac{28}{10}$
4. $\dfrac{7}{8}$ ● $\dfrac{12}{15}$

Write each ratio as a fraction in lowest terms.

5. 10:15
6. 8:6
7. $\dfrac{125}{100}$
8. $\dfrac{18}{20}$

Write = or ≠ for each ◉. Use cross products to decide.

1. $\frac{2}{5}$ ◉ $\frac{4}{10}$
2. $\frac{4}{8}$ ◉ $\frac{4}{16}$
3. $\frac{8}{12}$ ◉ $\frac{2}{3}$
4. $\frac{16}{8}$ ◉ $\frac{8}{3}$
5. $\frac{1}{3}$ ◉ $\frac{3}{9}$
6. $\frac{2}{7}$ ◉ $\frac{6}{21}$
7. $\frac{3}{5}$ ◉ $\frac{5}{9}$
8. $\frac{4}{10}$ ◉ $\frac{8}{25}$
9. $\frac{6}{10}$ ◉ $\frac{5}{9}$
10. $\frac{5}{2}$ ◉ $\frac{40}{16}$
11. $\frac{3}{7}$ ◉ $\frac{4}{9}$
12. $\frac{20}{16}$ ◉ $\frac{5}{4}$
13. $\frac{1}{3}$ ◉ $\frac{4}{11}$
14. $\frac{3}{21}$ ◉ $\frac{1}{7}$
15. $\frac{2}{3}$ ◉ $\frac{6}{12}$
16. $\frac{7}{8}$ ◉ $\frac{6}{7}$

Write each ratio as a fraction in lowest terms.

17. 4 to 6
18. 10 to 2
19. 12:16
20. 18:20
21. 6 out of 15
22. 20 out of 25
23. 9:3
24. 50:200
25. $\frac{16}{32}$
26. $\frac{75}{100}$
27. $\frac{40}{15}$
28. $\frac{36}{60}$
29. $\frac{10}{45}$
30. $\frac{18}{60}$
31. $\frac{15}{24}$
32. $\frac{20}{28}$

33. Cans of tomato paste were priced at 3 cans for 87¢. Another store had the same tomato paste priced at 1 can for 29¢. Are the ratios $\frac{87¢}{3}$ and $\frac{29¢}{1}$ equal?

34. Apples were priced at 96¢ for a dozen in one store. The same size apples were priced at 4 for 35¢ in another store. Are the ratios equal? At which price are the apples cheaper?

35. Three of the four ratios are equal. Which one does not belong with the equal ratios? Check the cross products.
   A $\frac{7.8}{13}$ B $\frac{19.2}{28.9}$ C $\frac{2.7}{4.5}$ D $\frac{0.45}{0.75}$

36. **DATA HUNT** Write ratios for the prices of loose potatoes and large bags of potatoes. Which price is more economical?

SKILLKEEPER

Solve each equation.

1. $7n = 7$
2. $6x = 36$
3. $13t = 65$
4. $20z = 100$
5. $9y = 108$
6. $\frac{h}{5} = 17$
7. $\frac{k}{9} = 21$
8. $\frac{r}{3} = 30$
9. $\frac{q}{12} = 8$
10. $\frac{j}{20} = 3$
11. $\frac{a}{17} = 1$
12. $\frac{c}{11} = 5$
13. $\frac{m}{4} = 22$
14. $\frac{p}{32} = 7$
15. $\frac{s}{2} = 24$

More Practice, page 434, Set A

Solving Proportions

A bottle capping machine is set to cap 6 bottles every 5 s.

At this rate, how many bottles would the machine cap in 60 s, or 1 min?

Since the rate of capping stays the same, we can use equal ratios to write a proportion.

Let b = the number of bottles capped in 60 s.

Write the proportion.

Bottles → $\dfrac{6}{5} = \dfrac{b}{60}$
Seconds →

Find the cross-product equation.

$$5b = 6 \cdot 60$$
$$5b = 360$$

Solve the equation.

$$\dfrac{5b}{5} = \dfrac{360}{5}$$
$$b = 72$$

The machine would cap 72 bottles in 60 s.

Other Examples

$\dfrac{4}{x} = \dfrac{18}{27}$ $\dfrac{3}{8} = \dfrac{15}{t}$ $\dfrac{n}{42} = \dfrac{15}{18}$

$18x = 27 \cdot 4$ $3t = 8 \cdot 15$ $18n = 15 \cdot 42$

$18x = 108$ $3t = 120$ $18n = 630$

$x = 6$ $t = 40$ $n = 35$

Warm Up

Solve each proportion.

1. $\dfrac{2}{x} = \dfrac{6}{9}$ 2. $\dfrac{5}{6} = \dfrac{15}{x}$ 3. $\dfrac{x}{10} = \dfrac{9}{30}$ 4. $\dfrac{20}{6} = \dfrac{x}{15}$

Solve each proportion.

1. $\frac{1}{3} = \frac{x}{18}$
2. $\frac{5}{6} = \frac{x}{24}$
3. $\frac{2}{3} = \frac{8}{x}$
4. $\frac{10}{16} = \frac{15}{x}$
5. $\frac{15}{3} = \frac{x}{8}$
6. $\frac{10}{x} = \frac{4}{8}$
7. $\frac{2}{5} = \frac{x}{250}$
8. $\frac{15}{1} = \frac{x}{4}$

Read the problem. Then solve the proportion.

9. A machine can seal 12 bottles in 45 s. How long would it take the machine to seal 60 bottles?
   Let t = the number of seconds to seal 60 bottles.

   Bottles → $\frac{12}{45} = \frac{60}{t}$
   Seconds →

10. If 1 out of 80 bottles had a defect, how many defective bottles were in a shipment of 1,200 bottles?
    Let b = the number of defective bottles in 1,200 bottles.

    Defective bottles → $\frac{1}{80} = \frac{b}{1,200}$
    All bottles →

Complete and solve the proportion for each problem.

11. A machine put labels on cans at a rate of 7 labels every 4 s. How many cans could the machine label in 28 s?
    Let n = the number of cans labeled in 28 s.

    Labels → $\frac{7}{4} = \frac{\rule{0.5cm}{0.15mm}}{\rule{0.5cm}{0.15mm}}$
    Seconds →

12. A movable belt for a can-labeling machine travels 115 cm in 5 s. How far does a point on the belt move in 1 s?
    Let d = the distance the point moves in 1 s.

    Centimeters → $\frac{\rule{0.5cm}{0.15mm}}{\rule{0.5cm}{0.15mm}} = \frac{\rule{0.5cm}{0.15mm}}{\rule{0.5cm}{0.15mm}}$
    Seconds →

Write and solve a proportion for each problem.

13. A machine can fill 3 bottles in 8 s. At this rate, how long will it take to fill 48 bottles?
    Let t = the time to fill 48 bottles.

14. A machine sealed 225 boxes in 3 h. At this rate, how many boxes would be sealed in an 8-hour shift?
    Let b = the number of boxes sealed in 8 h.

THINK MATH

Guess and Check

How many proportions can you make using only these numbers? Each number can be used only once in each proportion.

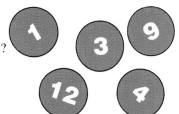

More Practice, page 434, Set B

PROBLEM SOLVING: Using the 5-Point Checklist

QUESTION
DATA
PLAN
ANSWER
CHECK

To Solve a Problem

1. Understand the QUESTION
2. Find the needed DATA
3. PLAN what to do
4. Find the ANSWER
5. CHECK back

Review the 5-Point Checklist for solving problems. Then study the problem below and its solution.

A tape recording of 42 min used 1,464 ft of tape. How many feet of tape were used to record a song that lasted 14 min?

1. **Understand the QUESTION**
   How many feet of tape are needed for 14 min of taping?

2. **Find the needed DATA**

minutes	42	14
feet	1,464	?

3. **PLAN what to do**
   Let n = the number of feet of tape needed for 14 min of taping. Write a proportion using the data and n.

   $$\frac{42}{1,464} = \frac{14}{n}$$

4. **Find the ANSWER**
   Solve the proportion for n.

   $$\frac{42}{1,464} = \frac{14}{n}$$
   $$42 \cdot n = 20,496$$
   $$n = 488$$

   488 ft of tape were used.

5. **CHECK back**
   The answer seems reasonable.

   $\frac{42}{1,464} = \frac{14}{488}$ because
   $14 \cdot 1,464 = 42 \cdot 488 = 20,496$

Solve. Use the 5-Point Checklist.

1. At a recording company, 2 out of every 7 employees are technicians. The company employs 21 people. How many are technicians?

2. A popular jazz group has averaged 2 hit singles for every 15 singles recorded. They have recorded 45 singles. How many were hits?

3. A group of musicians worked in a recording studio 6 h a day for 46 days to make an album. The studio charged $135 an hour. What was the total cost of using the studio?

4. In January, the ratio of records sold to the number of tapes sold was 5 to 2. There were 2,235 records sold. How many tapes were sold?

5. A 16-track tape uses 15 in. of tape per second. How many inches of tape are used to record a 7-minute song?

6. A salesperson sells about 11 record albums per hour. If he works 8 h a day, about how many record albums does he sell in 5 days?

7. To make a record, each singer and instrument is recorded on a different track of 24-track tape. The 24 tracks are then mixed into 2 tracks. A 35-minute recording takes about 300 h to mix. How long will it take to mix a 14-minute recording?

8. What fraction of the business was radio and TV advertisements?

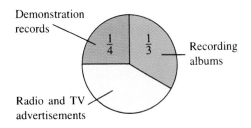

9. Lametra bought 3 cassette tapes that cost $2.79 each. She bought 2 eight-track tapes that cost $9.95 each. She paid for them with two $20 bills. How much money did she get back?

10. A pop band receives 78¢ in royalties for every album sold. Use the table to compute their royalties. What is the total amount they received?

Country	Number of Albums Sold
Japan	86,534
Germany	64,287
Australia	91,064
United States	123,250

11. **DATA BANK** If the same orchestra is heard on 3 out of 5 FM stations in Utah, how many FM stations is this? (See page 413.)

12. **Try This** The drummer in a band received her paycheck on Wednesday. She spent $\frac{1}{4}$ of it that day. She spent $\frac{1}{2}$ of what was left on Thursday, and still had $48. How much was the paycheck before any money was spent?

Similar Figures

A photograph for use in a book is 25 cm wide and 20 cm in height. A lithographer must make a reduced photograph that will be only 10 cm wide to fit the space in the book. What will be the height of the reduced photograph?

Original photograph

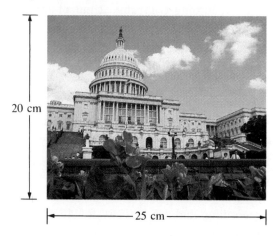

The two photographs are examples of a pair of **similar figures**. Similar figures have the same shape.

The lengths of the corresponding sides of similar figures have equal ratios. We can use these ratios to form a proportion.

Reduced photograph

Let $h =$ the height of the reduced photograph.

Original → $\dfrac{25}{10} = \dfrac{20}{h}$
Reduced →

$$25 \cdot h = 200$$
$$h = 8$$

The reduced photograph will have a height of 8 cm.

Warm Up

1. The two triangles are similar. Find the length x.

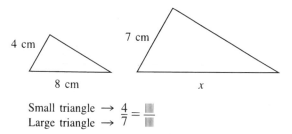

Small triangle → $\dfrac{4}{7} = \dfrac{\text{▥}}{\text{▥}}$
Large triangle →

2. The two parallelograms are similar. Find the length x.

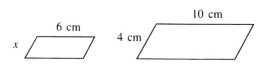

Small parallelogram → $\dfrac{6}{10} = \dfrac{\text{▥}}{\text{▥}}$
Large parallelogram →

Find the length *x* for each pair of similar figures.

1.

2.
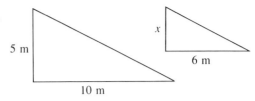

3.
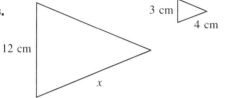

4.
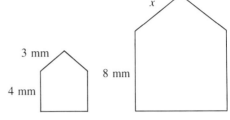

5.
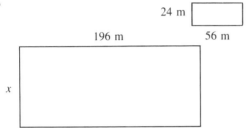

6.
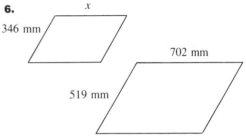

7. A small wallet-sized photograph is 4 cm wide and 5 cm in height. An enlarged copy of the photograph will be 12 cm wide. What will be the height of the enlarged photograph?

8. A rectangular picture that was 15 cm wide and 9 cm high was reduced to fit a space 5 cm wide and 3 cm high. What is the ratio of the lengths of the sides of the large picture to the small picture? Give the ratio in lowest terms.

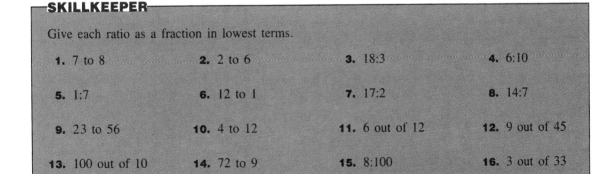

SKILLKEEPER

Give each ratio as a fraction in lowest terms.

1. 7 to 8   **2.** 2 to 6   **3.** 18:3   **4.** 6:10

5. 1:7   **6.** 12 to 1   **7.** 17:2   **8.** 14:7

9. 23 to 56   **10.** 4 to 12   **11.** 6 out of 12   **12.** 9 out of 45

13. 100 out of 10   **14.** 72 to 9   **15.** 8:100   **16.** 3 out of 33

PROBLEM SOLVING: Using Data from a Map

QUESTION
DATA
PLAN
ANSWER
CHECK

On this map of the island of Oahu, Hawaii, it is about 1.5 cm from Diamond Head to Koko Head. What is the actual distance in kilometers?

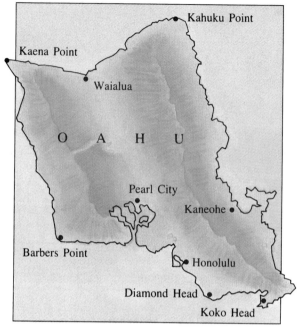

The map is a **scale drawing**. The **scale** gives the ratio of the map distance to the actual distance.

Map distance → $\frac{1}{7}$ cm
Actual distance → $\frac{1}{7}$ km

You can use this ratio to write and solve a proportion for the problem.

Let d = the actual distance from Diamond Head to Koko Head.

$\frac{1}{7} = \frac{1.5}{d}$

$1 \cdot d = 7 \cdot 1.5$

$d = 10.5$

The actual distance is about 10.5 km.

Scale: 1 cm : 7 km
↑ ↑
Map Actual
distance distance

Solve. Find all distances to the nearest tenth.

1. The map distance from Koko Head to Pearl City is 4.3 cm. What is the actual distance?

2. The map distance from Barbers Point to Kaena Point is about 5 cm. How many kilometers is the actual distance?

3. The distance from Honolulu to Barbers Point is 24 km. What is the map distance in centimeters?

4. On the map, it is 7.9 cm from Koko Head to Kahuku Point. How many kilometers is the actual distance?

5. The map distance from Honolulu to Kahuku Point is about 6.5 cm. The sailing distance is about twice this straight line distance. What is the actual sailing distance?

6. The map distance between Kaneohe and Waialua is 5.2 cm. What is the actual round-trip distance?

7. **Try This** Of 44 people on a sightseeing trip, 29 people had visited the island of Kauai, 30 had visited the island of Maui, and 6 had not visited either of these islands. How many people had visited both Kauai and Maui? Hint: Use logical reasoning.

PROBLEM SOLVING: Using Estimation

QUESTION
DATA
PLAN
ANSWER
CHECK

Choose an estimate for the answer. Then solve the problem. Compare your answer with your estimate.

1. At a marine park, about three out of four people visiting the park see the performing porpoises and whales. One day 2,416 people visited the park. How many persons saw the porpoises and whales? Hint: round 2,416 to 2,400.

   A 600   **B** 800   **C** 1,800

2. The 10 porpoises at the park eat 140 kg of fish each day. The park got 2 more porpoises. How many kilograms of fish are needed each day to feed the porpoises now?

   A 28 kg   **B** 100 kg   **C** 170 kg

3. A porpoise swam 114 km in 3 h. At this rate, what distance would the porpoise swim in 7 h?

   A 300 km   **B** 700 km   **C** 210 km

4. The ratio of the length of a common porpoise to the largest dolphin, called a killer whale, is 1 to 5. The length of a common porpoise is 1.8 m. What is the length of a killer whale?

   A 10 m   **B** 15 m   **C** 20 m

5. The ratio of the length of a bowhead whale's head to its total length is 2 to 5. Its head is 6.2 m long. What is its total length?

   A 10 m   **B** 15 m   **C** 20 m

6. **DATA BANK** What is the ratio of the length of a Baird's beaked whale to the length of a black right whale? (See page 416.)

7. **Try This** There are 25 animals that perform at the park. The park has a morning show in which 13 animals perform. There are 14 animals in the afternoon show. Six animals are in training and do not appear in either show. How many animals perform in both shows?

287

PROBLEM SOLVING: Using the Strategies

| QUESTION |
| DATA |
| PLAN |
| ANSWER |
| CHECK |

Problem Solving Strategies

- Guess and Check
- Make an Organized List
- Find a Pattern
- Draw a Picture
- Work Backward
- Solve a Simpler Problem
- Make a Table
- Choose the Operations
- Use Logical Reasoning

Choose one or more of the strategies to solve each problem below.

1. Rita ran the 100-meter dash in 13.2 s. Jessica's time was 0.3 s less than Maria's. Carla beat Rita by 0.4 s. Pat's time of 12.9 s was 0.8 s less than Maria's time. What was the order in which the five girls finished the race? What was the time for each girl?

2. Ann has $100 in a savings account and puts in $8 each week. Sid has $60 in his savings account but puts $10 each week into his account. How many weeks will it be until Sid and Ann have the same amount in their savings accounts?

3. A group of 6 tennis players are going to play in a "round robin" tournament. Each player must play 1 match with each of the other players. How many matches will be played in the tournament?

4. The Loporto family is planning a trip. They know they have to get back from their vacation on Sunday, July 5. They plan 10 days for camping in a national park. They want to spend 5 days visiting friends. They know it will take 4 days of driving in addition to the other days. What date and day of the week should they start on their trip?

CHAPTER REVIEW/TEST

Write each ratio as a fraction.

1. 4 out of 11
2. $21 for 2
3. 3 to 16
4. 7:10

Write each ratio as a fraction in lowest terms.

5. 5 to 15
6. 8:16
7. 16 out of 24
8. 20:15

Write = or ≠ for each ⬤. Use cross products to decide.

9. $\frac{3}{2}$ ⬤ $\frac{12}{8}$
10. $\frac{2}{15}$ ⬤ $\frac{6}{40}$
11. $\frac{3}{15}$ ⬤ $\frac{15}{75}$
12. $\frac{4}{3}$ ⬤ $\frac{20}{12}$

Solve each proportion.

13. $\frac{5}{9} = \frac{20}{x}$
14. $\frac{12}{18} = \frac{x}{21}$
15. $\frac{x}{12} = \frac{14}{6}$
16. $\frac{3}{7} = \frac{x}{35}$

Find the length x for each pair of similar figures.

17.

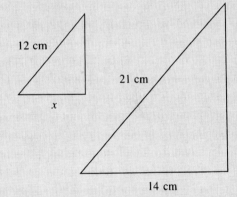

19. The scale on a map is 1 cm: 25 km. The distance between Hampton and Bellville on the map is 5 cm. What is the actual distance between the two towns?

20. The distance between two cities is 180 km. They are shown on a map with a scale of 1 cm representing 40 km. What is the map distance in centimeters?

21. There are 3 books for every 4 students. How many books are there for 24 students?

22. Kinuko drove 170 km in 2 h. At this rate, how far will she drive in 3 h?

ANOTHER LOOK

Write each ratio as a fraction.

1. 6 to 4
2. 3 out of 5
3. 9:7
4. 3 for $15
5. 5.4 to 1.8
6. 174 to 560

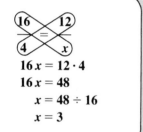

Write = or ≠ for each ●. Use cross products to decide.

7. $\frac{1}{2}$ ● $\frac{2}{4}$
8. $\frac{4}{16}$ ● $\frac{2}{8}$
9. $\frac{3}{5}$ ● $\frac{4}{7}$
10. $\frac{9}{10}$ ● $\frac{7}{8}$
11. $\frac{2}{3}$ ● $\frac{4}{6}$
12. $\frac{7}{6}$ ● $\frac{4}{3}$

Solve each proportion.

13. $\frac{x}{32} = \frac{5}{8}$
14. $\frac{12}{18} = \frac{x}{6}$
15. $\frac{6}{16} = \frac{18}{x}$
16. $\frac{x}{24} = \frac{2}{3}$
17. $\frac{21}{x} = \frac{7}{8}$
18. $\frac{15}{20} = \frac{x}{12}$

Find the missing lengths in each pair of similar figures.

19.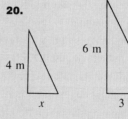

20.

21.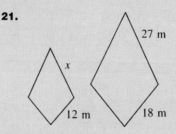

22.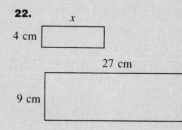

ENRICHMENT

Finding the Golden Ratio

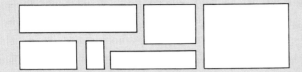

Which rectangle do you like best?

Some rectangles appear more pleasing to the eye than others.

The ancient Greeks believed that a certain shape of rectangle, called a **golden rectangle**, was the best shape. Golden rectangles have been found in Greek art and architecture and are still in use today.

The Erechtheion

Use graph paper and follow the steps below to make a golden rectangle.

1. Draw square ABCD.
2. Mark M, the midpoint of $\overline{AB}$, and draw $\overline{MC}$.
3. With a compass point at M, draw $\overset{\frown}{CP}$.
4. Draw $\overline{PN}$ and complete the rectangle APND.

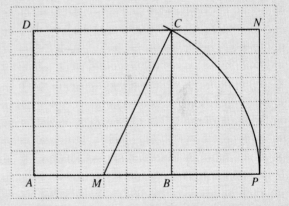

Rectangle APND is a golden rectangle. The **golden ratio** is the ratio of the length divided by the width of a golden rectangle.

What is the golden ratio of your golden rectangle? Give the ratio as a decimal rounded to the hundredths place.

A spiral drawn in a golden rectangle resembles the spiral seen in the shell of the nautilus.

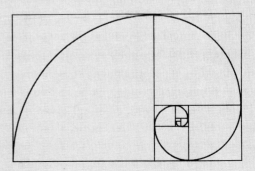

CUMULATIVE REVIEW

1. Which is correct?
   - A $1\frac{1}{2} > 1\frac{1}{3}$
   - B $5\frac{3}{4} < 5\frac{2}{3}$
   - C $2\frac{3}{4} > 2\frac{7}{8}$
   - D not given

2. Find the missing number.
   $\frac{12}{14} = \frac{\square}{28}$
   - A 12
   - B 6
   - C 24
   - D not given

3. What is $\frac{9}{54}$ in lowest terms?
   - A $\frac{1}{8}$
   - B $\frac{1}{6}$
   - C $\frac{3}{5}$
   - D not given

4. What is the mixed number for $\frac{52}{9}$?
   - A $6\frac{1}{9}$
   - B $5\frac{8}{9}$
   - C $5\frac{7}{9}$
   - D not given

Add.

5. $\frac{9}{16} + \frac{5}{8}$
   - A $\frac{7}{8}$
   - B $1\frac{3}{4}$
   - C $\frac{19}{16}$
   - D not given

6. $12\frac{3}{4}$
   $+ \ 5\frac{7}{8}$
   - A $17\frac{5}{8}$
   - B $18\frac{5}{8}$
   - C $18\frac{1}{2}$
   - D not given

7. Subtract.
   $\frac{3}{4} - \frac{7}{12}$
   - A $\frac{1}{6}$
   - B $\frac{1}{12}$
   - C $\frac{1}{4}$
   - D not given

Change each measurement to millimeters.

8. 124 cm
   - A 0.124 mm
   - B 12.4 mm
   - C 1,240 mm
   - D not given

9. 9.2 m
   - A 9,200 mm
   - B 0.092 mm
   - C 920 mm
   - D not given

10. What is the perimeter?

    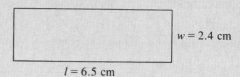

    - A 11.3 cm
    - B 17.8 cm
    - C 15.6 cm
    - D not given

11. What is the area?

    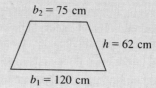

    - A 319 cm²
    - B 159.5 cm²
    - C 257 cm²
    - D not given

12. What is the volume?

    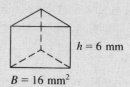

    - A 96 mm³
    - B 28 mm³
    - C 192 mm³
    - D not given

13. A nail is $2\frac{3}{8}$ in. long. A piece of wood is $1\frac{1}{4}$ in. thick. How much longer is the nail than the thickness of the wood?
    - A $1\frac{1}{4}$ in.
    - B $1\frac{1}{8}$ in.
    - C $1\frac{3}{4}$ in.
    - D not given

14. What is the total length of molding needed for 18 windows that each measure 85 cm by 105 cm?
    - A 6,840 cm
    - B 3,420 cm
    - C 8,925 cm
    - D not given

12 Percent

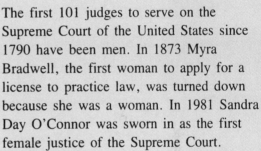

The first 101 judges to serve on the Supreme Court of the United States since 1790 have been men. In 1873 Myra Bradwell, the first woman to apply for a license to practice law, was turned down because she was a woman. In 1981 Sandra Day O'Connor was sworn in as the first female justice of the Supreme Court.

The number of women entering the law profession rises every year. The greatest changes have happened since 1970. From 1950 to 1970 the percentage of women lawyers was fairly constant, from 2.5 percent in 1950 to 2.8 percent in 1970. By 1980, the number of women in the profession had grown to 7.5 percent. At least one out of every three students entering law school was a woman. In the year 2000, women are expected to make up 35 to 40 percent of the law profession.

Percent

A full sheet of postage stamps had 100 stamps. There are 78 stamps left.

We can describe the part of the sheet of stamps in several ways.

Ratio: **78** out of **100** stamps are left.

Fraction: $\frac{78}{100}$ of the stamps are left.

Decimal: **0.78** of the stamps are left.

We can also use a **percent**. **Percent** means **per hundred**. The symbol for percent is **%**. 78 percent or 78% of the stamps are left.

What part of the large square is shaded?

There are 25 squares out of 100 squares shaded.

Fraction: $\frac{25}{100}$
Decimal: **0.25**
Percent: **25%**

Other Examples

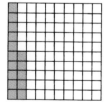

Fraction: $\frac{15}{100}$
Decimal: **0.15**
Percent: **15%**

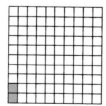

Fraction: $\frac{2}{100}$
Decimal: **0.02**
Percent: **2%**

Fraction: $\frac{100}{100}$
Decimal: **1.00**
Percent: **100%**

What percent of each square is shaded?

1.
2.
3.
4.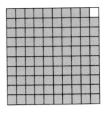

5.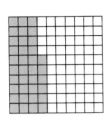
6.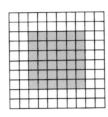
7.
8.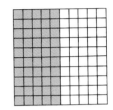

One dollar is 100 cents. What percent of a dollar is each amount?
Example: 35¢ is 35% of a dollar.

9. 15¢
10. 1¢
11. 50¢
12. 95¢
13. 38¢
14. $0.25
15. $0.64
16. $0.08
17. $0.41
18. $0.03

Copy and complete the table.

	Ratio	Fraction	Decimal	Percent
19.	8 to 100			
20.	17 to 100			
21.	6 per hundred			
22.	37 out of 100			
23.	39 to 100			
24.	95 to 100			
25.	100 to 100			

26. A sheet of 100 stamps had 39 stamps left. What percent of the stamps were left?

27. A sheet of 100 stamps had 27 stamps left. Then 18 of these stamps were used. What percent of 100 stamps are there now?

Percents and Fractions

A pullover is 80% cotton. What fractional part of the material in the pullover is cotton? Give the part as a lowest-terms fraction.

$$80\% = \frac{80}{100} = \frac{80 \div 20}{100 \div 20} = \frac{4}{5}$$

The lowest-terms fraction for 80% is $\frac{4}{5}$.

The pullover is $\frac{4}{5}$ cotton.

The pullover is marked "$\frac{1}{4}$ off." Give the fraction as a percent.

$$\frac{1}{4} = \frac{1 \times 25}{4 \times 25} = \frac{25}{100} = 25\%$$

The pullover is marked 25% off.

Other Examples

Percents to fractions

$$45\% = \frac{45}{100} = \frac{9}{20}$$

$$125\% = \frac{125}{100} = \frac{5}{4}$$

Fractions to percents

$$\frac{2}{5} = \frac{2 \times 20}{5 \times 20} = \frac{40}{100} = 40\%$$

$$3 = \frac{3}{1} = \frac{300}{100} = 300\%$$

Warm Up

Give the lowest-terms fraction for each percent.

1. 5%   2. 16%   3. 8%   4. 20%   5. 50%
6. 84%   7. 60%   8. 2%   9. 33%   10. 85%

Give the percent for each fraction.

11. $\frac{1}{10}$ 12. $\frac{3}{4}$ 13. $\frac{3}{20}$ 14. $\frac{4}{5}$ 15. $\frac{1}{100}$

16. $\frac{5}{20}$ 17. $\frac{99}{100}$ 18. $\frac{7}{1}$ 19. $\frac{8}{10}$ 20. $\frac{3}{2}$

Write the lowest-terms fraction for each percent.

	%	Fraction		%	Fraction		%	Fraction
1.	10%		5.	40%		9.	75%	
2.	20%		6.	50%		10.	80%	
3.	25%		7.	60%		11.	90%	
4.	30%		8.	70%		12.	100%	

Write each fraction as a percent.

13. $\frac{8}{25}$ 14. $\frac{9}{1}$ 15. $\frac{4}{20}$ 16. $\frac{39}{100}$ 17. $\frac{8}{5}$

18. $\frac{13}{50}$ 19. $\frac{3}{25}$ 20. $\frac{11}{25}$ 21. $\frac{5}{4}$ 22. $\frac{10}{1}$

Write each fraction in lowest terms. Then write the percent for the fraction.

23. $\frac{9}{12}$ 24. $\frac{7}{14}$ 25. $\frac{9}{36}$ 26. $\frac{21}{30}$ 27. $\frac{18}{40}$

28. $\frac{36}{60}$ 29. $\frac{99}{198}$ 30. $\frac{36}{18}$ 31. $\frac{29}{116}$ 32. $\frac{144}{1,600}$

33. A suit is 5% silk. What is the lowest-terms fraction for the part of the suit that is silk?

34. A dress is 60% polyester, 25% cotton, and the rest is silk. What percent of the dress is silk? What is the lowest-terms fraction for the part that is silk?

35. A coat is $\frac{4}{5}$ wool, $\frac{1}{10}$ linen, and the rest is polyester. What percent of the coat is made of linen and wool together?

36. **DATA HUNT** Find an article of clothing that is made of different materials. Find the percent of each material. What is the fraction in lowest terms for each percent?

THINK MATH

Guess and Check

What is the greatest amount of money you can have (using pennies, nickels, dimes, quarters, and half dollars) and still not be able to give someone change for a dollar?

More Practice, page 434, Set C

Percents and Decimals

To use percents in problem solving, we must often write a decimal for a percent.

What is the decimal for 13.3%? What is the decimal for 10.7%? Remember that percent means per hundred.

$13.3\% = \frac{13.3}{100} = 0.133$

$10.7\% = \frac{10.7}{100} = 0.107$

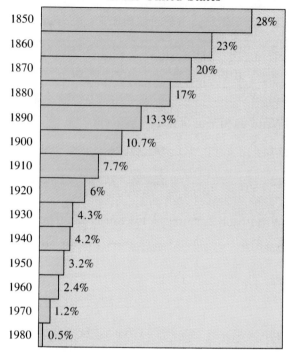

The Decreasing Percent of Illiteracy in the United States

Year	Percent
1850	28%
1860	23%
1870	20%
1880	17%
1890	13.3%
1900	10.7%
1910	7.7%
1920	6%
1930	4.3%
1940	4.2%
1950	3.2%
1960	2.4%
1970	1.2%
1980	0.5%

Percent of illiteracy

Dividing by 100 shifts the decimal point 2 places to the left. This is a shortcut.

26.7% = 0.267   125% = 1.25

7.8% = 0.078   0.5% = 0.005

63% = 0.63   200% = 2.00

To write a percent for a decimal, shift the decimal 2 places to the right.

0.287 = 28.7%   1.33 = 133%

0.033 = 3.3%   0.008 = 0.8%

0.07 = 7%   5.00 = 500%

Write each percent as a decimal.

1. 38%   2. 16%   3. 8%   4. 79%   5. 11%
6. 23.5%   7. 12.8%   8. 9.6%   9. 10.2%   10. 18.45%
11. 0.6%   12. 105%   13. 325%   14. 1.55%   15. 100%

Write each decimal as a percent.

16. 0.48   17. 0.95   18. 0.09   19. 0.19   20. 0.02
21. 0.064   22. 0.1667   23. 0.028   24. 0.505   25. 0.0975
26. 1.36   27. 3   28. 0.0025   29. 10.00   30. 0.0125

More Practice, page 435, Set A

Estimating Percents

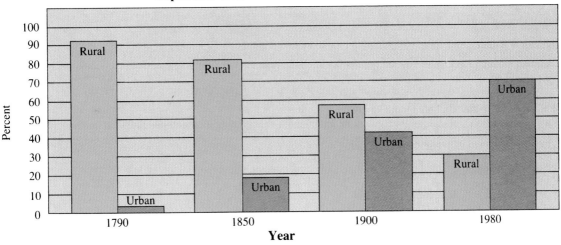

What is an estimate of the percent of the population in 1790 that was rural?

The graph shows more than 90% but less than 95%.

Estimate: About 93% of the population was rural.

Estimate the percents.

1. What percent of the population in 1850 was urban? What percent was rural?

2. What percent of the population in 1900 was urban? What percent was rural?

3. What percent of the population in 1980 was urban? What percent was rural?

4. What was the increase of the percent of urban population from 1790 to 1980?

Andrew made a circle graph showing how he uses his leisure time. Use the circle graph to estimate each percent.

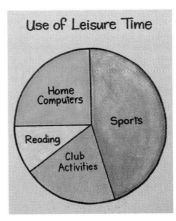

5. About what percent of the leisure time does the graph show for using the computer?

6. About what percent of the time is shown for sports?

7. About what percent of the time is shown for reading?

8. About what percent of the time is shown for club activities?

Finding Percents for Other Fractions

Serena Miller is a meteorologist. She found that in 7 out of 30 days of June it had rained. What percent of the days in June had it rained?

To find the percent for 7 out of 30, we can find a decimal for $\frac{7}{30}$ by dividing. Then we can write a percent for the decimal.

$$30\overline{)7.00}^{\,0.23\frac{10}{30}} = 0.23\frac{1}{3} = 23\frac{1}{3}\%$$

It rained on $23\frac{1}{3}\%$ of the days.

Sometimes percents are rounded to the **nearest whole percent**. $23\frac{1}{3}\%$ rounded to the nearest whole percent is 23%.

Other Examples

What percent is $\frac{6}{7}$?

$7\overline{)6.00}^{\,0.85\frac{5}{7}}$ $\frac{6}{7} = 85\frac{5}{7}\%$

To the nearest whole percent, this is 86%.

What percent is $\frac{1}{8}$?

$8\overline{)1.000}^{\,0.125} = 12.5\%$ $\frac{1}{8} = 12.5\% = 12\frac{1}{2}\%$

To the nearest whole percent, this is 13%.

Warm Up

Give the percent for each fraction. Round to the nearest whole percent.

1. $\frac{1}{3}$
2. $\frac{3}{8}$
3. $\frac{1}{6}$
4. $\frac{7}{8}$
5. $\frac{5}{8}$
6. $\frac{5}{9}$
7. $\frac{1}{8}$
8. $\frac{5}{12}$
9. $\frac{9}{16}$
10. $\frac{7}{12}$

Find the percent for each fraction. Round to the nearest whole percent.

1. $\frac{5}{6}$ 2. $\frac{8}{12}$ 3. $\frac{9}{11}$ 4. $\frac{8}{9}$ 5. $\frac{5}{11}$

6. $\frac{11}{16}$ 7. $\frac{10}{17}$ 8. $\frac{5}{18}$ 9. $\frac{11}{12}$ 10. $\frac{13}{16}$

11. $\frac{6}{9}$ 12. $\frac{13}{18}$ 13. $\frac{17}{24}$ 14. $\frac{33}{40}$ 15. $\frac{19}{35}$

16. $\frac{26}{27}$ 17. $\frac{17}{29}$ 18. $\frac{5}{18}$ 19. $\frac{29}{60}$ 20. $\frac{35}{45}$

The decimal for each fraction has been found using a calculator.
Write each decimal as a percent rounded to the nearest tenth.

Example: $\frac{84}{152} = 0.5526316$ 55.26316% rounded to the nearest tenth is 55.3%

21. $\frac{75}{160} = 0.46875$ 22. $\frac{33}{111} = 0.2972972$ 23. $\frac{71}{90} = 0.7888889$

24. $\frac{133}{87} = 1.5287356$ 25. $\frac{110}{245} = 0.4489795$ 26. $\frac{1,475}{1,525} = 0.9672131$

27. In Cleveland, Ohio, 212 out of 365 days were cloudy. What percent of the days were cloudy? Round the percent to the nearest whole percent.

28. In Albuquerque, New Mexico, only 84 out of 365 days were cloudy. What percent of the days were not cloudy? Round the percent to the nearest tenth of a percent.

★ 29. In Bridge City, about 66% of the days are cloudy. Estimate how many days a year are cloudy.

30. **DATA BANK** How many days of precipitation does Miami, Florida average during the month of August? What is the percent of days in August with precipitation? Round the percent to the nearest whole percent. (See page 414.)

SKILLKEEPER

Find the area of each triangle. Use $A = \frac{1}{2}bh$.

1. $b = 20$ cm
   $h = 12$ cm
2. $b = 32$ m
   $h = 16$ m
3. $b = 4.5$ cm
   $h = 2.6$ cm
4. $b = 80$ m
   $h = 62$ m
5. $b = 2.6$ cm
   $h = 0.8$ cm

Find the volume of each prism. Use $V = Bh$.

6. $B = 16$ cm²
   $h = 10$ cm
7. $B = 12.2$ cm²
   $h = 5.9$ cm
8. $B = 6$ cm²
   $h = 1.9$ cm
9. $B = 295$ mm²
   $h = 27$ mm
10. $B = 0.92$ m²
    $h = 0.002$ m

More Practice, page 435, Set B

Finding a Percent of a Number

In a class of 25 students, 24% of the students play a musical instrument. How many students play a musical instrument?

Since 24% of 25 means 24% × 25, we can change the percent to a decimal and multiply.

24% = 0.24

```
    2 5
  × 0.2 4
  ─────
    1 0 0
    5 0
  ─────
    6.0 0
```

There are 6 students who play a musical instrument.

Other Examples

Find 20% of 60.

Since 20% = $\frac{1}{5}$, it is easy to use the fraction for the percent.

$\frac{1}{5} \times 60 = 12$

What is 7% of 85?

```
    8 5
  × 0.0 7
  ─────
    5.9 5
```

Find 150% of 240.

```
    2 4 0
  × 1.5 0
  ─────
    1 2 0 0 0
    2 4 0
  ─────
    3 6 0.0 0
```

Warm Up

Give the percent of each number. Use a lowest-terms fraction for the percent.

1. 10% of 80
2. 50% of 38
3. 75% of 28
4. $33\frac{1}{3}$% of 15
5. 80% of 30
6. 25% of 108

Give the percent of each number. Use a decimal for the percent.

7. 37% of 26
8. 77% of 125
9. 67% of 130
10. 8.5% of 200
11. 123% of 94
12. 0.5% of 700

Find the percent of each number. Use a lowest-terms fraction for each number.

1. 25% of 80
2. 50% of 62
3. 80% of 75
4. 10% of 250
5. 30% of 30
6. 75% of 36
7. 40% of 45
8. 90% of 300

Find the percent of each number. Use a decimal for the percent.

9. 21% of 54
10. 18% of 325
11. 95% of 700
12. 3% of 66
13. 150% of 48
14. 110% of 60
15. 6.5% of 200
16. 12.5% of 124

Find the percent of each number.

17. 24% of 84
18. 50% of 214
19. 10.3% of 78
20. 25% of 200
21. 75% of 60
22. 13% of 620
23. 45% of 250
24. 200% of 800

Find 1% of each number.

Example: 1% of 800 = 0.01 × 800 = 8.00

25. 400
26. 1,000
27. 750
28. 80
29. 2,500
30. 260
31. 2,860
32. 50,000

Find 10% of each number.

Example: 10% of 450 = 0.1 × 450 = 45.0

33. 300
34. 870
35. 1,600
36. 25
37. 2,000
38. 8,000
39. 60,000
40. 1,000,000

41. In a class of 25 students, 8% of the students play the piano. How many students play the piano?

42. In a school of 1,640 students, 10% of the students either sing in the chorus or play in the band. There are 73 students in the chorus. How many students are in the band?

43. A school has a total enrollment of 480 students. One day 97.5% of the students were in school. How many students were in school? How many students were absent?

44. Make up a problem about finding the percent of a number. Use real data from your school or your class. Solve the problem.

More Practice, page 435, Set C

Finding Simple Interest

Darlene Fields put $800 in a savings account that will earn **interest** at a **rate** of 9% per year. She plans to keep the money in the savings account for 3 years. How much interest will be earned? What will be the **amount** after 3 years?

To find the interest, we can use the interest formula $I = PRT$.

$P = \$800$, $R = 9\%$, $T = 3$ years
$I = \$800 \times 0.09 \times 3$
$I = \$216.00$

The interest is $216.

To find the amount after 3 years, we add the **principal** to the interest.

$A = P + I$
$P = \$800$, $I = \$216$
$A = \$800 + \$216 = \$1{,}016$

The amount is $1,016.

Principal (P):	Amount loaned or borrowed
Interest (I):	Charge for the use of money
Rate (R):	A percent of the principal charged as interest
Time (T):	Length of time the principal is loaned or borrowed
Amount (A):	Principal plus interest

Interest = Principal · Rate · Time
$I = PRT$

Amount = Principal + Interest
$A = P + I$

Other Examples

Find the interest and amount on a principal of $2,000 at 10% per year for 18 months.

$P = \$2{,}000$ $I = PRT$ $A = P + I$
$R = 10\% = 0.10$ $I = \$2{,}000 \times 0.10 \times 1.5 = \$300$ $A = \$2{,}000 + \$300 = \$2{,}300$
$T = 18$ months $= 1.5$ years

The interest is $300. The amount is $2,300.

Find the interest.

1. $P = \$3{,}000$
   $R = 12\%$ per year
   $T = 2$ years

2. $P = \$20{,}000$
   $R = 16\%$ per year
   $T = 6$ months
   (Use 0.5 year.)

3. $P = \$2{,}500$
   $R = 1.5\%$ per month
   $T = 5$ months

Find the amount.

4. $P = \$650$
   $I = \$26$

5. $P = \$1{,}800$
   $R = 10\%$ per year
   $T = 5$ years

6. $P = \$9{,}000$
   $R = 18\%$ per year
   $T = 2.5$ years

PROBLEM SOLVING: Using Simple Interest Formulas

QUESTION
DATA
PLAN
ANSWER
CHECK

Solve.

1. If a rate of 8% per year were paid on a principal of $500, what would be the interest at the end of 1 year?

2. If a rate of 15% per year were paid on a principal of $6,600, what would be the interest for 3 months? (Use 0.25 of a year.)

3. Will Maxon borrowed $500 for 1 year. The rate was 19% per year. What was the interest on the principal?

4. Gwen Masoni put $5,000 into a savings account that paid an interest rate of 12% per year. What is the amount she will have at the end of 6 years?

5. Isabelle Wilson put $3,600 into a money-market fund that paid a 14% rate of interest per year. How much interest would the principal earn in 2 months? (Use $\frac{1}{6}$ year for 2 months.)

6. Kevin Hough had $600 in a savings account that paid an interest rate of 6% per year. At the end of 4 months Kevin decided to draw out all his money, including the interest. What was the total amount?

7. Janalee Hughes bought a used car. She borrowed $3,800 for 4 years at a rate of 17% per year. What was the amount that she had to repay?

8. Calvin Wong borrowed $2,000 for 1 year. At the end of the year he repaid his loan with $2,300. How much was the interest? What was the interest rate?

9. A business borrowed $32,000 for 1 year at an interest rate of 18%. What is the amount that must be repaid in 1 year?

★ 10. Sylvia can earn up to $2,000 tax-free interest in 1 year. How much principal must she invest at an interest rate of 16% per year in order to earn $2,000 interest?

11. **Try This** A bank has checkbook covers that are either pocket- or desk-size. The covers are orange, black, red, or tan. The customer's name will be stamped on the cover in gold or silver. How many different combinations of covers are possible? Hint: Make an organized list.

Finding the Percent One Number Is of Another

A campground has 80 camping spaces. One night, 52 of the spaces were in use. The manager had to record the percent of spaces in use. What percent of the spaces were in use?

To solve the problem, we can find the percent for the ratio of the two numbers.

Ratio: **52** out of **80** spaces → $\frac{52}{80}$

Percent: $\frac{52}{80} = 0.65 = 65\%$

65% of the spaces were in use.

Check: 65% of 80 = 0.65 × 80 = 52

Other Examples

What percent of 20 is 16?

Ratio: $\frac{16}{20}$

Percent: $\frac{16}{20} = 0.80 = 80\%$

12 is what percent of 8?

Ratio: $\frac{12}{8}$

Percent: $\frac{12}{8} = 1.50 = 150\%$

What percent is 9 out of 15?

Ratio: $\frac{9}{15}$

Percent: $\frac{9}{15} = 0.60 = 60\%$

What percent of 4 is 12?

Ratio: $\frac{12}{4}$

Percent: $\frac{12}{4} = 3.00 = 300\%$

Warm Up

1. What percent of 12 is 6?
2. What percent of 8 is 20?
3. 3 is what percent of 30?
4. What percent of 4 is 6?
5. What percent of 16 is 4?
6. 18 is what percent of 24?

Find the percents.

1. 12 is what percent of 60?
2. What percent is 8 out of 25?
3. 30 is what percent of 40?
4. What percent of 72 is 18?
5. What percent is 9 out of 12?
6. 24 is what percent of 12?
7. 3 is what percent of 8?
8. What percent of 15 is 45?
9. What percent of 35 is 14?
10. 8 is what percent of 400?
11. What percent of 2 is 5?
12. What percent is 20 out of 25?

Find the percents. Round each percent to the nearest whole percent.

Example: 2 out of 3 is what percent?

$\frac{2}{3} = 0.66\frac{2}{3} = 66\frac{2}{3}\%$ or 67%, rounded to the nearest whole percent.

13. What percent is 11 out of 33?
14. 18 is what percent of 66?
15. What percent of 9 is 12?
16. 15 is what percent of 16?
17. What percent is 21 out of 78?
18. What percent of 360 is 6?
19. 216 is what percent of 512?
20. What percent is 475 out of 1,800?
21. A campground had only 24 out of 80 camping spaces filled. What percent of the camping spaces were filled?
22. There were 120 requests for 80 camping spaces. To the nearest whole percent, what percent of the 120 requests could not be given a camping space?

THINK MATH

Magic Square Percents

Find the percent for each ratio. Write the percent in the matching square. The correct answers in the squares will form a Magic Square. The sum in each row, column, and diagonal should be the same.

A 63 to 175
D 3.875 to 12.5
G 0.4 to 1.25
B 7.25 ÷ 25
E 2,211 : 6,700
H $\frac{29 + 45}{2^3 \cdot 5^2}$
C $\frac{153}{450}$
F 91 out of 260
I $\frac{100 - 67}{4,267 - 4,157}$

What is the magic sum?

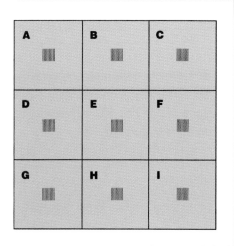

More Practice, page 436, Set A

PROBLEM SOLVING: Using Percents for Test Scores

QUESTION
DATA
PLAN
ANSWER
CHECK

Beverly got 27 out of 30 problems correct on a math test. What percent of the problems did she get correct?

Since you want to find a percent for a ratio, you can find a decimal and then a percent for the ratio.

Number correct → $\dfrac{27}{30} = \dfrac{9}{10} = 0.90 = 90\%$
Total number →

Beverly got 90% of the problems correct. Her test score was 90%.

Jake got 25 out of 30 problems correct on a spelling test. What was his percent score?

$\dfrac{25}{30} = 0.83\dfrac{1}{3} = 83\dfrac{1}{3}\%$

Jake's test score was $83\dfrac{1}{3}\%$. His score to the nearest whole percent is 83%.

Solve. Round all percents to the nearest whole percent.

1. Tish got 19 out of 20 questions correct on a science test. What was her test score?

2. Bud took a multiple-choice test. There were 100 items on the test. His test score was 83%. How many items did he answer correctly?

3. The lowest passing score for a test was 70%. Elise got 2 out of every 3 problems correct. What was her test score? Was it a passing score?

4. A teacher said that 68% would be the lowest passing test score. Kanisha got 19 out of 28 problems correct. What was her test score? Was it a passing score?

5. Akira got 48 out of 50 spelling words correct on a spelling test. What was his test score?

6. Valerie's teacher told her that her test score was 85%. If there were 40 problems on the test, how many problems did Valerie get correct?

10. On a semester test, Hanna got 65 out of 75 problems correct. What was her test score?

11. On a true-false quiz of 50 questions, Randy was sure he knew the correct answers to 36 of the questions. He guessed at the rest of the questions, expecting to get one half of them correct. What test score does Randy think he should make?

12. A driver's license test has 40 questions. To pass the test, no more than 6 questions can be missed. What is the lowest passing score?

13. Monica got 9 out of 10 problems correct for 90% on one test. She got 16 out of 20 problems correct for 80% on another test. What percent of the problems on both tests did she get correct? Is this the same as the average of 90% and 80%?

14. A passing grade on a test with 45 true-false questions is 72%. Gil got 36 questions correct. Is this a passing grade?

15. Check your answers to problems 1–14. What percent of the problems did you answer correctly?

16. Try This A class of 35 students took a math test and a science test. 12 students got 100% on the math test. 9 students got 100% on the science test. There were 19 students who made less than 100% on both tests. How many students made 100% on both tests?

7. Elliot's test score on a true-false test of 60 questions was 90%. How many questions did Elliot miss?

8. Ryan got 27 out of 32 questions correct on a social studies test. What was his test score?

9. Marquita got 17 out of 20 problems correct on Monday's quiz. She got 13 out of 15 problems correct on Wednesday's quiz. What was her test score for each quiz? On which day was her score the highest?

Finding a Number when a Percent of It Is Known

The volcano in Washington called Mount St. Helens erupted in 1980 and blew off part of its peak. It then had a height of 8,419 ft. This is only about 87% of its height before the eruption. What was its height before the eruption?

To find the height before eruption, we can write an equation.

Let h = the height before eruption.

Percent equation $\qquad 87\% \times h = 8{,}419$

Write the percent as a decimal. $\qquad 0.87 \times h = 8{,}419$

Divide both sides by 0.87. $\qquad \dfrac{0.87}{0.87} \times h = \dfrac{8{,}419}{0.87}$

Simplify and round to the nearest whole number. $\qquad h = 9{,}677$

The height before the eruption was about 9,677 ft.

Other Examples

15% of what number is 24?

$15\% \times n = 24$
$0.15 \times n = 24$
$\dfrac{0.15}{0.15} \times n = \dfrac{24}{0.15}$
$\qquad n = 160$

40% of what number is 1.8?

$40\% \times n = 1.8$
$0.40 \times n = 1.8$
$\dfrac{0.40}{0.40} \times n = \dfrac{1.8}{0.40}$
$\qquad n = 4.5$

Warm Up

Solve.

1. $8\% \times n = 4$
2. $50\% \times n = 52$
3. $1\% \times n = 19$
4. $13\% \times n = 52$
5. $200\% \times n = 84$
6. $80\% \times n = 15$

Solve.

1. $5\% \times n = 20$
2. $16\% \times n = 40$
3. $10\% \times n = 6$
4. $50\% \times n = 22$
5. $6\% \times n = 21$
6. $75\% \times n = 15$
7. $9\% \times n = 63$
8. $12\% \times n = 3.6$
9. $100\% \times n = 83$
10. $7\% \times n = 84$
11. $17\% \times n = 136$
12. $15\% \times n = 2.10$

Write and solve an equation for each question.

13. 20% of what number is 18?
14. 45% of what number is 36?
15. 56% of what number is 42?
16. 99% of what number is 495?
17. 9% of what number is 2.7?
18. 60% of what number is 333?
19. 150% of what number is 9?
20. 15% of what number is 48?

Find the number.

21. 31% of $n = 4.65$
22. $57\% \times n = 17.1$
23. $9\% \times n = 33.48$
24. $67\% \times n = 548.73$
25. $2.5\% \times n = 4.70$
26. $106\% \times n = 97.52$
27. $29\% \times n = 104.4$
28. $72\% \times n = 45.36$
29. $6.3\% \times n = 6224.4$

30. At one time Mt. Mazama, the mountain of Crater Lake, Oregon, was about 12,000 ft high. This is about 150% of its present height. What is the height of Mt. Mazama today?

31. About $4\frac{1}{2}$ million years ago, the Grand Canyon may have been about 3,960 ft deep. This is about 75% of its depth today. What is the depth of the Grand Canyon today?

SKILLKEEPER

Write each fraction as a percent.
Round to the nearest whole percent.

1. $\frac{7}{15}$
2. $\frac{10}{12}$
3. $\frac{5}{16}$
4. $\frac{9}{14}$
5. $\frac{2}{9}$
6. $\frac{7}{32}$
7. $\frac{17}{20}$
8. $\frac{9}{40}$
9. $\frac{5}{11}$
10. $\frac{12}{27}$
11. $\frac{7}{25}$
12. $\frac{8}{22}$
13. $\frac{45}{48}$
14. $\frac{9}{20}$
15. $\frac{32}{40}$

More Practice, page 436, Set B

Skills Practice

Write each percent as a decimal.

1. 15%
2. 39%
3. 75%
4. 8%
5. 20%
6. 3%
7. 400%
8. 33%
9. 62%
10. 125%

Write each decimal as a percent.

11. 0.35
12. 0.18
13. 0.39
14. 0.50
15. 0.07
16. 2.50
17. 0.01
18. 0.99
19. 0.002
20. 6.8

Find the percent for each fraction. Round to the nearest tenth of a percent.

21. $\frac{3}{5}$
22. $\frac{6}{10}$
23. $\frac{3}{4}$
24. $\frac{7}{20}$
25. $\frac{11}{25}$
26. $\frac{3}{8}$
27. $\frac{79}{100}$
28. $\frac{3}{2}$
29. $\frac{1}{3}$
30. $\frac{5}{6}$
31. $\frac{9}{10}$
32. $\frac{4}{9}$
33. $\frac{11}{12}$
34. $\frac{4}{15}$
35. $\frac{8}{5}$

Find the lowest-terms fraction for each percent.

36. 25%
37. 10%
38. 80%
39. 15%
40. 4%
41. 16%
42. 45%
43. 17%
44. 28%
45. 96%
46. 150%
47. 200%
48. 76%
49. 55%
50. 125%

Find the percent of each number.

51. 50% of 48
52. 10% of 300
53. 5% of 80
54. 18% of 2,500
55. 108% of 300
56. 6% of 75

Find the percents.

57. 7 is what percent of 10?
58. 16 is what percent of 20?
59. 9 is what percent of 15?
60. 45 is what percent of 150?

Find the numbers.

61. 10% of a number is 8.
62. 25% of a number is 40.
63. 90% of a number is 315.
64. 64% of a number is 16.

Discounts and Sale Prices

A garden shop reduced the price of a lawnmower by 20%. The regular price was $350. What was the sale price?

The amount that the price was reduced is called the **discount**. The **percent of discount** is 20%.

The diagram shows the relation between the regular price, the sale price, and the percent of discount.

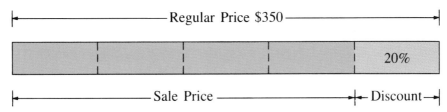

Discount percent × regular price = discount

20% × $350 = discount
0.20 × $350 = $70
The discount is $70.

Regular price − discount = sale price

$350 − $70 = sale price
$350 − $70 = $280
The sale price is $280.

Find the discount and the sale price.

1. Regular Price: $60
   Discount Percent: 15%

2. Regular Price: $75
   Discount Percent: 50%

3. Regular Price: $240
   Discount Percent: 20%

4. Regular Price: $27.50
   Discount Percent: 10%

5. Regular Price: $12.50
   Discount Percent: 40%

6. Regular Price: $6,500
   Discount Percent: 8%

Solve.

7. Potted citrus trees had a regular price of $16.00. They were put on sale at 30% off. What was the discount? What was the sale price?

8. Peat moss was regularly priced at $10 a bale. It was put on sale at $7.50 a bale. What was the discount? What was the discount percent?

Proportions and Percents

Percent problems can be solved by writing and solving a **proportion**. Each ratio in the proportion compares a **part** to a **whole**.

Find 15% of 80.

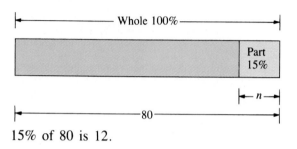

$\text{part} \rightarrow \dfrac{15}{100} = \dfrac{n}{80} \leftarrow \text{part}$
$\text{whole} \rightarrow \phantom{\dfrac{15}{100}} \phantom{\dfrac{n}{80}} \leftarrow \text{whole}$

$100\,n = 1{,}200$
$n = 12$

15% of 80 is 12.

30 is what percent of 50?

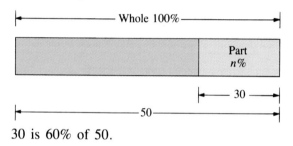

$\text{part} \rightarrow \dfrac{n}{100} = \dfrac{30}{50} \leftarrow \text{part}$
$\text{whole} \rightarrow \phantom{\dfrac{n}{100}} \phantom{\dfrac{30}{50}} \leftarrow \text{whole}$

$50\,n = 3{,}000$
$n = 60$

30 is 60% of 50.

25% of a number is 8. What is the number?

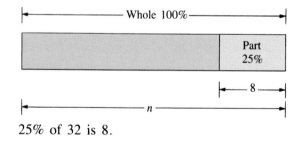

$\text{part} \rightarrow \dfrac{25}{100} = \dfrac{8}{n} \leftarrow \text{part}$
$\text{whole} \rightarrow \phantom{\dfrac{25}{100}} \phantom{\dfrac{8}{n}} \leftarrow \text{whole}$

$25\,n = 800$
$n = 32$

25% of 32 is 8.

Warm Up

Write and solve a proportion for each question.

1. What is 60% of 35?
2. 20% of a number is 12. What is the number?
3. 24 is what percent of 60?
4. What is 65% of 300?
5. 16 is what percent of 20?
6. 64% of a number is 16. What is the number?

Write and solve a proportion for each question.

1. What is 48% of 75?

2. 10% of a number is 24. What is the number?

3. 20 is what percent of 25?

4. What percent of 50 is 37?

5. What is 36% of 200?

6. 75% of a number is 45. What is the number?

7. What is 90% of 330?

8. What is 66% of 800?

9. 12.5% of a number is 18. What is the number?

10. 17 is what percent of 68?

11. What is 125% of 88?

12. What is 13% of 56?

13. 60% of a number is 213. What is the number?

14. 85% of a number is 17. What is the number?

15. 28 is what percent of 56?

16. What percent of 80 is 48?

17. Janelle is 144 cm tall. This is 90% of Kimberly's height. How tall is Kimberly?

18. Carlene is 171 cm tall. Justin is 190 cm tall. Carlene's height is what percent of Justin's height?

19. Lawan is 170 cm tall. Andy's height is 85% of Lawan's height. What is Andy's height?

20. **DATA HUNT** Find your height. Find a classmate's height. Your height is what percent of your classmate's height?

THINK MATH

Space Perception

An exercise course in a park is on both sides of a stream and on two small islands in the stream. There are seven bridges that cross the stream and connect the islands. Is it possible to go through each exercise station and cross each bridge exactly one time?

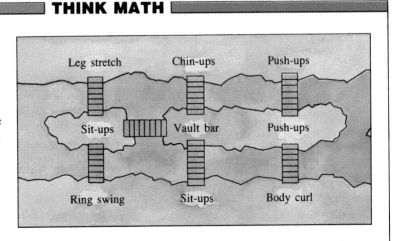

More Practice, page 436, Set C

APPLIED PROBLEM SOLVING

QUESTION
DATA
PLAN
ANSWER
CHECK

You have a neighbor who is unable to go out to shop for groceries. You have volunteered to do the shopping. Your neighbor gave you a grocery list. You need to decide where you will do the shopping.

Grocery List	Main Market	Lesso's Super	Broadway Food Store
1 dozen eggs	$0.77	$0.79	$0.69
laundry soap	$1.69	$1.78	$1.73
2 lb chicken *or*	$0.99 / lb	$0.79 / lb	$0.89 / lb
1 lb ground beef	$1.99 / lb	$1.89 / lb	$1.69 / lb
cat food	$0.99	$0.99	$0.99
loaf of bread	$0.79	$0.95	$0.69
$\frac{1}{2}$ gal milk	$1.05	$1.04	$1.05
1 lb apples	$0.59	$0.69	$0.49
cereal	$2.29	$2.33	$2.19

Some Things to Consider

- You can shop at any one of three stores.
- You can use the newspaper advertisements to compare prices at the three stores.
- Your neighbor gave you $10 to buy the groceries.
- Your neighbor is on a strict budget and wants to save as much money as possible.

Some Questions to Answer

1. What is the most expensive item on the list?

2. What is the price of 2 lb of chicken at each store? How do these prices compare with the price of 1 lb of ground beef at each store?

3. Will you have enough money to do all your shopping at Main Market?

4. Will you have enough money to do all your shopping at Broadway Food Store?

5. Will you have enough money to do all your shopping at Lesso's Super?

What Is Your Decision?

In which store will you buy the groceries on the list?

CHAPTER REVIEW/TEST

Write each ratio or number as a percent.

1. 7 to 100
2. 45 out of 100
3. $\frac{3}{100}$
4. 0.17

Write each fraction as a percent.

5. $\frac{1}{4}$
6. $\frac{3}{20}$
7. $\frac{5}{50}$
8. $\frac{3}{5}$

Write each percent as a lowest-terms fraction.

9. 10%
10. 75%
11. 16%
12. 90%

Write each decimal or fraction as a percent.

13. 0.1886
14. 0.0251
15. $\frac{3}{8}$
16. $\frac{7}{8}$

17. Find 25% of 74.
18. Find 40% of 90.

19. Find 13.7% of 300.
20. 15 is what percent of 60?

21. What percent is 18 of 48?
22. What percent of 80 is 15?

23. 90 is 3% of what number?
24. 75% of what number is 30?

Write and solve a proportion for each question.

25. What is 42% of 80?
26. 20% of a number is 24. What is the number?
27. 30 is what percent of 60?

Solve.

28. June has 30 days. During that month, 40% of the days were sunny. How many days were sunny?
29. In one class, 9 out of 30 students play musical instruments. What percent of the class plays musical instruments?

ANOTHER LOOK

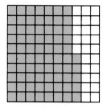

100 squares   75 shaded

$\frac{75}{100} = \frac{3}{4}$ ← fraction

$\frac{75}{100} = 0.75$ ← decimal

$\frac{75}{100} = 75\%$ ← percent

What percent of each square is shaded?

1. 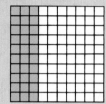   **2.**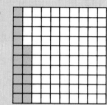

Write each fraction as a percent.

3. $\frac{27}{100}$ **4.** $\frac{3}{10}$ **5.** $\frac{4}{5}$ **6.** $\frac{17}{25}$

Write the lowest-terms fraction for each percent.

7. 20%   **8.** 50%   **9.** 44%   **10.** 65%

Write each percent as a decimal.

57.3% = 0.573
shift 2 places left

0.333 = 33.3%
shift 2 places right

11. 28%   **12.** 160%   **13.** 62.5%   **14.** 9%

Write each decimal as a percent.

15. 0.23   **16.** 0.06   **17.** 0.315   **18.** 1.80

$\frac{5}{6} = 6\overline{)5.00}$ 0.833 → 83%

Find the percent for each fraction. Round to the nearest whole percent.

19. $\frac{3}{8}$ **20.** $\frac{2}{3}$ **21.** $\frac{11}{17}$ **22.** $\frac{7}{12}$

5% of 80
$0.05 \times 80 = 4.00$

3 out of 5
$\frac{3}{5} = \frac{60}{100} = 60\%$

20% of n = 10
$n = \frac{10}{0.20} = 50$

23. Find 20% of 250.   **24.** Find 7% of 900.

25. What percent is 12 out of 20?   **26.** What percent of 25 is 15?

27. 80% of a number is 24. What is the number?   **28.** 5% of a number is 30. What is the number?

ENRICHMENT

Calculating Percents

1980 Census Data on Metropolitan Growth

Standard Metropolitan Area	1970 Population	1980 Population	Percent Change
Fort Lauderdale-Hollywood, FL	620,000	1,005,000	62
Phoenix, AZ	971,000	1,505,000	
Tampa-Saint Petersburg, FL	1,013,000	1,552,000	
Houston, TX	1,985,000	2,887,000	
Atlanta, GA	1,390,000	2,004,000	
Riverside-San Bernardino-Ontario, CA	1,139,000	1,537,000	
Denver-Boulder, CO	1,228,000	1,613,000	
Dallas-Fort Worth, TX	2,318,000	2,961,000	
Knoxville, TN	409,000	477,000	
Portland, OR	1,007,000	1,236,000	
Tulsa, OK	549,000	690,000	
San Jose, CA	1,065,000	1,283,000	

The Fort Lauderdale-Hollywood area increased in population by 62% between 1970 and 1980. You can use a calculator to find the percent of change in population for the other areas. Follow the steps shown below for Fort Lauderdale-Hollywood.

	Enter	Press	Read
Find the difference between the 1980 and 1970 populations.	1,005,000	−	1005000
	620,000	÷	385000
Divide by the 1970 population. Round to the nearest hundredth. Write a percent for the decimal.	620,000	=	0.6209677 0.62 62%

1. Find the percent of change for each of the other areas in the chart.

2. Find the percent of change in population between 1970 and 1980 for your metropolitan area or for your local community.

319

CUMULATIVE REVIEW

1. What is $\frac{7}{8}$ of 72?

   A 67   B 63
   C 56   D not given

Find the products.

2. $6 \times \frac{5}{6}$ A 41   B 40
      C 5   D not given

3. $4\frac{2}{3} \times \frac{3}{7}$ A 12   B $2\frac{2}{7}$
      C 2   D not given

4. Find the quotient.

   $1\frac{5}{6} \div 3\frac{1}{2}$

   A $\frac{11}{21}$ B $\frac{11}{12}$
   C $\frac{5}{21}$ D not given

5. What is the decimal for $\frac{6}{17}$?

   A 0.36   B 0.32
   C 0.28   D not given

6. What is the lowest-terms fraction for 0.68?

   A $\frac{17}{25}$ B $\frac{12}{25}$
   C $\frac{24}{50}$ D not given

Find the missing numbers.

7. 60 km = ▨ m

   A 6,000   B 60,000
   C 600   D not given

8. 1,675 cm = ▨ m

   A 0.1675   B 167.5
   C 16.75   D not given

9. 0.967 kg = ▨ g

   A 967   B 9.67
   C 9,670   D not given

10. 720 mg = ▨ g

    A 7,200   B 0.72
    C 7.2   D not given

11. What is the perimeter?

    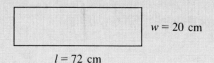

    $w = 20$ cm $l = 72$ cm

    A 92 cm   B 184 cm
    C 1,440 cm   D not given

12. What is the area?

    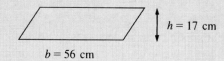

    $h = 17$ cm $b = 56$ cm

    A 952 cm²   B 9,520 cm²
    C 146 cm²   D not given

13. There were 39 players trying out for a field hockey team. The coach chose $\frac{1}{3}$ of the players for the team. How many players did she choose?

    A 13   B 11
    C 16   D not given

14. One face of a cube has an area of 16.24 cm². What is the surface area of the cube?

    A 129.92 cm²   B 64.96 cm²
    C 97.44 cm²   D not given

Circles and Cylinders

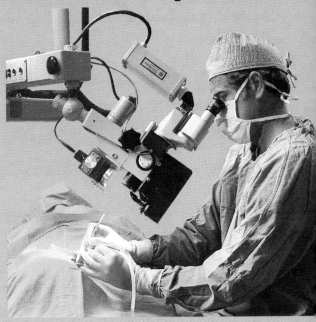

A laser beam has special characteristics. It is narrow and spreads out very little, so it can travel much farther than ordinary light. When a laser beam was bounced off the moon, the pencil-thin beam of light spread out only about 3 km in diameter after traveling for approximately 380,000 km. The light of a laser is only one color and it is much brighter than sunlight.

Lasers are used in industry to drill tiny holes and cut thick steel. Surgeons can use lasers because the laser light can be focused down to a spot 0.0005 mm in diameter. This narrow beam of light becomes a laser knife that can seal off blood vessels as it cuts. With the help of lasers, photographers can take three-dimensional pictures and scientists can measure pollution in the air. The storage capacity of computers has been greatly increased using laser technology. Television broadcasting companies hope to one day use laser beams as transmitters.

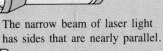

Light from a flashlight spreads out in all directions.

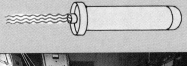

The narrow beam of laser light has sides that are nearly parallel.

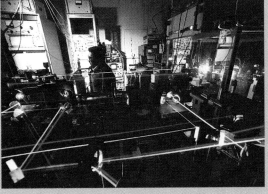

Circumference

Maurice Thompson is a mineralogist. He measured the **diameter (d)** of some petrified logs and then measured the distance around the logs. The distance around a circle is called the **circumference (C)**.

Diameter (d)	Circumference (C)	$\frac{C}{d}$
0.8 m	2.5 m	3.1
2.1 m	6.6 m	3.1
1.4 m	4.4 m	3.1
1.6 m	5.0 m	3.1

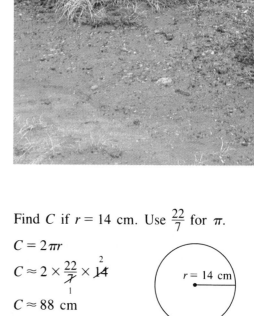

For every circle, the ratio $\frac{C}{d}$ $\left(\frac{\text{Circumference}}{\text{diameter}}\right)$ is the same number. We use the Greek letter π **(pi)** for this ratio. The decimal for π is unending and does not repeat.

$\pi = 3.14159265358979323846 2643\ldots$

We will use 3.14 or $\frac{22}{7}$ for π.

Circumference $= \pi \times$ **diameter**
$C = \pi d$ or $C = 2\pi r$

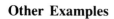
$d = 2r$

Other Examples

Find C if $d = 8$ cm. Use 3.14 for π.

$C = \pi d$
$C \approx 3.14 \times 8$
$C \approx 25.12$ cm
$\approx$ means "is approximately equal to"

$d = 8$ cm

Find C if $r = 14$ cm. Use $\frac{22}{7}$ for π.

$C = 2\pi r$
$C \approx 2 \times \frac{22}{\underset{1}{\cancel{7}}} \times \overset{2}{\cancel{14}}$
$C \approx 88$ cm

$r = 14$ cm

Warm Up

Find the circumference of each circle.

Use 3.14 for π.

1. $d = 3$ cm   **2.** $d = 0.75$ m

Use $\frac{22}{7}$ for π.

3. $d = 42$ cm   **4.** $d = 70$ m

Find the circumference of each circle. Use 3.14 for π.

1.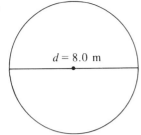
$d = 8.0$ m

2.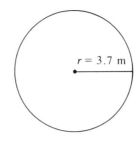
$r = 3.7$ m

3.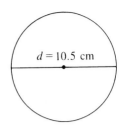
$d = 10.5$ cm

4. $r = 3.6$ m
5. $d = 6.0$ cm
6. $r = 15$ km

Find the circumference of each circle. Use $\frac{22}{7}$ for π.

7. $d = 21$ mm
8. $r = 7$ m
9. $d = 10.5$ m
10. $r = 0.7$ m
11. $r = 28$ cm
12. $d = 3.5$ km

Since $C = \pi d$, $d = \frac{C}{\pi}$. Use this formula to find the diameter of each circle with the given circumference. Use 3.14 for π.

13. $C = 18.84$ cm
14. $C = 28.26$ m
15. $C = 37.68$ mm
16. $C = 4.71$ m
17. $C = 25.12$ cm
18. $C = 78.5$ m

19. A petrified conifer has a diameter of 3.4 m. What is the circumference of the petrified tree?

20. A petrified tree found in Michigan has a diameter of 1.2 m. How much greater is the circumference of a petrified tree with a diameter of 2.9 m?

★ 21. The "Queen Tree," a petrified tree in Calistoga, California, has a circumference of 11.46 m. What is the radius of the tree, to the nearest tenth of a meter?

22. Find a decimal for the fraction $\frac{355}{113}$. To what decimal place is this a correct value for π?

More Practice, page 437, Set A

Area of Circles

A department store prices its rugs according to the area of the rugs. What is the area of a large circular rug that has a radius of 2 m?

We can estimate the area by counting the squares and parts of squares inside the circle on the grid below.

Whole squares → 4
Estimated part squares → +8
Total squares → 12

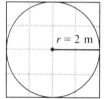

We can use multiplication to find area.

Area of a circle = π × square of radius
$A = \pi r^2$

Use 3.14 for π.
$A \approx 3.14 \times 2^2$
$A \approx 3.14 \times 4$
$A \approx 12.56$

The area of the rug is about 12.56 m².

Other Examples

Find A if $r = 5$ cm.

$A = \pi r^2$
$A \approx 3.14 \times 5^2$
$A \approx 3.14 \times 25$
$A \approx 78.50$

The area is about 78.50 cm².

Find A if $d = 2.4$ m.

$A = \pi r^2$
$r = 2.4 \div 2 = 1.2$
$A \approx 3.14 \times (1.2)^2$
$A \approx 3.14 \times 1.44$
$A \approx 4.5216$

The area is about 4.5216 m².

Warm Up

Find the area of each circle. Use 3.14 for π.

1. $r = 10$ cm
2. $d = 18$ m
3. $r = 2.4$ m
4. $d = 40$ cm
5. $r = 0.5$ m
6. $r = 7$ mm

Find the area of each circle. Use 3.14 for π.

1.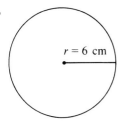
2.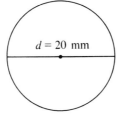
3.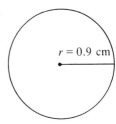

4. $r = 12$ cm
5. $d = 100$ m
6. $r = 0.1$ cm
7. $r = 2$ cm
8. $d = 8$ cm
9. $r = 11$ cm
10. $r = 1$ m
11. $r = 1.5$ m
12. $d = 4.4$ m
13. $d = 0.6$ cm
14. $r = 75$ cm
15. $d = 9$ cm

16. A small circular throw rug has a radius of 0.4 m. What is the area of the rug, to the nearest tenth of a square meter?

17. One circular throw rug has a diameter of 1 m. Another circular rug has a diameter of 2 m. Is the area of the larger rug 2 times or 4 times the area of the smaller rug?

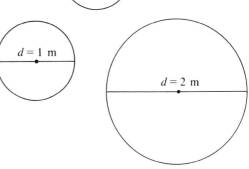

SKILLKEEPER

Write each percent as a decimal.

1. 29%
2. 12.6%
3. 2.3%
4. 16.5%
5. 0.7%
6. 125%
7. 19.96%
8. 2%
9. 1.16%
10. 179%

Find the percent of each number.

11. 30% of 52
12. 95% of 41
13. 45% of 200
14. 80% of 23
15. 5% of 45
16. 35% of 26
17. 75% of 41
18. 50% of 15
19. 7% of 100
20. 40% of 81

More Practice, page 437, Set B

PROBLEM SOLVING: Practice

Solve. Use 3.14 for π.

1. Robin's bicycle wheels each have a radius of 33 cm. For each revolution of a wheel, the bicycle will roll forward the circumference of the wheel. What is this distance?

2. Giles estimated that his bicycle travels about 2 m for each revolution of the wheels. How many revolutions would one of the wheels make in traveling 5 km?

3. Alvin bought a new bicycle tire for $8.99 and a bicycle light for $6.50. If he paid for these items with a $20 bill, how much change should he get back?

4. On a bicycle trip, Charity and three friends traveled 54 km in 2.5 h. How many kilometers did they average in 1 h?

5. A reflector lens on Nora's bicycle has a radius of 4 cm. What is the area of the reflector lens?

6. The circular headlight lens on Tom's bicycle has a diameter of 6 cm. What is the area of the lens?

7. A bicycle store has an antique "high-wheeler" bicycle on display. The front wheel of the bicycle has a diameter of 1.5 m. What is the circumference of the front wheel?

8. The back wheel of the high-wheeler has a diameter of 0.4 m. What is the circumference of the back wheel?

9. How many times would the back wheel of the high-wheeler have to turn while the front wheel is making one complete turn?

10. Molly would like to buy a new bicycle that costs $188. She has already saved $60 and is planning to save $8 a week until she has enough money to pay for the bicycle. How many weeks must she plan to save $8 a week?

11. Carlotta rode her bicycle for 1.5 hours, averaging 32 km/h. She has 9 more kilometers to ride. How far will she ride in all?

12. **Try This** Ralph rode his bicycle from his house to a state park. He rode $\frac{1}{2}$ of the way in the first half-hour. In the next half-hour he rode $\frac{2}{5}$ of the distance. At this point he was still 3 km from the park. What is the total distance from Ralph's house to the park? Hint: Choose the operations.

Area of Cylinders

Moriko covered the curved part of a can with a rectangular piece of paper. The area of this part of a cylinder is called the **lateral area**. What is the lateral area of the can?

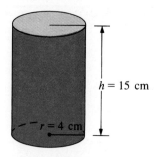

To find the lateral area, we can find the area of the rectangle.

Length = circumference of the cylinder ($2\pi r$)
Width = height of the cylinder (h)
Area   = circumference × height

Lateral area = circumference × height
$A = 2\pi rh$

$A \approx 2 \times 3.14 \times 4 \times 15$
$A \approx 376.8$

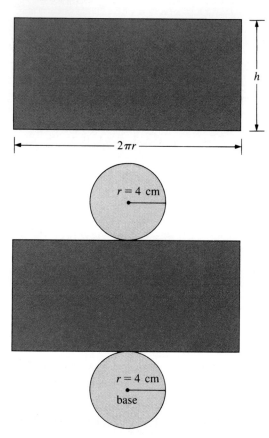

The lateral area of the can is about 376.8 cm².

To find the **total surface area** of the cylinder, we must add the area of the top and bottom circular bases to the lateral area.

Area of each base: $A \approx 3.14 \times 4^2 = 50.24$
Lateral area: $A \approx 376.8$
Total surface area: $\approx 376.8 + 50.24 + 50.24$
                      ≈ 477.28

The total surface area is about 477.28 cm².

Warm Up

Find the lateral area of each cylinder. Use 3.14 for π.

1. $r = 10$ cm
   $h = 12$ cm

2. $r = 1$ cm
   $h = 15$ cm

3. $r = 5$ cm
   $h = 7$ cm

Find the total area of each cylinder. Use 3.14 for π.

4. $r = 3$ cm
   $h = 5$ cm

5. $r = 2$ cm
   $h = 4$ cm

6. $r = 0.5$ cm
   $h = 5$ cm

Find the lateral area of each cylinder. Use 3.14 for π.

1.
$r = 3.3$ cm
$h = 4$ cm

2.
$r = 4$ cm
$h = 10$ cm

3.
$r = 3.3$ cm
$h = 7$ cm

4.
$r = 2.8$ cm
$h = 13.5$ cm

5. $r = 5$ cm
$h = 10$ cm

6. $r = 6.4$ cm
$h = 12$ cm

7. $r = 10$ cm
$h = 15$ cm

8. $r = 0.3$ cm
$h = 8$ cm

Find the total surface area of each cylinder. Use 3.14 for π.

9. $r = 2$ m
$h = 5$ m

10. $r = 2$ mm
$h = 10$ mm

11. $r = 9$ cm
$h = 3$ cm

12. $r = 0.8$ m
$h = 2.5$ m

13. $r = 1$ cm
$h = 1$ cm

14. $r = 6$ cm
$h = 4.5$ cm

15. $r = 10$ m
$h = 1$ m

16. $r = 11$ cm
$h = 20$ cm

17. A rolled-oats carton is a cylinder with a radius of 5 cm and a height of 18 cm. What is the area of the label around the carton?

18. A container of salt is a cylinder with a radius of 4 cm and a height of 13.5 cm. What is the total surface area of the container?

19. Use this information to write and solve your own problem.
Cylinder: $r = 12$ cm
$h = 30$ cm

20. **DATA HUNT** Measure the radius and height of a cylinder, such as a can of food. What is the lateral area and the total surface area of the can?

THINK MATH

Guess and Check

Dave has only $7.00. Which choice below will give him the most pizza for his money?

Two 24 cm pizzas
One 24 cm and one 32 cm pizza
One 40 cm pizza

Diameter	Cost
24 cm pizza	$3.00
32 cm pizza	$4.00
40 cm pizza	$6.00

Guess which is the best choice, then find the area of each pizza to check your guess.

More Practice, page 437, Set C

Volume of Cylinders

If a cylinder and a prism have equal heights and the areas of their bases are equal, then their volumes are equal.

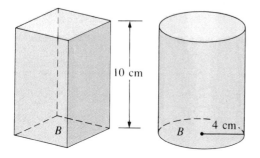

To find the volume of a cylinder, we can use this formula.

Volume = area of Base × height
$V = Bh$ or $V = \pi r^2 h$

If $r = 4$ cm and $h = 10$ cm, then
$V \approx 3.14 \times 4^2 \times 10$ Use 3.14 for π.
$V \approx 3.14 \times 16 \times 10$
$V \approx 502.40$

The volume of the cylinder is about 502.40 cm³.

Other Examples

$r = 2$ cm $V = \pi r^2 h$
$h = 8$ cm $V \approx 3.14 \times 2^2 \times 8$
$V \approx 3.14 \times 4 \times 8$
$V \approx 100.48$

Volume ≈ 100.48 cm³

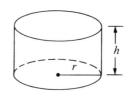

$r = 6$ m $V = \pi r^2 h$
$h = 6$ m $V \approx 3.14 \times 6^2 \times 6$
$V \approx 3.14 \times 36 \times 6$
$V \approx 678.24$

Volume ≈ 678.24 m³

Warm Up

Find the volume of each cylinder. Use 3.14 for π.

1. $r = 10$ cm
   $h = 20$ cm
2. $r = 8$ dm
   $h = 12$ dm
3. $r = 20.8$ cm
   $h = 4$ cm
4. $r = 2$ m
   $h = 6$ m

Find the volume of each cylinder. Use 3.14 for π.

1.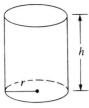
$r = 5$ cm
$h = 12$ cm

2.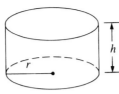
$r = 7$ dm
$h = 10$ dm

3.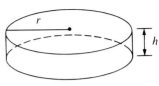
$r = 10$ cm
$h = 4$ cm

4.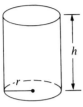
$r = 6$ mm
$h = 16$ mm

5.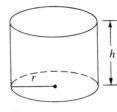
$r = 3$ m
$h = 5$ m

6.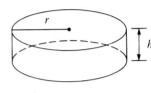
$r = 4$ m
$h = 2.5$ m

7. Which has the greater volume, the rectangular prism or the cylinder? How much greater is the volume?

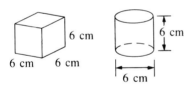

8. A mug has a base with an area of 44 cm². The height of the mug is 7.5 cm. How high should it be filled so that it contains 250 cm³ or 250 mL of water? Round your answer to the nearest tenth.

SKILLKEEPER

Find the circumference of each circle. Use 3.14 for π.

1. $r = 2.5$ cm
2. $r = 17.2$ mm
3. $d = 19$ m
4. $d = 0.6$ mm
5. $r = 32$ cm
6. $d = 607$ mm
7. $r = 20.25$ cm
8. $d = 66$ m
9. $r = 2.5$ km
10. $r = 0.008$ mm

Find the area of each circle. Use 3.14 for π.

11. $r = 29$ km
12. $d = 15$ cm
13. $d = 12$ mm
14. $r = 0.12$ cm
15. $d = 2.4$ mm
16. $r = 62$ cm
17. $r = 2.6$ mm
18. $d = 24$ km
19. $d = 12.2$ mm
20. $d = 0.3$ cm

More Practice, page 437, Set D

PROBLEM SOLVING: Practice

QUESTION
DATA
PLAN
ANSWER
CHECK

Wood is a valuable resource. Wood and wood products are used in many different ways. Much of the wood milled in the United States is used to make paper. Each person in this country uses about 300 kg of paper each year.

Modern sawmills are careful not to waste wood. Some of them use computers to help decide how to cut a log so that it will yield the greatest amount of lumber.

Solve. Use 3.14 for π.

1. A large redwood log has the shape of a cylinder. The log is 12 m long and has a radius of 1 m. What is the volume of the log in cubic meters?

2. A sawmill uses water under high pressure to remove bark from logs. The machine uses 3,700 L of water per minute. The water travels at a rate of 8 km/min. How many liters of water does the machine use per hour?

3. During a 16-hour period, a machine removed the bark from 600 logs. About how many minutes per log does it take to remove the bark?

4. To record the diameter of a log which is not exactly round, a sawmill worker makes two measurements of the diameter. The **recorded diameter** is the average of the two measurements. For one log, the first measurement was 92 cm and the second measurement was 74 cm. What was the recorded diameter of the log?

5. A water-storage tank at a sawmill has the shape of a cylinder. The radius of the tank is 2 m and the height is 5 m. What is the volume of the storage tank?

6. A painting contractor must find the lateral area of the water tank in order to estimate the cost of painting the tank. What is the lateral area?

7. In a wooded area, a forest worker counted 18 Douglas firs, 30 western hemlocks, and 12 Sitka spruces. What percent of the total number of trees were Douglas firs?

8. The diameter of a log was 1.8 m. After the bark was removed, the diameter was 1.6 m. How much less was the circumference of the log after the bark was removed?

9. A log had a diameter of 1 m and was 6 m long. A large square post, 70 cm on each side and 6 m long, was sawed from the log. What was the volume of the square post? What was the volume of the part of the log that was sawed off to form the post?

10. A log had a diameter of 1.5 m and was 8 m long. Which measurement is not needed to find the circumference of the log?

11. The dogwood tree is native to 76% of the states in the United States. The dogwood is native to how many states?

12. **Try This** A tree was 48 m tall. During a storm it was broken off, and the part broken off was 3 times as long as the part that remained standing. What were the lengths of each part?

333

Cross Sections

Linda Wahnee works in the drafting department of a company that makes machine parts. Linda makes blueprints that sometimes show **cross sections** of parts.

Linda thinks of a plane cutting the figure with a straight cut. The cross section is the intersection of the cutting plane and the figure.

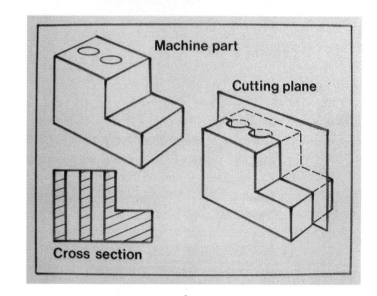

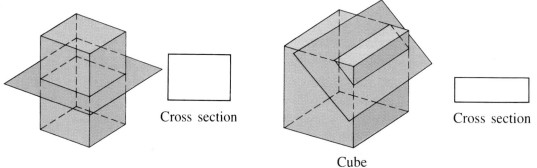

Choose the correct cross section for each figure.

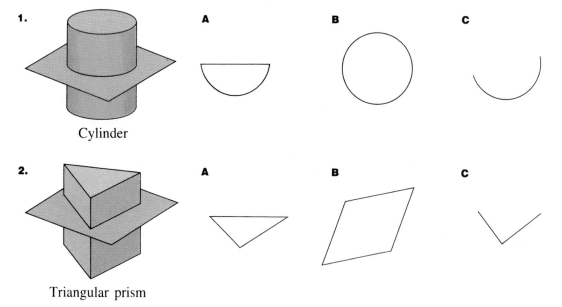

Draw a picture of each cross-section view.

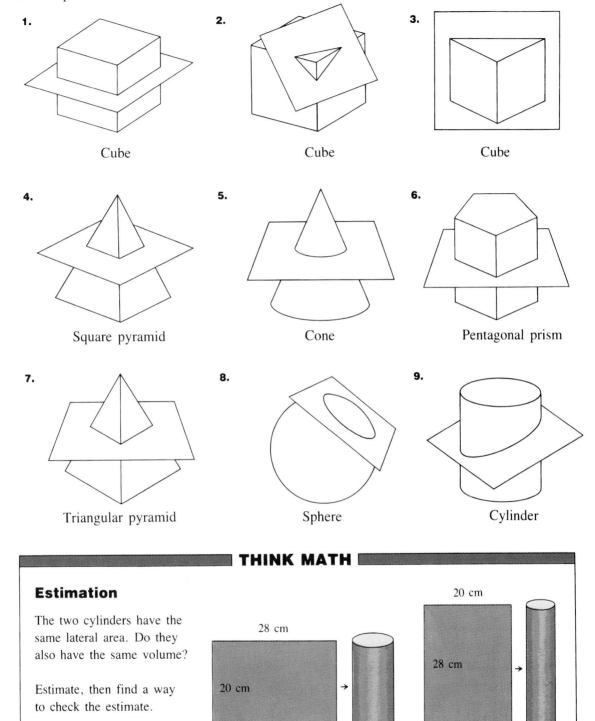

1. Cube
2. Cube
3. Cube
4. Square pyramid
5. Cone
6. Pentagonal prism
7. Triangular pyramid
8. Sphere
9. Cylinder

THINK MATH

Estimation

The two cylinders have the same lateral area. Do they also have the same volume?

Estimate, then find a way to check the estimate.

28 cm
20 cm

20 cm
28 cm

335

APPLIED PROBLEM SOLVING

QUESTION
DATA
PLAN
ANSWER
CHECK

You have an old TV set and you want to buy a new one. You could try to sell the old set through a newspaper advertisement before you buy a new TV set, or you could trade in your old set on a new set.

Some Things to Consider

- The old TV set is in fairly good condition.
- You want to get a new TV set within two weeks.
- A newspaper advertisement costs $2.75 a day.
- You expect to sell the old set for $50.
- A new TV set costs $389.
- You can trade in your old set on a new set and get $25 off the cost of the new set.

Some Questions to Answer

1. How much money will you have to pay for the new TV set with a trade-in?

2. What is the cost of an advertisement for one week? What is the cost of an advertisement for two weeks?

3. If the old TV does not sell after the advertisement runs for two weeks, you must use the old TV as a trade-in. What is the total cost of the new TV set and the advertisement?

4. How much money must you pay for the new TV set if someone buys the old TV after only one day?

What Is Your Decision?

How will you plan to use your old TV set to help you buy a new TV set?

CHAPTER REVIEW/TEST

Find the circumference of each circle. Use 3.14 for π.

1.
2.
3.

4. $r = 2.5$ m
5. $d = 20$ km
6. $d = 12$ mm

Find the circumference of each circle. Use $\frac{22}{7}$ for π.

7. $r = 28$ cm
8. $r = 21$ m
9. $d = 98$ cm

Find the area of each circle. Use 3.14 for π.

10. $r = 4$ cm
11. $r = 7$ cm
12. $d = 2$ cm

13. $r = 10$ km
14. $d = 4$ m
15. $d = 30$ m

Find the lateral area and the total surface area of each cylinder. Use 3.14 for π.

16. $r = 2$ cm
    $h = 10$ cm
17. $r = 5$ cm
    $h = 25$ cm
18. $r = 3$ cm
    $h = 8$ cm

Find the volume of each cylinder.

19. $r = 1$ cm
    $h = 2$ cm
20. $r = 3$ cm
    $h = 5$ cm
21. $r = 5$ cm
    $h = 2$ cm

Draw a picture of each cross-section view.

22.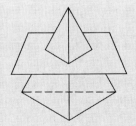
23.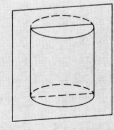
24.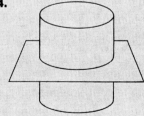

25. An irrigated field is a circle with a radius of 80 m. What is the area of the field?

26. An irrigation tank is a cylinder with a diameter of 8 m and a depth of 2 m. What is the volume of the tank?

ANOTHER LOOK

Circumference is the distance around a circle.

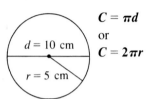

$C = \pi d$
or
$C = 2\pi r$

Use 3.14 or $\frac{22}{7}$ for π.

Find the circumference of each circle. Use 3.14 for π.

1. $d = 20$ m
2. $r = 8$ cm

Find the circumference of each circle. Use $\frac{22}{7}$ for π.

3. $d = 35$ m
4. $r = 10.5$ m

$A = \pi r^2 = \pi \times r \times r$

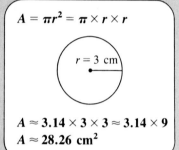

$A \approx 3.14 \times 3 \times 3 \approx 3.14 \times 9$
$A \approx 28.26$ cm²

Find the area of each circle. Use 3.14 for π.

5.
   $r = 5$ cm

6.
   $r = 2$ km

7. $r = 0.8$ cm
8. $r = 30$ cm

Lateral area of a cylinder equals circumference of the base times the height.

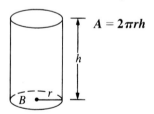

$A = 2\pi rh$

Find the lateral area of each cylinder.

9.
   $r = 2$ cm
   $h = 6$ cm

10.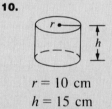
    $r = 10$ cm
    $h = 15$ cm

Total surface area = lateral area + area of two bases

Find the total surface area of each cylinder.

11. $r = 1$ cm
    $h = 2$ cm

12. $r = 4$ cm
    $h = 5$ cm

Volume = Bh
$V = \pi r^2 h$

Find the volume of each cylinder.

13. $r = 3$ cm
    $h = 9$ cm

14. $r = 2.5$ mm
    $h = 4$ mm

ENRICHMENT

Hex Numeration System

Some computers and microcomputers use a numeration system based on sixteen instead of base ten.

This system is called a **hexadecimal**, or **hex**, numeration system. The hex system has sixteen digits.

Hex	0	1	2	3	4	5	6	7	8	9	A	B	C	D	E	F
Base ten	0	1	2	3	4	5	6	7	8	9	10	11	12	13	14	15

To use the hex system, we must think about grouping by **sixteens** instead of by tens.

2 sixteens and 3 ones
We write: 23_{hex}
We read: "two three base hex"

Since 2 sixteens and 3 ones in base ten is 35, we can write $23_{hex} = 35_{ten}$.

We can use the expanded notation of a hex numeral to find the base-ten numeral.

$A8_{hex} = 10 \cdot 16 + 8 = 160 + 8 = 168_{ten}$

$13C_{hex} = 1 \cdot 16^2 + 3 \cdot 16 + 12 = 256 + 48 + 12 = 316_{ten}$

$3FD_{hex} = 3 \cdot 16^2 + 15 \cdot 16 + 13 = 768 + 240 + 13 = 1{,}021_{ten}$

Write a hex numeral for each figure.

1.

2.

3.

Write a base-ten numeral for each hex numeral.

4. 21_{hex}
5. $2B_{hex}$
6. 89_{hex}
7. AA_{hex}

8. $5C_{hex}$
9. $A2_{hex}$
10. $E0_{hex}$
11. FF_{hex}

12. 111_{hex}
13. $1AB_{hex}$
14. $20D_{hex}$
15. 615_{hex}

16. ACE_{hex}
17. $B0B_{hex}$
18. FED_{hex}
19. DAB_{hex}

TECHNOLOGY

Loops in Computer Programs

A computer can be programmed to keep track of items by counting. To do this, the computer must be told how far to count. In the program below, the computer is told to count and print the first five multiples of 3. The program uses a **loop**. A loop is a command that causes the computer to go back to an earlier step in the program and repeat the step.

A Universal Product Code symbol

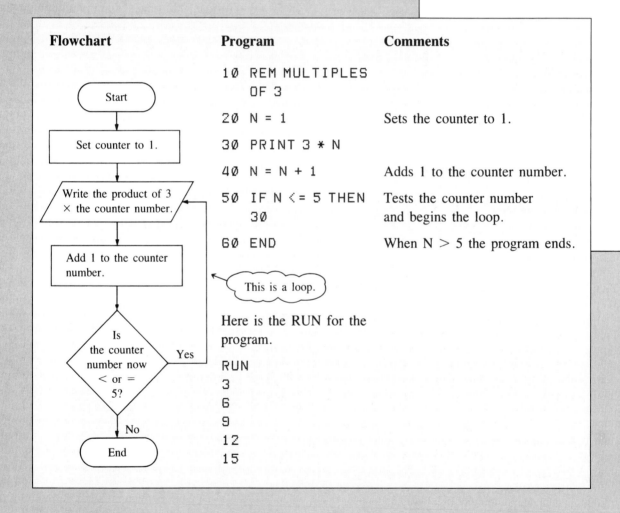

1. How could the program be changed so that the RUN would be the first ten multiples of 3?

2. How could the program be changed so that the RUN would be multiples of 4?

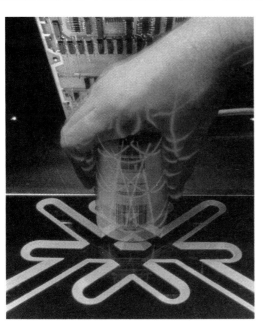

The bar code scanner used in supermarkets reads the Universal Product Code symbols to give an accurate tally of purchases and manages the store inventory.

Write a RUN for each program.

1.
```
10  REM COUNT TO 10
20  N = 1
30  PRINT N
40  N = N + 1
50  IF N <= 10 THEN 30
60  END
```

2.
```
10  REM ODD NUMBERS
20  N = 0
30  PRINT 2 * N + 1
40  N = N + 1
50  IF N <= 5 THEN 30
60  END
```

3.
```
10  REM COUNTING BACKWARD
20  N = 6
30  PRINT N
40  N = N - 1
50  IF N > 0 THEN 30
60  END
```

4.
```
10  REM NUMBER PATTERN
20  N = 1
30  B = 4 * N - 3
40  PRINT B
50  N = N + 1
60  IF N <= 6 THEN 30
70  PRINT "FIND MORE NUMBERS"
80  PRINT "IN THIS PATTERN."
90  END
```

5.
```
10  REM SUM OF 10 NUMBERS
20  N = 1
30  T = 0
40  T = T + N
50  N = N + 1
60  IF N <= 10 THEN 40
70  PRINT T
80  END
```

6. Write a computer program that has a loop in it. Write a RUN of your program.

CUMULATIVE REVIEW

Find the missing numbers.

1. 12.5 cm = ▓ m

   A 1.25 B 125
   C 0.125 D not given

2. 6,250 m = ▓ km

   A 6.250 B 625
   C 62.50 D not given

3. What is the perimeter?

   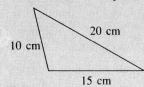

   A 40 cm
   B 45 cm
   C 3,750 cm
   D not given

4. What is the area?

   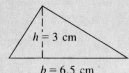

   A 9.75 cm²
   B 9.5 cm²
   C 19.5 cm²
   D not given

5. What is the volume?

   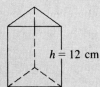

   A 64 cm³
   B 240 cm³
   C 52 cm³
   D not given

6. What is the capacity?

   Volume: 6 cm³

   A 12 mL
   B 36 mL
   C 24 mL
   D not given

7. What is the percent for $\frac{3}{5}$?

   A 16% B 60%
   C 35% D not given

8. What is the decimal for 129%?

   A 1.29 B 12.9
   C 0.129 D not given

9. What is the percent for 0.076?

   A 76% B 760%
   C 7.6% D not given

10. What is the lowest-terms fraction for 18%?

    A $\frac{3}{25}$ B $\frac{9}{25}$
    C $\frac{6}{50}$ D not given

11. What is 25% of 9?

    A 2.25 B 3.2
    C 4 D not given

12. What percent of 5 is 9?

    A 180% B 55%
    C 18% D not given

13. Which metric unit would you use to measure the length of a pencil?

    A meter B kilometer
    C centimeter D not given

14. If $500 earns 6% interest per year, how much interest will be earned in 7 years?

    A $350 B $210
    C $30 D not given

14

Probability, Statistics, and Graphs

The bald eagle was chosen as the national bird of the United States in 1782. The eagle has long been thought of as a symbol of strength and bravery.

In the wild, eagles generally live from 20 to 30 years. The nests of bald eagles are built near water in tall trees. Eagles first breed when they are about four years old. Bald eagles do not defend a large territory. There may be an average of three nests within 1.6 km of each other.

Bald eagles are not actually bald. Their heads are covered with white feathers. The length of bald eagles ranges from 76 cm to 89 cm. Their weights range from about 3.6 kg to 6 kg. Their wingspans measure from 1.8 m to 2.3 m.

At one time bald eagles were found throughout North America. Today there are less than 15,000 bald eagles in the United States, and all but about 5,000 of them are in Alaska. Federal laws and wildlife refuges now protect the bald eagle.

Equally Likely Outcomes

Before playing a game of tennis, Christy spun her tennis racquet to see who would serve first.

The letter U on the racquet handle could come "Up" or "Down." These are the only possible **outcomes**.

Each outcome is **equally likely** to happen.

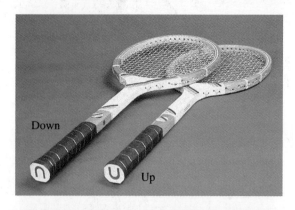

If a tennis ball can were dropped, there would be three possible outcomes.

These three outcomes are **not equally likely**. The can is most likely to fall on the side.

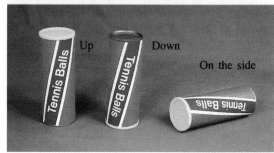

Warm Up

For each experiment, state if the outcomes are equally likely or not equally likely. If the outcomes are not equally likely, state which outcome is most likely.

1. Toss a coin.
   Outcomes: Heads, Tails

2. Roll a die.
   Outcomes: 1, 2, 3, 4, 5, 6

3. Draw a ball from the box while blindfolded.
   Outcomes:
   Red ball,
   Green ball,
   Blue ball

4. Spin the spinner.
   Outcomes: Yellow, Green

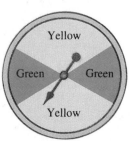

344

List all of the possible outcomes for each experiment.
State if the outcomes are equally likely or not equally likely.

1. Spin the spinner.

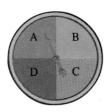

2. Draw a name from the hat while blindfolded.

3. Spin the spinner.

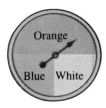

4. Roll a cube with faces lettered **A, B, C, D, E,** and **F**.

5. Drop a plastic cup on the floor.

6. Drop a thumbtack on a hard floor.

Write the number of possible outcomes for each experiment.
State if the outcomes are equally likely or not equally likely.

7. Spin the spinner.

8. Spin the spinner.

9. Guess the answer to a multiple-choice question.

   A 16.2   **B** 1.62
   C 0.162   **D** not given

10. Pick one of the days of June without looking.

11. Turn up a card from a deck of playing cards.

12. Draw a door-prize ticket from tickets numbered 1 to 100.

345

Chance and Probability

Sid has a stack of 5 records. There are vocal recordings (V) on 3 records and there is band music (B) on 2 records. Sid chose a record without looking and put it on the record player. What is the probability that it is a vocal recording?

Each of the 5 records is equally likely to be chosen.

There are **3 chances in 5** that Sid chose a vocal recording.

The **probability** that it is a vocal recording is $\frac{3}{5}$.

We write: $P(V) = \frac{3}{5}$ ← Vocal records / Total records

There are **2** chances in **5** that Sid chose a band record.

The probability of it being a band record is $\frac{2}{5}$.

$P(B) = \frac{2}{5}$ ← Band records / Total records

If an **outcome** cannot occur, its probability is 0.

$P(\text{jazz record}) = \frac{0}{5}$ ← Jazz records / Total records

If an **outcome** is certain, its probability is 1.

$P(\text{non-jazz record}) = \frac{5}{5}$ ← Non-jazz records / Total records

Warm Up

Give the missing numbers. Then give the probability.

1. Toss a coin. There is 1 chance in it will be heads.
   $P(\text{heads}) =$ ▒

2. Spin the spinner. There are 2 chances in ▒ it will be even.
   $P(\text{even number}) =$ ▒

Give each probability for a spinner with the numbers **1, 2, 3, 4, 5, 6, 7,** and **8**.

1. $P(2)$
2. $P(6)$
3. $P(4)$
4. P(an odd number)
5. P(a number less than 6)
6. P(a prime number)
7. P(a number greater than 10)
8. P(a number less than 8)
9. P(a number less than 1)
10. P(a number less than 3)

Give each probability for a die with the numbers **1, 2, 3, 4, 5,** and **6**.

11. $P(4)$
12. $P(1)$
13. $P(7)$
14. P(a number greater than 4)
15. P(a number less than 1)
16. P(a prime number)
17. $P(2)$
18. P(a number less than 7)

19. Laurie put a record on the stereo without looking to see if it was side 1 or side 2. What is P(side 2)?

20. Yuki has 3 classical records, 4 jazz records, and 1 rock'n'roll record. What kind of record has a probability of $\frac{1}{2}$ of being selected?

21. Write and solve a word problem for this data. P(an outcome) $= \frac{4}{9}$

22. **DATA HUNT** Toss a coin 100 times. Keep a tally of the results. Find P(heads). Is it about $\frac{1}{2}$?

SKILLKEEPER

Find the volume of each prism. Use $V = lwh$.

1. $l = 7$ m
   $w = 4.5$ m
   $h = 9$ m

2. $l = 25$ cm
   $w = 20$ cm
   $h = 30$ cm

3. $l = 15$ cm
   $w = 12$ cm
   $h = 20$ cm

4. $l = 13$ m
   $w = 9.5$ m
   $h = 4$ m

5. $l = 7$ m
   $w = 4.5$ m
   $h = 9$ m

Find the volume of each prism. Use $V = Bh$.

6. $B = 42$ m^2
   $h = 1.5$ m

7. $B = 135.9$ cm^2
   $h = 9.5$ cm

8. $B = 150$ m^2
   $h = 4$ m

9. $B = 12$ m^2
   $h = 9$ m

10. $B = 1,959$ cm^2
    $h = 14$ cm

Ordered Pairs in Probability

Using a red die and a green die from a game of backgammon, what is the probability of rolling a sum of 10?

Each outcome is an **ordered pair** of numbers: the number on the red die first and the number on the green die second.

We must find the total number of outcomes *and* the number of outcomes which have a sum of 10.

The grid shows there are 36 ordered-pair outcomes.

Only 3 of the outcomes have a sum of 10.

$P(\text{sum of } 10) = \dfrac{3}{36} = \dfrac{1}{12}$

There are 3 chances in 36 or 1 chance in 12 of rolling a sum of 10. The probability is $\dfrac{1}{12}$.

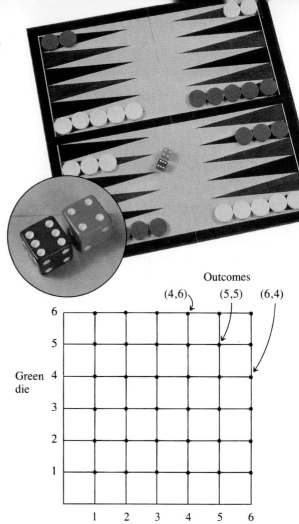

Other Examples

$P(2,4) = \dfrac{1}{36}$

$P(\text{sum of } 11) = \dfrac{2}{36} = \dfrac{1}{18}$
The outcomes are **(5,6)** and **(6,5)**.

$P(\text{sum} < 5) = \dfrac{6}{36} = \dfrac{1}{6}$
The outcomes are **(1,1),(1,2),(2,1), (3,1),(2,2),(1,3)**.

Warm Up

1. Use the grid above and list the ordered pairs that have a sum of 4. What is $P(\text{sum of } 4)$?

2. Use the grid above and list the ordered pairs that have a sum of 9. What is $P(\text{sum of } 9)$?

3. $P(\text{sum of } 8)$
4. $P(\text{sum of } 7)$
5. $P(\text{sum of } 6)$
6. $P(\text{sum} > 10)$
7. $P(1,1)$
8. $P(\text{sum} < 7)$

Give each probability for tossing a penny and a nickel.

1. List the possible outcomes as ordered pairs. How many outcomes are possible?

2. What is $P(H,T)$?

3. What is $P(H,H)$?

4. What is $P(T,T)$?

5. Which probability is greater, that both coins will be the same or that they will be different?

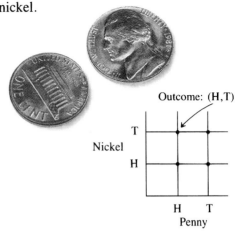

Give each probability for a coin with heads (H) and tails (T) and a spinner with the numbers **1, 2, 3, 4,** and **5**.

6. List the possible outcomes as ordered pairs. How many outcomes are possible?

7. What is $P(T,5)$?

8. What is the probability of getting heads and a number greater than 3?

9. What is the probability of getting tails and an odd number?

10. What is the probability of getting heads and an even number?

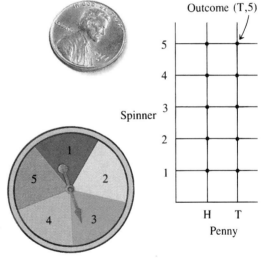

THINK MATH

Estimating Probability

Can you estimate how many times out of 25 you would get 3 heads if you tossed 3 coins at once? Make an estimate. Then try the experiment and record the results.

Tally Sheet

3H	2H,1T	1H,2T	3T
l	lll	l l l l l	ll

PROBLEM SOLVING: Practice

QUESTION
DATA
PLAN
ANSWER
CHECK

Weather records for New York City show that the probability of rain on any day in June is $33\frac{1}{3}\%$ or $\frac{1}{3}$. About how many rainy days can be expected in New York City during the month of June?

There are 30 days in June. Since $P(\text{rain}) = 33\frac{1}{3}\%$ or $\frac{1}{3}$, this means that 1 day in every 3 days is expected to be a rainy day.

Expected rainy days = $P(\text{rain}) \times$ number of days
$= \frac{1}{3} \times 30 = 10$

About 10 rainy days are expected in June.

Remember, if 10 rainy days are expected, this does not mean that 10 rainy days will occur. There may be more or less than 10 rainy days.

Solve.

1. A meteorologist said the probability of snow in the mountains on any of the next 30 days was 30%. How many snowy days were expected?

2. The probability of a cloudy day in Phoenix, Arizona is about 14%. About how many cloudy days are expected in Phoenix during the year? About how many non-cloudy days a year are expected?

3. An air traffic controller knew that the probability of a clear day in Juneau, Alaska is 32%. About how many clear days a year are expected in Juneau?

4. A travel agent read that the probability of rain for any day of the year in San Juan, Puerto Rico is about 55%. About how many rainy days a year are expected in San Juan?

5. Weather records in Reno, Nevada show that there was rainfall on 51 days out of 365 days over a period of several years. What is the probability of rainfall in Reno? Write the probability as a percent to the nearest whole percent.

6. The probability of rainfall in Los Angeles for any day in a year is about 9%. How many days with rainfall would be expected in Los Angeles during a year?

7. Milly tossed a coin 150 times and kept a tally of the heads and tails. About how many times should heads be expected to come up?

8. Fayetta plans to toss a die with numbers from one to six 30 times and to keep a tally of the outcomes. About how many times can she expect to get an outcome of 5?

9. Casey tossed a pair of dice 180 times. About how many times did he get a sum of 7? Hint: What is P(sum of 7) with two dice?

10. If Danielle tosses a penny and a nickel 20 times, about how many times should both coins be expected to come up heads? Hint: What is $P(H,H)$ when two coins are tossed?

11. Luis tossed two coins 100 times. He got an outcome of "one head, one tail" 57 times. How close was this to the expected number of outcomes?

12. Holly made two spinners like the ones below. She spun both spinners and recorded the pairs of letters for an outcome. About how many times should the outcome (B,B) be expected in 60 trials?

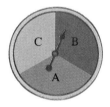 

13. The probability of guessing the correct answer on a true-false test without reading the question is $\frac{1}{2}$. On a true-false test with 40 questions, students must answer 28 questions correctly to pass the test. About how many questions might a student who guesses without reading the questions get right? Would this be a passing score?

14. **Try This** Tim, Ada, Carol, and Betsy each used a different coin for a probability experiment. They used a penny, a nickel, a dime, and a quarter. Ada's coin had Lincoln's face on the heads side. Carol did not use a quarter. Betsy's coin had Monticello on the tails side. Which coin did each person use?

351

Frequency, Range, and Mode

The city of Bridgeport has some young people working part time.

The employment manager made this **frequency chart** to show the number of people of each age that work for the city.

The **mode** is the age that occurs most often. Twelve people are 17 years old. The mode of the ages is 17 years. Some data may have more than one mode.

The ages of the part-time employees are listed from 16 years to 23 years. The **range** is the difference between the greatest number of years and the least number of years.

Range = 23 − 16 = 7

The range of ages is 7 years.

Part-time Employees

Age in years	Number
16	6
17	12
18	10
19	7
20	5
21	4
22	3
23	3
	Total 50

Warm Up

1. What is the mode of this list of ages of senior citizens? What is the range?

   72, 74, 74, 78, 78, 78,
   83, 85, 85, 90, 92, 96

2. This list of ages has more than one mode. What are the modes? What is the range?

   68, 70, 72, 80, 82, 82, 82, 82, 85,
   87, 87, 87, 87, 93, 94, 94, 96, 98

Use the frequency chart at the right to answer questions 1–5.

Hours Worked by Part-time Employees

Hours worked per week	Number of employees
3	1
5	4
6	5
8	10
10	16
12	14
15	11

1. How many people worked 15 h per week?

2. What is the mode of the hours worked per week?

3. What is the range of hours worked per week?

4. How many employees are shown in the frequency chart?

5. Which group worked the greatest total number of hours during the week, the 14 employees who worked 12 h each or the 11 employees who worked 15 h each?

Use the frequency chart at the right to answer questions 6–10.

Earnings per Week by Part-time Employees

Earnings per week	Number of employees
$10	2
$15	4
$20	5
$24	8
$30	15
$36	7
$45	9

6. How many employees earned $45 a week?

7. How many people earned $20 per week?

8. What is the mode of the earnings?

9. What is the range of the earnings?

10. What are the total earnings of the 50 employees?

SKILLKEEPER

Give each probability for tossing a die with numbers from **1** to **6**.

1. $P(5)$ 2. $P(2)$ 3. $P(7)$ 4. $P(6)$ 5. $P(4)$

6. P(a number less than 4)   7. P(a number greater than 5)

8. P(an odd number)   9. P(an even number)

10. P(a prime number)   11. P(a number less than 6)

More Practice, page 438, Set A

Arithmetic Mean and Median

Freshwater fish are a valuable food source and are also caught for recreational sport.

The table shows the adult length of some common freshwater fish.

The **arithmetic mean** or **average** of the lengths of the 5 fish is the sum of their lengths divided by the number of fish.

Sum = **75 + 66 + 50 + 30 + 20 = 241**

Arithmetic mean = **241 ÷ 5 = 48.2**

The average length is 48.2 cm.

The **median** is the middle number in a list of numbers given in order.

Median length = **50 cm**

When there are two middle numbers, find the mean of those middle numbers.

22, 23, 25, 28, 33, 37
         ↑   ↑
     middle numbers

Median = $\dfrac{25 + 28}{2} = \dfrac{53}{2} =$ **26.5**

Freshwater Fish
Adult length (cm)

Catfish	75
Walleye	66
Bass	50
Yellow perch	30
Bluegill	20

Find the arithmetic mean and median of each list of numbers.

1. 18
   28
   41

2. 30
   30
   40
   80

3. 93
   104
   109
   110

4. 1.8
   2.7
   4.5

Find the arithmetic mean and median of each list of numbers. Round to the nearest tenth if necessary.

5. 78, 78, 82, 88, 90

6. 22, 23, 28, 31, 35, 37

7. 7, 29, 18, 11, 16, 20, 14

8. 320, 195, 222, 307

PROBLEM SOLVING: Using Data from a Table

Solve. Use the table of tropical fish for exercises 1–4.

1. List the lengths of the fish in the table from greatest to least.

2. What is the median length of the seven fish? Which fish has the median length?

3. What is the arithmetic mean of the lengths of the fish?

4. Which fish is nearest the mean length? How much more or less than the mean length is it?

Use the table of freshwater fish for exercises 5–9.

5. List the lengths of the eight freshwater fish in order from greatest to least. What are the two middle lengths?

6. What is the median length of the freshwater fish?

7. What is the arithmetic mean (to the nearest tenth) of the lengths of the fish?

8. Which fish is nearest the mean length? How much more or less than the mean length is it?

9. How many adult sunfish would it take to have a total length equal to the length of an alligator gar?

Tropical Fish
Adult length (cm)

Clown anemone	5
Bluehead	15
Butterfly	13
Flamefish	11
Moorish idol	18
Neon goby	5
Trunkfish	10

Freshwater Fish
Adult length (cm)

Alligator gar	250
Carp	100
Chinook salmon	130
Sunfish	30
Drum	65
Inconnu	125
Lake sturgeon	140
Muskellunge	110

10. **DATA BANK** Four kinds of trout are the rainbow, the lake, the cutthroat, and the brook. What is their mean adult length to the nearest tenth of a centimeter? Which kind of trout is nearest the mean length? (See page 413.)

11. **Try This** Mara caught three catfish. The first two fish were the same length. The other fish was 12 cm shorter than the combined length of the first two fish. The total length of all three fish was 124 cm. What was the length of each fish?

Bar Graphs

This **double bar graph** shows data from a coin-tossing experiment.

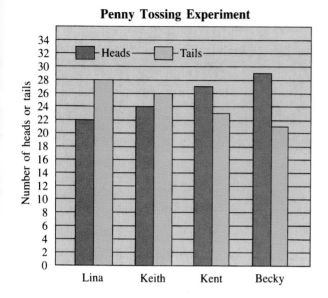

1. How many heads did Lina get? How many tails did she get?

2. How many tosses did Lina make in all?

3. Which person got the least number of tails? How many was this?

4. What is the total number of tosses made by all four people?

5. How many of the total tosses were heads? How many were tails?

This **histogram**, or bar graph of frequencies, shows the outcomes for tossing a pair of dice.

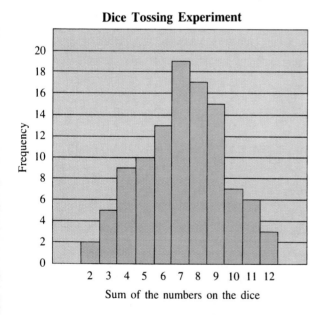

6. Which outcome had the greatest frequency? What was its frequency?

7. Which outcome had the least frequency? What was its frequency?

8. What is the total number of tosses shown by the graph?

9. What percent of the total number of tosses had the outcome of 7? Round your answer to the nearest whole percent.

Coin Toss

Outcome	4H	3H,1T	2H,2T	1H,3T	4T
Frequency	3	17	24	15	5

10. Make a histogram that will show the data in the table for tossing four coins at a time.

Line Segment Graphs

An exercise machine records pulse rates and the amount of time the machine is used. This **line segment graph** shows the results.

1. What was the pulse rate at the start?

2. How much did the pulse rate rise by the end of the first 2 min of exercise?

3. Estimate the highest pulse rate shown by the line graph. After how many minutes did this rate occur?

4. What was the pulse rate at the end of 10 min?

5. What is the range of the pulse rates?

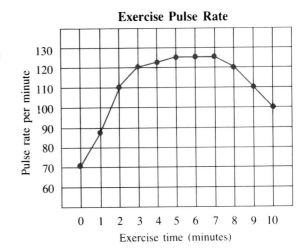

A record of pulse rates after stopping an exercise is shown on this line segment graph.

6. What was the pulse rate at 0 min?

7. How much did the pulse rate drop after 1 min?

8. Estimate the pulse rate at the end of 2 min.

9. Estimate the pulse rate at the end of 8 min.

10. What is the range of the pulse rates?

11. Make a line segment graph of the data in the table.

Exercise Pulse Rate

Time (minutes)	0	1	2	3	4	5	6
Pulse beats per minute	60	75	90	120	115	95	80

357

Pictographs

This **pictograph** or **picture graph** shows the number of farms in the United States from 1940 to 1980.

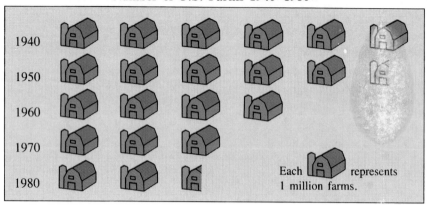

1. Each 🏠 represents how many farms?

2. How many farms does the symbol 🏠 represent?

3. Give the number of farms for each year shown by the pictograph.

4. How many fewer farms were there in 1980 than in 1940?

This pictograph shows how the farm population has changed from 1940 to 1980.

5. Each figure represents how many people?

6. What was the farm population in 1940?

7. What was the farm population in 1950?

8. What was the farm population in 1960, in 1970, and in 1980?

9. How much less was the farm population in 1980 than in 1940?

10. Make a pictograph that shows the average size of farms for the years shown in the table.

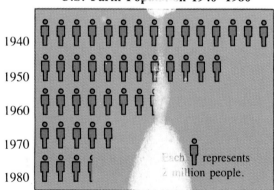

Farm Size in the U.S.

Year	1940	1950	1960	1970	1980
Number of acres	167	213	297	374	450

Circle Graphs

This circle graph shows how land in the United States is used.

The sector for grazing represents 30% of the 9 million km².

$0.30 \times 9 = 2.70$

The graph shows 2.70 million km² of land for grazing.

The central angle for the grazing sector is 30% of 360°.

$0.30 \times 360° = 108°$

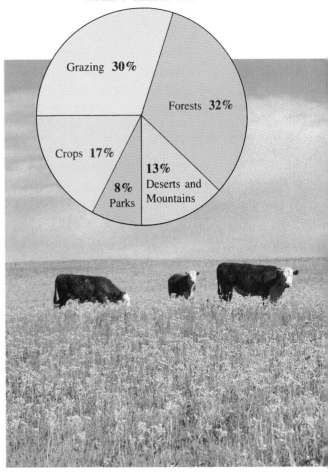

Land Use in the U.S.
Total: 9 million km²

Use the circle graph to answer these questions.

1. How many square kilometers does the graph show for forests?

2. What is the central angle of the sector for forests?

3. How many square kilometers does the graph show for crops?

4. What is the central angle of the sector for crops?

5. How many more square kilometers of land are used for grazing than for crops?

6. How many square kilometers does the graph show for deserts and mountains?

7. What is the central angle of the sector for deserts and mountains?

8. How many square kilometers does the graph show for parks?

9. What is the central angle of the sector for parks?

10. Make a circle graph using the data in the table below.

Kinds of Farms in U.S.
Total: 1,700,000 farms

Grain	34%	Cotton and	
Livestock	29%	tobacco	7%
Dairy	12%	Poultry	3%
Fruit and vegetable	4%	Other	11%

PROBLEM SOLVING: Using Data from a Graph

QUESTION
DATA
PLAN
ANSWER
CHECK

Use the double bar graph for questions 1–5.

1. What is the range of elevations for the state of Ohio?

2. What is the arithmetic mean of the highest and lowest elevations in Ohio?

3. What is the median of the highest elevations of the four states?

4. What is the median of the lowest elevations of the four states?

5. Which state has the greatest range of elevations?

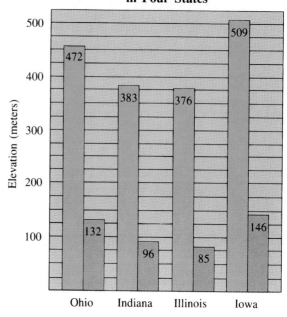

Michael tossed four coins at a time and recorded how many coins came up heads and how many coins came up tails. The graph shows the frequency of the outcomes.

6. What is the total number of tosses shown by the graph?

7. What fractional part of the tosses resulted in an outcome of 4 heads?

8. What fractional part of the tosses resulted in 3 heads, 1 tail?

9. Did more than $\frac{1}{2}$ of the tosses result in 2 heads, 2 tails? How much more or less than $\frac{1}{2}$ of the tosses gave this outcome?

10. Michael found that the probability of tossing 3 heads, 1 tail with four coins was $\frac{1}{4}$. About how many times should Michael expect this outcome in 80 tosses?

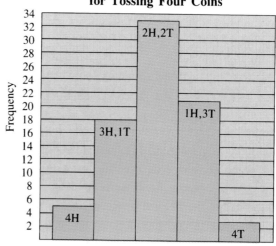

Use the line segment graph for problems 11–15.

11. Which day had the greatest range of temperatures? What was the range?

12. Which day had the smallest range of temperatures? What was the range?

13. The **mean daily temperature** is the average of the daily high and low temperatures. What was the mean daily temperature for Saturday?

14. List the daily high temperatures from lowest to highest. What is the median of the daily high temperatures?

15. What is the median of the low temperatures?

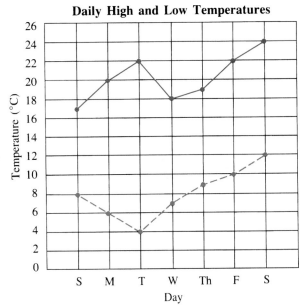

Yolanda made a circle graph to show how she uses her time.

16. How many hours for school are shown by the graph?

17. How many more hours are shown for sleep than for school?

18. What percent of the circular region is "Hobbies?"

19. If 37.5% of each day for a year is spent sleeping, how many hours is this? How many days is this?

20. **Try This** In 24 h, Arthur sleeps 30% of the time, works 35% of the time, eats for 2 h, and spends half as much time for exercise as for sleeping. How many hours does Arthur have left for all other activities during one day?

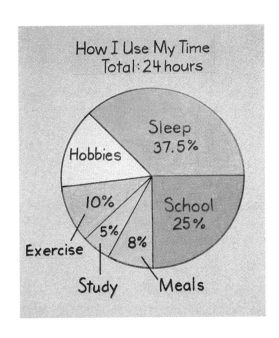

361

APPLIED PROBLEM SOLVING

QUESTION
DATA
PLAN
ANSWER
CHECK

There are about 150 deer who get much of their food from the acorns of an oak forest that they inhabit. You are helping a naturalist who needs to decide if there is enough food for the deer. To find the amount of acorns available, you take several samples from plots of one square meter and count the acorns in these samples.

Sample Plots (1 m²)	Number of Acorns
A	4
B	1
C	10
D	3
E	7
F	0
G	2
H	9

Some Things to Consider

- The oak forest covers about 4 km².
- About 450 kg of acorns are needed for each deer.
- Acorns weigh about 5 g each.
- The oak trees are fairly evenly distributed throughout the forest.

Some Questions to Answer

1. What is the number of square meters in the oak forest?

2. Which sample plot has the greatest number of acorns? Which sample plot has the least number of acorns?

3. Use sample plot D to estimate the number of acorns in the forest. Number of acorns = number in sample × area of forest in square meters.

4. Using sample plot D, about how many grams of acorns are there in the forest? How many deer would this feed?

5. What is the average number of acorns in all 8 samples?

What Is Your Decision?

Are there enough acorns in the forest for the deer population, or should some of the deer be resettled in other areas where there is a better supply of food?

CHAPTER REVIEW/TEST

Give the probability of each outcome when a die with faces numbered **1, 2, 3, 4, 5,** and **6** is tossed.

1. What is $P(3)$?
2. What is $P(\text{odd number})$?
3. What is $P(\text{a number less than 4})$?

Give each probability for a penny and a dime tossed together.

4. List all possible outcomes as ordered pairs of heads and tails.
5. What is the probability of getting a head with the penny and a tail with the dime?
6. What is the probability of getting one head and one tail?
7. What is the probability of getting two heads?

Use the line segment graph to answer questions 8–12.

8. How many days had a low temperature of 6°C?
9. What is the range of temperatures?
10. What is the mode of the temperatures?
11. What is the median temperature?
12. Find the arithmetic mean of the temperatures.

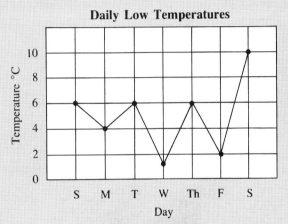

Use the circle graph to answer questions 13–17.

13. How many hours are shown for school?
14. How many hours are shown for sleep?
15. How many hours are shown for recreation?
16. What is the central angle of the sector for meals?
17. How many more hours are shown for sleep than for school?

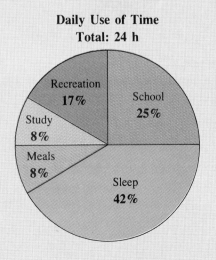

ANOTHER LOOK

There is 1 **chance** in 4 that the spinner will land on 3. The **probability** of getting 3 is $\frac{1}{4}$.

$P(3) = \frac{1}{4}$

Think of spinning the spinner.

1. List all of the possible outcomes.
2. Is each outcome equally likely?
3. What is $P(1)$?
4. What is $P(\text{even number})$?

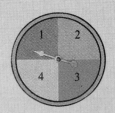

The outcome shown is (A,3). There is 1 chance in 6 that this outcome will occur.

$P(A,3) = \frac{1}{6}$

Think of spinning both spinners.

5. List all of the possible outcomes.
6. What is $P(A,1)$?
7. What is $P(C,4)$?

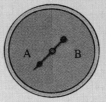 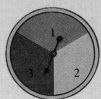

Range = highest−lowest
Median = middle
Mode = most often
Mean = average

Use the table to answer questions 8–10.

Daily Temperature (°C)

28	22	30	34	28	31	25
S	M	T	W	Th	F	S

8. Give the highest and lowest temperatures and the range of temperatures.
9. Give the mode and the median of the temperatures.
10. Find the mean temperature to the nearest whole degree.

8% for clothing
percent × total

8% × $1,200 = 0.08 × 1,200
           = $96.00

Use the circle graph to answer questions 11 and 12.

11. How much money is shown for housing?
12. How much more money is shown for food than for savings?

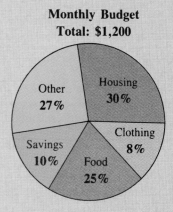

Monthly Budget Total: $1,200

Other 27%
Housing 30%
Clothing 8%
Food 25%
Savings 10%

ENRICHMENT

Probability

You can find an approximate value of $\pi = 3.14159\ldots$ by doing this experiment.

You will need a penny and a large grid of squares drawn on paper or posterboard.

Make the sides of each square of the grid the same length as the diameter of a penny (1.8 cm).

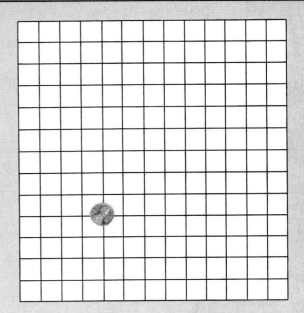

Toss a penny on the grid. The outcome is a "hit" if the penny covers a point of intersection on the grid. Otherwise, it is a "miss."

Toss a penny on the grid 100 times. Keep a tally of the number of hits and misses.

Tosses = hits + misses

Use this formula to compute an approximate value for π.

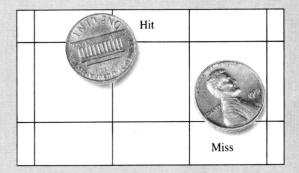

$$\pi \approx \frac{4 \times \text{number of hits}}{\text{number of tosses}}$$

Combine your numbers of hits and misses with those of several classmates. Use the formula for π again to see if the combined results are a better approximation of π.

CUMULATIVE REVIEW

1. Which proportion is correct?

   A $\frac{7}{10} = \frac{14}{22}$ B $\frac{2}{3} = \frac{1}{6}$
   C $\frac{3}{8} = \frac{6}{16}$ D not given

2. What is 12 to 20 as a fraction in lowest terms?

   A $\frac{6}{5}$ B $\frac{3}{5}$
   C $\frac{6}{10}$ D not given

Solve each proportion.

3. $\frac{3}{12} = \frac{x}{28}$ A $x = 7$ B $x = 19$
                                    C $x = 2$ D not given

4. $\frac{6}{10} = \frac{x}{15}$ A $x = 4$ B $x = 9$
                                    C $x = 5$ D not given

5. $\frac{15}{20} = \frac{21}{x}$ A $x = 40$ B $x = 26$
                                    C $x = 14$ D not given

6. The rectangles are similar. What is length x?

   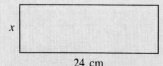

   A $x = 12$ B $x = 7$
   C $x = 8$ D not given

For 7–13, use 3.14 for π.

7. What is the circumference?

   A 26.376 m
   B 8.4 m
   C 13.188 m
   D not given

Find the area of each circle.

8. $r = 0.7$ m   A 4.396 m^2 B 2.198 m^2
                  C 1.5386 m^2 D not given

9. $d = 5$ cm   A 15.7 cm^2 B 19.625 cm^2
                 C 8.5 cm^2 D not given

10. What is the lateral area of a cylinder with $r = 0.2$ cm and $h = 6$ cm?

    A 1.2 cm^2 B 7.536 cm^2
    C 3.768 cm^2 D not given

11. What is the total surface area of a cylinder with $r = 4$ cm and $h = 11$ cm?

    A 276.32 cm^2 B 1,532.32 cm^2
    C 2,788.32 cm^2 D not given

12. What is the volume of a cylinder with $r = 0.4$ cm and $h = 6$ cm?

    A 6.0286 cm^3 B 15.072 cm^3
    C 7.536 cm^3 D not given

13. A mug has a base with an area of 35 cm^2. The height of the mug is 6.5 cm. About how much liquid will the mug hold?

    A 210 mL   B 113.75 mL
    C 227.5 mL D not given

14. A case packer can pack 16 boxes in 8 min. How long would it take the machine to pack 12 boxes?

    A 5 min   B 6 min
    C 4.7 min D not given

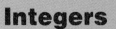

Integers

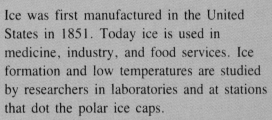

Ice was first manufactured in the United States in 1851. Today ice is used in medicine, industry, and food services. Ice formation and low temperatures are studied by researchers in laboratories and at stations that dot the polar ice caps.

Some scientists study very low temperatures. This field of study is called *cryogenics* (krī′ə jen′ iks). The field of cryogenics deals with temperatures starting at ⁻30°C. When scientists experiment at temperatures nine times colder than this, strange things happen. For example, liquid helium will flow up the side of a container and pour itself out.

An atomic clock has been made that keeps accurate time when cooled to ⁻248°C. "Absolute zero," the temperature at which all heat is removed, is only 25°C colder.

Positive and Negative Integers

To describe distances **above** and **below** sea level, we can use **positive** and **negative integers**.

$$\ldots\ \underbrace{{}^-3,\ {}^-2,\ {}^-1,}_{\text{Negative integers}}\ 0,\ \underbrace{{}^+1,\ {}^+2,\ {}^+3}_{\text{Positive integers}}\ \ldots$$

The integer 0 is neither positive nor negative.

$^+2$ is read "**positive** two."
$^-7$ is read "**negative** seven."

Object	Location	Integer
Boat	At sea level	0
Bird	40 m **above** sea level	$^+40$
Diver	10 m **below** sea level	$^-10$
Kite	70 m **above** sea level	$^+70$
Treasure	100 m **below** sea level	$^-100$

Each integer has an **opposite**.

The **opposite** of $^+3$ is $^-3$.
The **opposite** of $^-6$ is $^+6$.
The **opposite** of 0 is 0.

Give an integer for the location of each object.

1. Turtle **2.** Cliff height **3.** Helicopter

State if each integer is positive or negative.

4. $^+7$ **5.** $^-8$ **6.** $^+15$ **7.** $^-100$ **8.** 0   **9.** $^+1$

Give the opposite of each integer.

10. $^+10$ **11.** $^-1$ **12.** $^-19$ **13.** 0   **14.** $^+20$ **15.** $^+8$

Write each integer.

16. Positive six   **17.** Negative four   **18.** Negative eighteen

⁻8 ⁻7 ⁻6 ⁻5 ⁻4 ⁻3 ⁻2 ⁻1 0 ⁺1 ⁺2 ⁺3 ⁺4 ⁺5 ⁺6 ⁺7 ⁺8

We can use the horizontal number line above to help think of the order of integers and to compare two integers.

The integers increase in size from left to right on the number line.

⁻2 **is less than** ⁺7   ⁻2 < ⁺7

⁻1 **is greater than** ⁻4   ⁻1 > ⁻4

⁻2 **is less than** ⁺1   ⁻2 < ⁺1

0 **is greater than** ⁻5   0 > ⁻5

Write < or > for each ●.

1. ⁺8 ● ⁻2
2. ⁻4 ● ⁺4
3. ⁻1 ● 0
4. ⁻3 ● ⁻7
5. ⁺10 ● ⁻10
6. ⁺9 ● ⁺19
7. 0 ● ⁻8
8. ⁻5 ● ⁻6
9. ⁺1 ● ⁻100
10. ⁻6 ● ⁻5
11. ⁺12 ● ⁺1
12. ⁻13 ● 0
13. ⁻7 ● 8
14. ⁺9 ● ⁻10
15. ⁻32 ● ⁻20

Write the missing integers.

16. 10 degrees above zero is ⁺10°C.
    15 degrees below zero is ●.

17. A loss of 25¢ is ⁻25¢.
    A gain of 50¢ is ●

18. Six seconds before liftoff is ⁻6 s.
    Five seconds after liftoff is ●.

19. 3 km north is ⁺3 km.
    2 km south is ●.

20. Four years from now is ⁺4 years.
    Two years ago is ●.

21. 8 km east is ⁺8 km.
    6 km west is ●.

22. Write the integers in order from least to greatest.
    ⁻7, ⁻3, ⁻10, ⁺2, ⁻5

23. Write the integers in order from greatest to least.
    ⁻11, 0, 7, ⁻5, ⁻3

24. List all the negative integers that are greater than ⁻5.

25. List all the integers that are less than ⁻2 but greater than ⁻7.

More Practice, page 438, Set C

Basic Properties for Integers

Opposites Property	The sum of any integer and its opposite is zero.	$^+7 + {}^-7 = 0$
Zero Property	For every integer n, $n + 0 = n$	$^-6 + 0 = {}^-6$
One Property	For every integer n, $n \cdot {}^+1 = n$	$^-3 \cdot {}^+1 = {}^-3$
Commutative Properties	For any integers x and y, $x + y = y + x$ and $x \cdot y = y \cdot x$	$^-10 + {}^+4 = {}^+4 + {}^-10$ $^+9 \cdot {}^-3 = {}^-3 \cdot {}^+9$
Associative Properties	For any three integers, x, y, and z, $(x + y) + z = x + (y + z)$ and $(x \cdot y) \cdot z = x \cdot (y \cdot z)$	$(^+2 + {}^+3) + {}^-4 = {}^+2 + ({}^+3 + {}^-4)$ $(^-5 \cdot {}^+2) \cdot {}^+6 = {}^-5 \cdot ({}^+2 \cdot {}^+6)$
Distributive Property	For any three integers, x, y, and z, $x \cdot (y + z) = x \cdot y + x \cdot z$	$^+4 \cdot ({}^-5 + {}^+7) = {}^+4 \cdot {}^-5 + {}^+4 \cdot {}^+7$

Because of the commutative and associative properties for addition and multiplication, we can *group* or *order* addends or factors any way we choose. When only one operation is used, we can omit the parentheses.

$(^+5 + {}^-2) + {}^-3 = {}^+5 + {}^-2 + {}^-3$ $^+8 \cdot ({}^-10 \cdot {}^+2) = {}^+8 \cdot {}^-10 \cdot {}^+2$

Warm Up

Name the property used.

1. $^-5 \cdot {}^+1 = {}^-5$
2. $^+7 \cdot {}^+6 = {}^+6 \cdot {}^+7$
3. $^-2 + {}^-8 = {}^-8 + {}^-2$
4. $(^+6 + {}^+4) + {}^-4 = {}^+6 + ({}^+4 + {}^-4)$
5. $^+10 \cdot ({}^+5 + {}^-1) = {}^+10 \cdot {}^+5 + {}^+10 \cdot {}^-1$
6. $^+23 + 0 = {}^+23$
7. $(^-4 \cdot {}^-7) \cdot {}^-3 = {}^-4 \cdot ({}^-7 \cdot {}^-3)$
8. $^-26 + {}^+26 = 0$

Use the opposites property to find the missing integers.

1. $^+4 + {^-4} = \rule{1cm}{0.15mm}$
2. $^-14 + {^+14} = \rule{1cm}{0.15mm}$
3. $^+50 + {^-50} = \rule{1cm}{0.15mm}$
4. $^+2 + \rule{1cm}{0.15mm} = 0$
5. $^-11 + \rule{1cm}{0.15mm} = 0$
6. $\rule{1cm}{0.15mm} + {^+15} = 0$

Use the zero property or the one property to find the missing integers.

7. $^+12 + 0 = \rule{1cm}{0.15mm}$
8. $^-3 + 0 = \rule{1cm}{0.15mm}$
9. $^+16 \cdot {^+1} = \rule{1cm}{0.15mm}$
10. $^-4 + \rule{1cm}{0.15mm} = {^-4}$
11. $\rule{1cm}{0.15mm} \cdot {^+1} = {^-18}$
12. $^-25 \cdot \rule{1cm}{0.15mm} = {^-25}$
13. $\rule{1cm}{0.15mm} + {^-7} = {^-7}$
14. $\rule{1cm}{0.15mm} \cdot {^-9} = {^-9}$
15. $0 + \rule{1cm}{0.15mm} = {^+64}$

Use the commutative properties to write the addends or factors in a different way.

16. $^+7 \cdot {^+3}$
17. $^-6 \cdot {^+6}$
18. $^-7 + {^-1}$
19. $^+24 \cdot {^-1}$
20. $0 + {^+11}$
21. $^-3 \cdot {^-27}$

Use the associative properties to write the addends or factors in a different way.

22. $(^+9 + {^+7}) + {^-7}$
23. $^-1 \cdot ({^-1} \cdot {^+5})$
24. $^+12 + ({^-12} + {^+19})$
25. $(^-3 \cdot {^-2}) \cdot {^+2}$
26. $(^-7 + 0) + {^+7}$
27. $^-24 \cdot ({^-5} \cdot {^+5})$

Use the distributive property to write the addends or factors in a different way.

28. $^+4 \cdot ({^+2} + {^+3})$
29. $^+3 \cdot {^+5} + {^+3} \cdot {^-5}$
30. $^-8 \cdot ({^+4} + {^+6})$

THINK MATH

Logical Reasoning

Find the missing integers. Use the basic properties.

1. $^+8 + ({^+5} + \rule{1cm}{0.15mm}) = {^+8}$
2. $^-6 + ({^-3} + \rule{1cm}{0.15mm}) = {^-6}$
3. $^+10 + (\rule{1cm}{0.15mm} + {^-2}) = {^-2}$
4. $^-4 \cdot ({^+1} + \rule{1cm}{0.15mm}) = {^-4}$
5. $^-2 \cdot (\rule{1cm}{0.15mm} + {^-8}) = 0$
6. $\rule{1cm}{0.15mm} \cdot (0 + {^+1}) = {^+5}$

More Practice, page 439, Set A

Adding Integers

Eric rode his bicycle 3 km west from his home. Then he rode 2 km farther west. How far and in what direction was he from his home?

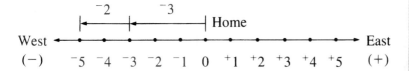

We can use the number line above to add the integers.

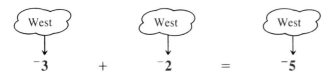

Eric was 5 km west of his home.

The number line below shows the addition of two positive integers.

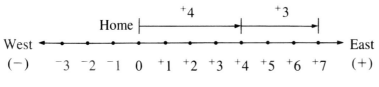

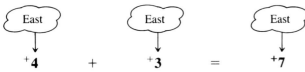

Positive integers are the same as whole numbers greater than 0. From now on we will omit the raised plus sign (⁺) for positive integers.

Other Examples

$4 + 5 = 9$ $\qquad$ $^-3 + {^-3} = {^-6}$ $\qquad$ $^-4 + {^-5} + {^-3} = {^-12}$

Warm Up
Add.

1. $^-2 + {^-2}$
2. $2 + 4$
3. $^-3 + {^-5} + {^-6}$
4. $^-4 + {^-1}$
5. $5 + 2$
6. $7 + 8 + 3$

Find the sums.

1. ⁻2 + ⁻3
2. 4 + 6
3. ⁻5 + ⁻4
4. ⁻1 + ⁻9
5. 4 + 3
6. ⁻8 + ⁻1
7. ⁻3 + ⁻7
8. 5 + 3
9. ⁻8 + ⁻4
10. ⁻5 + ⁻5
11. 4 + 1
12. ⁻6 + ⁻2
13. 2 + 6
14. ⁻2 + ⁻1
15. 7 + 2
16. 6 + 6
17. ⁻3 + ⁻8
18. ⁻1 + ⁻1
19. 6 + 7
20. ⁻5 + ⁻9
21. ⁻9 + ⁻9
22. 8 + 5
23. ⁻4 + ⁻7
24. ⁻9 + ⁻6

25. Negative four plus negative two
26. Positive seven plus positive three
27. Negative 8 plus negative 6
28. Positive 3 plus positive 9

Find the sums. Add the numbers inside the parentheses first.

29. (2 + 3) + 4
30. (⁻4 + ⁻1) + ⁻2
31. (⁻5 + ⁻3) + ⁻3
32. ⁻7 + (⁻2 + ⁻4)
33. ⁻6 + (⁻8 + ⁻4)
34. 9 + (2 + 5)
35. (⁻1 + ⁻2) + ⁻3
36. ⁻5 + (⁻5 + ⁻5)
37. (6 + 8) + 2

38. Is the sum of two positive integers a positive integer or a negative integer?

39. Is the sum of two negative integers a positive integer or a negative integer?

40. Dora rode her bike 7 km west from her home. Then she rode 4 km farther west. How far and in what direction was she from her home?

41. Clarence rode his bike 12 km east in one hour. The second hour he rode 9 km farther east. The third hour he rode 5 km east. How far and in what direction was he from his starting point?

SKILLKEEPER

Find the arithmetic mean and the median of each list of numbers. Round the answers to the nearest tenth if necessary.

1. 25, 29, 32, 36, 37, 39, 42
2. 5, 12, 16, 25, 7, 20
3. 16, 20, 25, 40
4. 252, 275, 310, 325, 350
5. 46, 19, 65, 22, 73
6. 172, 179, 185, 187, 195, 200
7. 2, 7, 12, 52, 3, 16, 25, 42, 49
8. 39, 45, 53, 65, 72, 84, 92

More Practice, page 439, Set B

Adding Positive and Negative Integers

Ricardo rode the train 4 km west from his home. Then he rode a return train 6 km east. How far and in what direction was he from his home?

We can use the number line to help think about the problem.

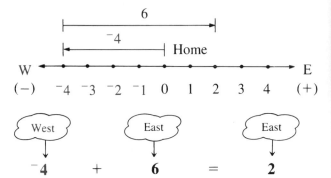

Ricardo was 2 km east of his home.

The number line below shows the addition of a positive integer and a negative integer.

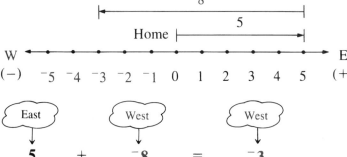

Warm Up

Add. Use a number line if necessary.

1. $^-4 + 9$
2. $6 + {^-6}$
3. $^-2 + {^-2}$
4. $^-3 + 0$
5. $4 + {^-7}$
6. $^-8 + 6$
7. $10 + {^-1}$
8. $5 + 11$

Find the sums. Use the number line if necessary.

```
←—•—•—•—•—•—•—•—•—•—•—•—•—•—•—•—•—•—•—•—•—•→
  ⁻10         ⁻5          0           5          10
```

1. ⁻4 + 7
2. 6 + ⁻4
3. 5 + ⁻6
4. 9 + ⁻7
5. ⁻10 + 2
6. 1 + ⁻9
7. 3 + ⁻3
8. ⁻7 + 3
9. ⁻8 + 1
10. ⁻7 + ⁻12
11. 8 + ⁻15
12. ⁻5 + 10
13. ⁻10 + 14
14. 6 + ⁻8
15. ⁻9 + 9
16. ⁻4 + ⁻7
17. 2 + ⁻6
18. ⁻8 + 16
19. ⁻5 + ⁻2
20. ⁻4 + 13

Find the sums. Add the integers inside the parentheses first.

21. (4 + ⁻2) + ⁻5
22. ⁻1 + (⁻3 + 5)
23. (2 + 5) + ⁻4
24. 6 + (⁻8 + 2)
25. (⁻3 + ⁻4) + 5
26. 6 + (⁻7 + ⁻6)
27. (⁻2 + ⁻3) + ⁻4
28. 9 + (⁻6 + ⁻4)
29. (⁻8 + 10) + ⁻7

30. Marcia rode 8 km east. Then she rode 13 km west. How far and in what direction was she from her starting point? Write an integer addition equation.

31. Rob rode 9 km west. Then he rode 6 km east, and 5 km west. How far and in what direction was he from his starting point? Write an integer equation.

THINK MATH

Using a Calculator

The [+/−] key on a calculator can be used to add integers. Enter any number, then press the [+/−] key. This enters the opposite of the number. To find ⁻6 + 4 on the calculator, follow these steps.

The answer displayed should be ⁻2.

Find each of these sums on a calculator.

1. ⁻10 + 7
2. ⁻8 + ⁻9
3. 15 + ⁻25
4. ⁻31 + 19
5. 57 + ⁻75
6. ⁻94 + 165
7. ⁻53 + ⁻49
8. ⁻217 + 528
9. 537 + ⁻880
10. ⁻2,647 + 1,999
11. 1,000 + ⁻3,775
12. ⁻2,364 + ⁻5,938

More Practice, page 439, Set C

Subtracting Integers

The air temperature outside a plane was ⁻10°C at an altitude of 1,500 m. The temperature at ground level was 2°C. What is the difference between the ground level temperature and the temperature at 1,500 m?

To find the difference, we subtract.

2 − ⁻10   *(What number added to ⁻10 equals 2?)*

Since **12 + ⁻10 = 2**, then **2 − ⁻10 = 12**.

The difference in the temperature is 12°C.

To *subtract* any integer, we *add* its *opposite*.

2 − ⁻10 = 2 + 10 = 12
    ↑   ↑
   opposites

3 − 8 = 3 + ⁻8 = ⁻5
↑   ↑
opposites

Other Examples

0 − ⁻6 = 0 + 6 = 6
 ↑   ↑
opposites

3 − 5 = 3 + ⁻5 = ⁻2
   ↑   ↑
opposites

⁻1 − ⁻3 = ⁻1 + 3 = 2
    ↑   ↑
  opposites

Warm Up

Subtract.

1. 9 − 2
2. 8 − ⁻3
3. ⁻1 − 3
4. ⁻6 − ⁻2
5. 5 − 3
6. 0 − ⁻4
7. ⁻2 − ⁻5
8. ⁻4 − ⁻4
9. ⁻11 − 4
10. 1 − ⁻1
11. 5 − ⁻6
12. 1 − ⁻5

Subtract.

1. ⁻6 − 2
2. 2 − ⁻7
3. 0 − 4
4. 5 − ⁻2
5. 2 − 8
6. ⁻5 − 8
7. ⁻7 − ⁻9
8. ⁻7 − ⁻1
9. 14 − ⁻6
10. ⁻7 − 13
11. ⁻9 − 3
12. ⁻4 − 10
13. 0 − ⁻5
14. ⁻6 − ⁻8
15. ⁻11 − ⁻6
16. 8 − 12
17. 12 − ⁻4
18. 17 − ⁻8
19. 4 − ⁻11
20. ⁻8 − ⁻10
21. ⁻4 − 5
22. ⁻3 − 7
23. 15 − ⁻6
24. 2 − 8
25. ⁻8 − 2
26. ⁻2 − 8
27. 5 − ⁻3
28. ⁻6 − 9
29. ⁻8 − 12
30. ⁻3 − ⁻5
31. 4 − 0
32. 27 − ⁻5
33. 19 − ⁻4
34. ⁻17 − ⁻17
35. ⁻10 − ⁻9
36. 27 − 27

37. When negative 4 is subtracted from positive 7, what is the difference?

38. If positive 10 is subtracted from negative 8, what is the difference?

39. The temperature at 2,000 m is ⁻15°C. At ground level the temperature is 3°C. What is the difference between the temperature at ground level and the temperature at 2,000 m?

40. The outside air temperature at an altitude of 6,100 m was ⁻26°C. When a pilot increased his altitude to 9,100 m, the temperature dropped to ⁻44°C. What is the difference between the temperatures at 6,100 m and 9,100 m?

THINK MATH

Guess and Check

Place the nine integers 4, 3, 2, 1, 0, ⁻1, ⁻2, ⁻3, ⁻4 in the circles so that the sum of the four integers along each side of the triangle is 0.

Then try to get a sum of ⁻1 along each side of the triangle.

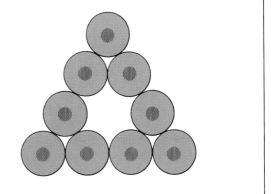

More Practice, page 440, Set A

PROBLEM SOLVING: Using Data from a Graph

QUESTION
DATA
PLAN
ANSWER
CHECK

The difference between the temperature of still air and the **apparent temperature** when the wind blows is called the **chill factor** of the wind.

The graph shows the chill factors for a still-air temperature of ⁻7°C.

When the wind speed is 24 km/h, the apparent temperature is ⁻23°C. The chill factor is 16°. The air feels 16° colder than if the air were still.

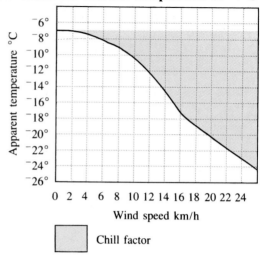

Solve. Use the data from the graph.

1. The still-air temperature is ⁻7°. What is the apparent temperature when the wind speed is 10 km/h? What is the chill factor?

2. What is the wind speed when the apparent temperature is ⁻22°C? What is the chill factor for this wind speed?

3. What is the apparent temperature when the wind speed is 12 km/h? What is the chill factor for this wind speed?

4. What is the wind speed when the chill factor is about 8°? What is the apparent temperature?

5. How much does the chill factor increase when the wind speed increases from 8 km/h to 16 km/h? Is this chill factor more or less than when the wind speed increases from 0 to 8 km/h?

6. Estimate the wind speed when the chill factor of the wind is about 10°C. What is the apparent air temperature at this wind speed?

7. **DATA BANK** If the lowest temperature of record for Wichita, Kansas was the still-air temperature and the apparent temperature was ⁻32°C, what was the chill factor? (See page 416.)

8. **Try This** Cynthia, Lee, and Bill guessed at the temperature on a cold winter morning. Cynthia's guess was 4° too high. Bill guessed ⁻10°C. Lee's guess was 8° below Bill's guess and 2° above Cynthia's guess. What was the actual temperature? Hint: Draw a picture.

378

PROBLEM SOLVING: Practice

QUESTION
DATA
PLAN
ANSWER
CHECK

Solve.

1. An **anemometer** is a device that is used to measure wind speed. In 2.5 h, the cups on an anemometer made 1,500 revolutions. How many revolutions is this per minute?

2. When the wind blows at a speed of 16 km/h, an anemometer will turn 600 times in 1 h. How many revolutions would the anemometer make in 1 h in a 20 km/h wind?

3. A wind-powered generator that is 9 m high produces about 125 kWh of electricity per month. If the generator were twice as tall, the electricity produced would increase by about 40% per month. What would be the increase? What would be the total amount of electricity produced?

Anemometer

4. John Herrington's yearly electricity costs were about $1,100 per year. With a windpowered generator, he now produces about 90% of the electricity he uses. How much money does he save per year on electricity?

5. A windmill with a blade diameter of 1.5 m produces 24 kWh of electricity per month. A windmill with a blade diameter of 4.5 m would produce about 6.5 times as much electricity per month. How much electricity would this be?

6. A jet stream of wind had a speed that was 204 km/h. This is about 0.17 of the speed of a supersonic wind. What is the speed of a supersonic wind?

7. **Try This** In a certain location, the air temperature at ground level was 20°C. The air temperature dropped about 6.4° for each 1,000 m above ground. What would be the temperature outside an airplane flying at 10,000 m?

Windpowered generator

379

Multiplying Integers

We can use the number line to help understand the rules we use for multiplication of integers.

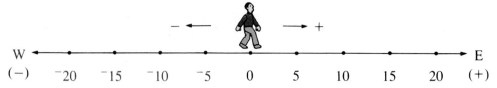

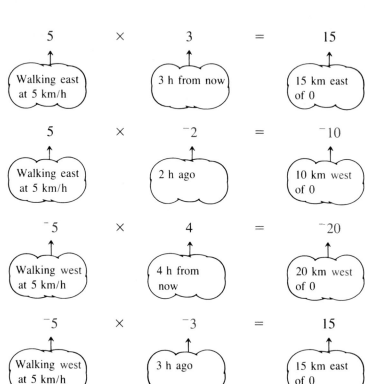

You are now at 0 walking **east** at 5 km/h. Where will you be **3 h from now**?

You are now at 0 walking **east** at 5 km/h. Where were you **2 h ago**?

You are now at 0 walking **west** at 5 km/h. Where will you be **4 h from now**?

You are now at 0 walking **west** at 5 km/h. Where were you **3 h ago**?

The product of a **positive integer** and a **negative integer** is a **negative integer**.

The product of **two positive integers** or **two negative integers** is a **positive integer**.

Warm Up

Multiply.

1. $4 \cdot {}^-4$
2. ${}^-7 \cdot {}^-1$
3. $0 \cdot {}^-8$
4. $21 \cdot 4$
5. $({}^-2 \cdot 4) \cdot 3$
6. $5 \cdot ({}^-5 \cdot {}^-5)$
7. ${}^-2 \cdot (3 \cdot {}^-1)$
8. $4 \cdot (8 \cdot 4)$

Multiply.

1. $7 \cdot {}^-6$
2. ${}^-5 \cdot 4$
3. ${}^-4 \cdot 3$
4. $8 \cdot 7$
5. ${}^-9 \cdot 6$
6. $7 \cdot {}^-8$
7. ${}^-6 \cdot {}^-4$
8. $9 \cdot {}^-9$
9. ${}^-8 \cdot 6$
10. $4 \cdot {}^-9$
11. ${}^-8 \cdot {}^-5$
12. ${}^-7 \cdot 3$
13. $6 \cdot {}^-10$
14. ${}^-7 \cdot 0$
15. ${}^-6 \cdot {}^-8$
16. ${}^-8 \cdot 1$
17. ${}^-18 \cdot 6$
18. $7 \cdot 13$
19. ${}^-12 \cdot 3$
20. ${}^-24 \cdot {}^-4$
21. ${}^-7 \cdot 15$
22. $42 \cdot {}^-9$
23. ${}^-21 \cdot {}^-5$
24. ${}^-36 \cdot {}^-4$
25. ${}^-95 \cdot {}^-4$
26. ${}^-46 \cdot {}^-2$
27. ${}^-9 \cdot {}^-28$
28. ${}^-54 \cdot 3$
29. $12 \cdot {}^-9$
30. ${}^-15 \cdot 11$
31. ${}^-8 \cdot 25$
32. ${}^-10 \cdot {}^-20$

Find the products.

33. $({}^-2 \cdot 3) \cdot {}^-2$
34. ${}^-4 \cdot ({}^-1 \cdot {}^-5)$
35. $(6 \cdot {}^-2) \cdot {}^-3$
36. $8 \cdot ({}^-5 \cdot {}^-2)$
37. $({}^-8 \cdot {}^-3) \cdot {}^-1$
38. $({}^-2 \cdot {}^-2) \cdot {}^-2$
39. $(10 \cdot 10) \cdot {}^-10$
40. ${}^-5 \cdot ({}^-4 \cdot 5)$
41. ${}^-7 \cdot ({}^-2 \cdot {}^-7)$

Solve.

42. $5 \cdot (2 + {}^-3)$
43. $({}^-4 \cdot 2) + ({}^-4 \cdot {}^-5)$
44. $(3 \cdot 7) + (3 \cdot {}^-4)$
45. ${}^-6 + (3 \cdot {}^-4)$
46. ${}^-2 \cdot (6 - 10)$
47. $(5 \cdot {}^-2) - {}^-3$
48. $({}^-1 \cdot {}^-5) - 7$
49. ${}^-6 \cdot ({}^-9 + 5)$
50. ${}^-10 - ({}^-8 \cdot {}^-2)$

THINK MATH

Integer Patterns

Find the pattern for each sequence of integers.
Give the next three integers in the sequence.

1. $1, {}^-1, 3, {}^-3, 5, {}^-5, \_\_\_, \_\_\_, \_\_\_$
2. $2, {}^-4, 8, {}^-16, 32, \_\_\_, \_\_\_, \_\_\_$
3. $10, 7, 4, 1, {}^-2, \_\_\_, \_\_\_, \_\_\_$
4. ${}^-2, {}^-3, {}^-5, {}^-9, {}^-17, \_\_\_, \_\_\_, \_\_\_$

More Practice, page 440, Set B

Dividing Integers

A clock lost 20 min in 5 h. How many minutes per hour did it lose?

To find the number of minutes per hour, we need to divide. To find a quotient, we can use the relation between multiplication and division.

$^-20 \div 5 = ?$ Since $^-4 \cdot 5 = {}^-20$, then $^-20 \div 5 = {}^-4$.

The clock lost 4 min per hour.

$12 \div {}^-4 = ?$ Since $^-3 \cdot {}^-4 = 12$, then $12 \div {}^-4 = {}^-3$.

$^-14 \div {}^-2 = ?$ Since $7 \cdot {}^-2 = {}^-14$, then $^-14 \div {}^-2 = 7$.

$15 \div 3 = ?$ Since $5 \cdot 3 = 15$, then $15 \div 3 = 5$.

The table at the right shows the kind of quotients we will get by using positive or negative dividends and divisors.

Remember, we cannot divide by 0.

Dividend	÷	Divisor	=	Quotient
+	÷	+	=	+
+	÷	−	=	−
−	÷	−	=	+
−	÷	+	=	−

Warm Up

Use the related multiplication fact to find the quotients.

1. $6 \cdot {}^-3 = {}^-18$
   $^-18 \div 6 = $ ▓
   $^-18 \div {}^-3 = $ ▓

2. $^-8 \cdot 5 = {}^-40$
   $^-40 \div 8 = $ ▓
   $^-40 \div {}^-8 = $ ▓

3. $^-7 \cdot {}^-1 = 7$
   $7 \div {}^-7 = $ ▓
   $7 \div {}^-1 = $ ▓

Divide.

4. $\frac{48}{^-6}$

5. $\frac{^-72}{^-9}$

6. $\frac{^-21}{3}$

7. $^-10 \overline{)70}$

8. $12 \overline{)^-108}$

9. $^-15 \overline{)^-300}$

Find the quotients.

1. $81 \div {}^-9$
2. $28 \div {}^-4$
3. $54 \div {}^-6$
4. $24 \div {}^-8$
5. ${}^-21 \div {}^-7$
6. ${}^-24 \div 8$
7. $56 \div {}^-7$
8. $72 \div {}^-9$
9. $63 \div 9$
10. ${}^-28 \div 4$
11. $21 \div {}^-7$
12. $45 \div {}^-5$
13. ${}^-9 \div {}^-9$
14. $49 \div {}^-7$
15. ${}^-72 \div 8$
16. $63 \div {}^-9$

Divide.

17. $\dfrac{{}^-35}{7}$
18. $\dfrac{40}{{}^-8}$
19. $\dfrac{{}^-18}{{}^-3}$
20. $\dfrac{{}^-42}{{}^-7}$
21. $\dfrac{64}{{}^-8}$
22. $\dfrac{48}{{}^-8}$
23. $\dfrac{{}^-12}{{}^-12}$
24. $\dfrac{{}^-9}{{}^-1}$
25. $\dfrac{27}{{}^-9}$
26. $\dfrac{{}^-40}{{}^-8}$
27. $\dfrac{48}{6}$
28. $\dfrac{{}^-1}{1}$

29. $19 \overline{){}^-285}$
30. ${}^-74 \overline){}^-1{,}998$
31. ${}^-27 \overline){729}$
32. ${}^-52 \overline){}^-1{,}976$

33. Divide ${}^-27$ by ${}^-3$.

34. Divide 54 by the opposite of 9.

35. Divide the opposite of 70 by ${}^-10$.

36. Divide the opposite of ${}^-24$ by the opposite of 6.

Solve.

37. $({}^-24 \div {}^-6) \div 2$
38. ${}^-24 \div ({}^-6 \div 2)$
39. $({}^-36 + 24) \div {}^-4$
40. $({}^-36 \div {}^-4) + (24 \div {}^-4)$

SKILLKEEPER

Add.

1. $17 + {}^-6$
2. $5 + {}^-1$
3. ${}^-5 + 2$
4. $25 + {}^-9$
5. ${}^-12 + 7$
6. ${}^-7 + 23$
7. $6 + {}^-16$
8. ${}^-3 + {}^-2$
9. ${}^-11 + 9$
10. ${}^-5 + {}^-6$
11. $12 + {}^-8$
12. ${}^-17 + 6$
13. $8 + {}^-5$
14. $19 + {}^-4$
15. ${}^-15 + 7$

Subtract.

16. $0 - {}^-3$
17. ${}^-12 - {}^-8$
18. ${}^-4 - 7$
19. ${}^-8 - {}^-3$
20. ${}^-11 - 2$
21. ${}^-6 - 7$
22. $9 - 14$
23. ${}^-12 - {}^-15$
24. $4 - 9$
25. ${}^-6 - {}^-2$
26. ${}^-5 - 8$
27. $4 - {}^-11$
28. $14 - {}^-6$
29. ${}^-9 - 3$
30. $12 - {}^-4$

More Practice, page 440, Set C

Integer Coordinates

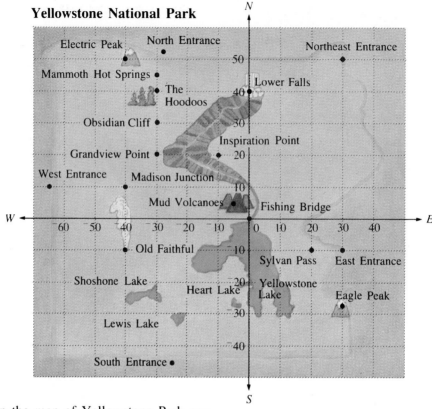

Locations on the map of Yellowstone Park can be given by ordered pairs of numbers called the **coordinates** of the points.

Fishing Bridge is at the **origin** (0,0).

East Entrance is 30 km east and 10 km south of Fishing Bridge. The coordinates of East Entrance are (30, ⁻10).

Give the coordinates of each location.

1. Lower Falls
2. Madison Junction
3. Electric Peak
4. West Entrance
5. Grandview Point
6. Sylvan Pass
7. Northeast Entrance
8. Obsidian Cliff
9. Inspiration Point
10. Mammoth Hot Springs
11. The Hoodoos
12. Old Faithful

Estimate the coordinates of each location.

13. South Entrance
14. Mud Volcanoes
15. Eagle Peak
16. North Entrance
17. Lewis Lake
18. Heart Lake

Give the coordinates of each point.

1. A
2. B
3. C
4. D
5. E
6. F
7. G
8. H
9. I
10. J
11. K
12. L
13. M
14. N
15. P

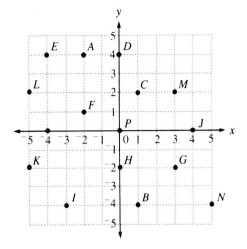

16. Give the coordinates of the three vertices of the triangle.

17. Give the coordinates of the four vertices of the square.

18. Give the coordinates of the four vertices of the parallelogram.

19. Give the coordinates of the four vertices of the rectangle.

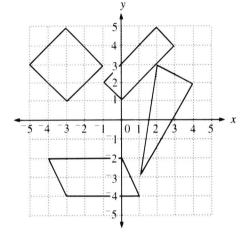

The coordinates of the vertices of six polygons are given below.
Graph the points and then draw each polygon. Name each polygon.

20. (⁻4,4), (2,4), (2,⁻2)
21. (1,2), (⁻3,3), (⁻4,⁻1), (0,⁻2)
22. (⁻2,⁻3), (2,⁻3), (4,0), (2,3), (⁻2,3), (⁻4,0)
23. (4,0), (0,3), (⁻4,0), (⁻2,⁻3), (2,⁻3)
24. (0,0), (⁻2,⁻1), (⁻3,⁻4), (3,⁻1)
25. (3,4), (⁻3,3), (⁻4,⁻3), (2,⁻2)

THINK MATH

Logical Reasoning

Graciela said, "I was 12 years old 2 days ago, but next year I will be 15!" What is the date of Graciela's birthday?

Graphing Equations

The points on a road on this map have coordinates given by the equation $y = 2x + 1$. What is the graph of the equation?

To graph the equation, first make a table of values for x and y.

Then choose any number for x. Multiply by 2 and add 1 to find the corresponding value for y.

Mark the points for the ordered pairs (x,y) found in the table, and connect the points.

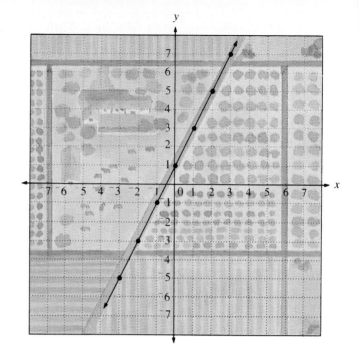

Table of values for the equation $y = 2x + 1$

x	3	2	1	0	⁻1	⁻2	⁻3
$2x + 1$	2·3+1	2·2+1	2·1+1	2·0+1	2·⁻1+1	2·⁻2+1	2·⁻3+1
y	7	5	3	1	⁻1	⁻3	⁻5
(x, y)	(3,7)	(2,5)	(1,3)	(0,1)	(⁻1,⁻1)	(⁻2,⁻3)	(⁻3,⁻5)

Warm Up

1. Copy and complete the table of values for the equation $y = x + 3$.

x	⁻5	⁻3	⁻1	0	1	3	5
y	⁻2						
(x,y)	(⁻5,⁻2)						

2. Use the table of values from exercise 1. Graph the equation $y = x + 3$.

386

Copy and complete each table of values. Then draw the graph of the equation.

1. Equation: $y = 5 - x$

x	8	5	3	1	$^-1$	$^-3$
y	$^-3$					
(x,y)	$(8,^-3)$					

2. Equation: $y = \frac{1}{2}x$

x	8	4	2	0	$^-2$	$^-4$	$^-8$
y	4						
(x,y)	$(8,4)$						

3. Equation: $y = x - 4$

x	7	5	3	1	$^-2$	$^-4$
y	3					
(x,y)	$(7,3)$					

4. Equation: $y = 2x - 2$

x	5	3	1	$^-1$	$^-3$
y	8				
(x,y)	$(5,8)$				

Make a table of values for each equation. Then draw the graph of each equation.

5. $y = x + 1$

6. $y = 3x - 1$

7. $y = 7 - x$

8. $y = x$

9. $y = \frac{1}{4}x + 2$

10. $y = 2x - 4$

11. On the same coordinate grid, draw the graphs of the equations $y = 3 - x$ and $y = 2x - 3$. What are the coordinates of the point where the two graphs meet?

12. DATA HUNT What are the coordinates of your town or city? Use more than one map. Report the different ways the coordinates are labeled.

APPLIED PROBLEM SOLVING

You want to learn to play the clarinet and become a member of the school band. To play in the band, you must have your own musical instrument and at least 24 hours of lessons.

Some Things to Consider

- You cannot pay cash for the clarinet.
- Music lessons are given once a week at $9 per hour.
- A music store sells new clarinets for $376 cash. They have three other plans for buying or renting instruments.

12-month Installment Plan to Buy	Rental Plan	Rent-to-Buy Plan
$34.50 per month	$45.00 for the first 3 months $27.00 per month thereafter	$45.00 for the first 3 months $41.00 per month for 9 months thereafter if you decide to buy

Some Questions to Answer

1. Using the 12-month installment plan, what is the cost of the clarinet? How much more is this than the cash price?

2. What does it cost to rent a clarinet for 12 months? How much more or less is the rental price for 12 months than the intallment price to buy a clarinet?

3. What is the cost of 24 hours of music lessons?

4. If you rent a clarinet while you are taking the lessons, what is the total amount you will spend by the time you finish the lessons?

5. If you use the rent-to-buy plan, how much will you pay for the instrument?

What Is Your Decision

What plan will you choose to be able to play a clarinet in the band?

CHAPTER REVIEW/TEST

Write < or > for each ●.

1. ⁻3 ● 1
2. ⁻7 ● ⁻9
3. 0 ● ⁻6
4. 7 ● ⁻50

Add.

5. ⁻7 + ⁻3
6. ⁻6 + ⁻8
7. 2 + 10
8. ⁻4 + ⁻4
9. ⁻16 + 7
10. 3 + ⁻10
11. ⁻9 + 9
12. 14 + ⁻17

Subtract.

13. 8 − ⁻5
14. ⁻1 − 6
15. 7 − 12
16. ⁻9 − ⁻3
17. ⁻4 − ⁻6
18. 0 − ⁻5
19. ⁻6 − 2
20. 12 − 7

Multiply.

21. 3 · ⁻8
22. ⁻7 · ⁻4
23. ⁻6 · ⁻6
24. ⁻9 · 5
25. 8 · ⁻7
26. ⁻6 · ⁻9
27. ⁻1 · 10
28. ⁻2 · 15

Divide.

29. 42 ÷ ⁻6
30. ⁻28 ÷ 7
31. ⁻18 ÷ ⁻3
32. ⁻45 ÷ 9
33. ⁻60 ÷ ⁻10
34. 36 ÷ ⁻3
35. ⁻54 ÷ ⁻6
36. 32 ÷ ⁻4

Give the coordinates of each point.

37. A
38. B
39. C
40. D
41. E
42. F

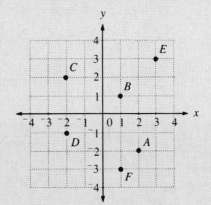

43. The temperature at ground level is 5°C. It is 20° colder at an altitude of 2,500 m. What is the temperature at 2,500 m?

44. The temperature due to the chill factor is ⁻13°C. In still air, the temperature would be 8° higher. What is the still-air temperature?

ANOTHER LOOK

$^-3\ ^-2\ ^-1\ 0\ 1\ 2\ 3$

$^-3$ is to the left of $^-2$.
$^-3 < ^-2$

2 is to the right of $^-2$.
$2 > ^-2$

Write < or > for each ●.

1. 6 ● $^-4$ 2. $^-10$ ● $^-4$ 3. 0 ● $^-7$
4. $^-3$ ● $^-1$ 5. $^-4$ ● 0   6. $^-8$ ● $^-9$

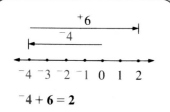

$^-4 + 6 = 2$

Add.

7. $^-3 + ^-4$ 8. $7 + ^-5$ 9. $^-8 + 4$
10. $^-11 + ^-3$ 11. $4 + ^-9$ 12. $0 + ^-6$
13. $^-6 + ^-5$ 14. $^-7 + 10$ 15. $^-8 + ^-7$

$7 - ^-3 = 7 + 3 = 10$

To subtract, *add* the opposite.

$^-9 - 4 = ^-9 + ^-4 = ^-13$

Subtract.

16. $6 - ^-2$ 17. $^-9 - 2$ 18. $^-4 - ^-7$
19. $8 - 4$ 20. $10 - 12$ 21. $^-1 - ^-6$
22. $9 - 15$ 23. $^-4 - ^-4$ 24. $0 - 8$

negative × positive = negative
positive × negative = negative
negative × negative = positive
positive × positive = positive

Multiply.

25. $^-7 \cdot 2$ 26. $8 \cdot ^-3$ 27. $^-5 \cdot ^-2$
28. $6 \cdot 3$ 29. $^-4 \cdot ^-5$ 30. $9 \cdot ^-8$
31. $^-1 \cdot ^-1$ 32. $6 \cdot ^-8$ 33. $^-7 \cdot ^-6$

negative ÷ positive = negative
positive ÷ negative = negative
negative ÷ negative = positive
positive ÷ positive = positive

Divide.

34. $^-8 \div 2$ 35. $^-9 \div ^-3$ 36. $21 \div ^-7$
37. $^-40 \div 5$ 38. $^-16 \div ^-8$ 39. $36 \div ^-4$
40. $^-24 \div ^-6$ 41. $10 \div ^-10$ 42. $^-56 \div ^-8$

ENRICHMENT

Function Rules

This machine operates on **input numbers** using a **rule** to give **output numbers**.

Rule: Multiply by 2 and add 1.

Input n	Output $2 \cdot n + 1$
5	11
3	7
$^-2$	$^-3$
$^-5$	$^-9$

Give the output numbers for each table.

1. Rule: Subtract 2 and multiply by 3.

Input n	Output $(n-2) \cdot 3$
4	
2	
0	
$^-1$	
$^-3$	

2. Rule: Divide by 3, then add 1.

Input n	Output $(n \div 3) + 1$
18	
6	
0	
$^-3$	
$^-15$	

3. Rule: Add $^-10$, then double.

Input n	Output $(n + {^-10}) \cdot 2$
20	
10	
5	
$^-5$	
$^-10$	

Give the rule for finding the output numbers in each table.

4. Rule: ?

Input n	Output ?
5	1
4	0
3	$^-1$
2	$^-2$
1	$^-3$

5. Rule: ?

Input n	Output ?
3	$^-15$
2	$^-10$
7	$^-35$
$^-1$	5
$^-8$	40

6. Rule: ?

Input n	Output ?
2	4
3	9
$^-6$	36
5	25
$^-10$	100

CUMULATIVE REVIEW

1. What is the percent for $\frac{7}{8}$?
   - A 87.5%
   - B 56%
   - C 114%
   - D not given

2. What is the lowest-terms fraction for 72%?
   - A $\frac{7}{25}$
   - B $\frac{35}{50}$
   - C $\frac{18}{25}$
   - D not given

3. What is 12% of 27?
   - A 5.4
   - B 3.24
   - C 6.2
   - D not given

4. What percent of 16 is 5?
   - A 3.2%
   - B 1.2%
   - C 31.25%
   - D not given

5. 20% of what number is 17?
   - A 54
   - B 340
   - C 85
   - D not given

6. What would be the amount of interest charged on a $700 loan for 3 years at 12%?
   - A $252
   - B $84
   - C $168
   - D not given

7. What is $P(4)$?

   - A $\frac{1}{2}$
   - B $\frac{1}{5}$
   - C $\frac{1}{4}$
   - D not given

8. What is P(odd number)?

   - A $\frac{1}{2}$
   - B $\frac{1}{4}$
   - C $\frac{1}{3}$
   - D not given

9. What is the arithmetic mean of 23, 32, 41, 56, and 63?
   - A 54.25
   - B 41
   - C 43.4
   - D not given

10. What is the median of 2, 4, 5, 7, 9, 12, 13, and 14?
    - A 7.5
    - B 8
    - C 8.5
    - D not given

11. What is the mode of 23, 24, 25, 27, 28, 28, 28, 30, 32, 34, 34?
    - A 28.5
    - B 34
    - C 28
    - D not given

12. What is the name of a graph that uses drawings to report information?
    - A pictograph
    - B line segment graph
    - C circle graph
    - D not given

13. Justine got 27 out of 30 problems correct on a spelling test. What was her percent score?
    - A 85%
    - B 89%
    - C 81%
    - D not given

14. Steve has 4 pencils—red, green, yellow, and black—in his desk drawer. If he reaches into the drawer without looking, what is the probability of getting the green pencil?
    - A $\frac{1}{4}$
    - B $\frac{1}{3}$
    - C $\frac{1}{2}$
    - D not given

16

Measurement: Customary Units

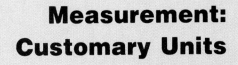

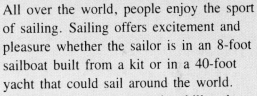

All over the world, people enjoy the sport of sailing. Sailing offers excitement and pleasure whether the sailor is in an 8-foot sailboat built from a kit or in a 40-foot yacht that could sail around the world.

Sailing also challenges the skill and experience of all sailors. They must learn how to adjust the sails to gain the most power from the wind. This is called trimming. A sailor must know the wind direction in order to trim the sails correctly.

When sailing with the wind, racing boats may use huge sails called spinnakers. Sailing across the wind is usually the fastest direction, although sailboats do not usually travel faster than 20 mph. Sailing into the wind requires a skill called tacking. In this basic maneuver, the boat zigzags back and forth across the wind.

Units of Length

Stephanie is building a planter box. She needs a nail that is $1\frac{7}{8}$ in. long. Which nail has this length?

Nail B has a length of $1\frac{7}{8}$ in.

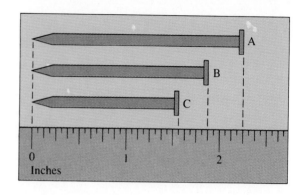

The tables below show some commonly used units of length.

Unit	Symbol
inch	in. or "
foot	ft or '
yard	yd
mile	mi

Relations Between Units
12 in. = 1 ft
3 ft = 1 yd
1,760 yd = 1 mi
5,280 ft = 1 mi

How many inches are there in 5 ft 7 in.?
There are 12 inches in 1 ft.

$5 \times 12 = 60$ $60 + 7 = 67$

5 ft 7 in. = 67 in.

How many yards are there in 27 ft?
There are 3 ft in 1 yd.

$27 \div 3 = 9$

27 ft = 9 yd

$1\frac{1}{4}$ ft is how many inches?

$1\frac{1}{4} \times 12 = \frac{5}{4} \times 12 = 15$

$1\frac{1}{4}$ ft = **15** in.

52 in. is how many feet?

$52 \div 12 = 4\frac{4}{12} = 4\frac{1}{3}$

52 in. = **$4\frac{1}{3}$** ft

Give the missing numbers.

1. 2 ft = ▒ in.
2. 5 yd = ▒ ft
3. 60 in. = ▒ ft
4. 2 mi = ▒ yd
5. 1 ft 6 in. = ▒ in.
6. 2 yd 1 ft = ▒ ft
7. 1 yd = ▒ in.
8. $2\frac{1}{4}$ ft = ▒ in.
9. $1\frac{1}{3}$ yd = ▒ ft
10. $\frac{1}{3}$ ft = ▒ in.
11. 2.5 mi = ▒ ft
12. 1 mi = ▒ in.
13. How many yards are there in $\frac{1}{4}$ mi?
14. How many yards are there in 3 mi?
15. How many inches are there in $3\frac{1}{4}$ ft?
16. Jeff is $58\frac{1}{4}$ in. tall. What is Jeff's height in feet and inches?

When measures of length are added or subtracted, we may need to regroup.

$$\begin{array}{r} \overset{1}{}7\text{ ft }9\text{ in.}\\ +\,3\text{ ft }6\text{ in.}\\ \hline 11\text{ ft }3\text{ in.}\end{array}$$
{ 9 in. + 6 in. = 15 in.
15 in. = 1 ft 3 in. }

$$\begin{array}{r}\overset{2}{\cancel{3}}\text{ yd }\overset{\overset{3}{\cancel{0}}}{\cancel{1}}\text{ ft }\overset{16}{\cancel{4}}\text{ in.}\\ -\,1\text{ yd }2\text{ ft }7\text{ in.}\\ \hline 1\text{ yd }1\text{ ft }9\text{ in.}\end{array}$$

Add the lengths.

1. 2 ft 7 in.
   +5 ft 4 in.

2. 4 ft 10 in.
   +2 ft 8 in.

3. 4 yd 2 ft
   +3 yd 2 ft

4. 12 ft 11 in.
   + 9 ft 5 in.

5. 4 ft 8 in.
   +3 ft 9 in.

6. 1 yd 1 ft 7 in.
   +3 yd 2 ft 6 in.

7. 1 yd 2 ft 10 in.
   +   11 in.

8. 1 ft $7\frac{1}{2}$ in.
   +1 ft $9\frac{1}{4}$ in.

Subtract the lengths.

9. 6 ft 9 in.
   −1 ft 7 in.

10. 5 ft 6 in.
    −2 ft 8 in.

11. 10 ft
    − 2 ft 4 in.

12. 2 yd
    −1 yd 2 ft

13. 2 ft
    −   10 in.

14. 6 ft 5 in.
    −3 ft 9 in.

15. 8 yd 1 ft 1 in.
    −2 yd 2 ft 2 in.

16. 1 yd 1 ft 8 in.
    −   2 ft 11 in.

17. How much longer is a nail that is $2\frac{1}{4}$ in. long than a nail that is $1\frac{7}{8}$ in. long?

18. Two boards of equal thickness are put on top of each other. A nail that is $2\frac{1}{2}''$ long is nailed through the boards. There is $\frac{1}{4}''$ of the nail sticking through the boards. How thick is each board?

19. **DATA HUNT** The **cubit** is an ancient unit of length. What is your height in cubits?

20. **DATA BANK** How many **rods** are there in one mile? (See page 414.)

SKILLKEEPER

1. $^{-}7 \times 2$
2. $7 \times {}^{-}7$
3. $12 \times {}^{-}3$
4. $9 \times {}^{-}6$
5. $2 \times {}^{-}5$
6. $^{-}4 \times {}^{-}9$
7. $0 \times {}^{-}4$
8. $^{-}9 \times 7$
9. $^{-}10 \times {}^{-}2$
10. $^{-}6 \times 5$
11. $12 \div {}^{-}4$
12. $30 \div {}^{-}5$
13. $20 \div {}^{-}4$
14. $^{-}14 \div {}^{-}2$
15. $^{-}24 \div 8$
16. $35 \div {}^{-}5$
17. $^{-}18 \div {}^{-}6$
18. $81 \div {}^{-}9$
19. $^{-}15 \div {}^{-}3$
20. $^{-}9 \div 3$

More Practice, page 441, Set A

Area

A rectangular section of a parade float is 16 ft long and 3 ft wide. What is the **area** of the section?

To find the area of a rectangle, we can use the formula for the area of a rectangle.

$A = lw$
$l = 16$ ft $w = 3$ ft
$A = 3 \times 16 = 48$

The area is 48 square feet (ft^2).

Area is often given in square yards (yd^2). How many square yards are there in 48 ft^2?

1 yd^2 contains 9 ft^2.

$48 \div 9 = 5\frac{1}{3}$

There are $5\frac{1}{3}$ yd^2 in 48 ft^2.

A circular section of a float has a diameter of 16 in. What is the area of that section?

To find the area, we can use the formula for the area of a circle.

$A = \pi r^2$
$r = 8$ in. $\pi \approx 3.14$
$A \approx 3.14 \times 8^2$
$A \approx 3.14 \times 64$
$A \approx 200.96$

The area is about 200.96 in.^{2}.

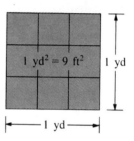

1 $yd^2 = 9$ ft^2 1 yd

1 yd

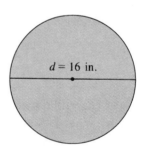

$d = 16$ in.

Warm Up

Give the area of each rectangle.

1. $l = 10$ in.
   $w = 8$ in.

2. $l = 6.5$ ft
   $w = 4$ ft

3. $l = 3\frac{1}{2}$ in.
   $w = 1\frac{1}{2}$ in.

4. $l = 3.1$ mi
   $w = 1.8$ mi

Give the area of each circle. Use 3.14 for π.

5. $r = 10$ ft

6. $d = 8$ in.

7. $r = 2.5$ in.

8. $r = 0.5$ mi

Find the area of each shaded region.

1.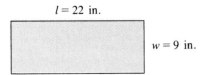

2.

3.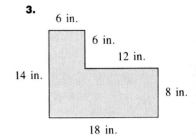

4.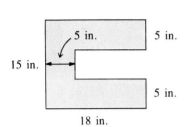

5.

6.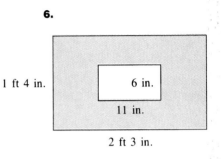

7.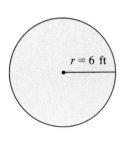

8.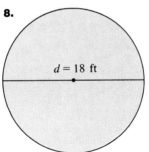

9.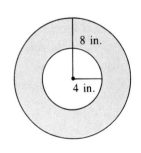

10. Find the area of a rectangle that is 12 ft long and 9 ft wide.

11. Find the area of a rectangle that is 15 ft long and 12 ft wide. Give the area in square yards.

★ **12.** A stairway is 36 in. wide. Each step is 12 in. deep and each riser is 6 in. high. There are 8 steps in all. What would it cost to cover the stairway with carpet costing $18.95 per square yard?

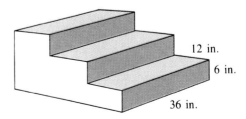

More Practice, page 441, Set B

Volume

A book company packs books in a box 1 ft 4 in. long, 10.5 in. wide, and 8 in. high. What is the **volume** of the box?

Use the volume formula $V = lwh$. All the measurements for l, w, and h must be in the same unit.

$l = 1$ ft 4 in. $= 16$ in.
$w = 10.5$ in.
$h = 8$ in.

$V = 16 \times 10.5 \times 8 = 1{,}344.0$

The volume is 1,344 cubic inches (in.^{3}).

A kitchen canister is a cylinder 8 inches in height with a diameter of 6 in. What is the volume of the canister?

Use the formula $V = \pi r^2 h$.

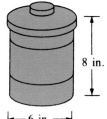

$\pi \approx 3.14$
$r = 3$ in.
$h = 8$ in.

$V \approx 3.14 \times 3^2 \times 8$
$V \approx 3.14 \times 9 \times 8$
$V \approx 226.08$

The volume is about 226.08 in.^{3}.

Warm Up

Give the volume of each figure. Use 3.14 for π.

1.

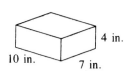

10 in., 4 in., 7 in.

2.

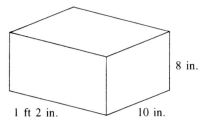

1 ft 2 in., 10 in., 8 in.

3.

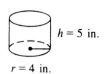

$h = 5$ in.
$r = 4$ in.

4.

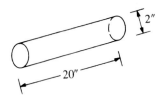

2″, 20″

Find the volume of each rectangular prism. Use $V = lwh$.

1.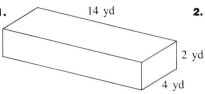
2.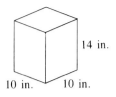
3.

4. $l = 4$ in.
   $w = 4$ in.
   $h = 11$ in.

5. $l = 10$ in.
   $w = 8\frac{1}{2}$ in.
   $h = 1\frac{3}{4}$ in.

6. $l = 1$ ft 7 in.
   $w = 1$ ft 2 in.
   $h = 8$ in.

Find the volume of each cylinder. Use the formula $V = \pi r^2 h$. Use 3.14 for π.

7.
8.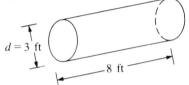
9.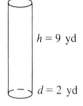

10. $r = 10$ ft
    $h = 10$ ft

11. $d = 24$ in.
    $h = 1$ ft 3 in.

12. $d = 18$ in.
    $h = 5$ in.

13. A cubic foot is a cube that measures 1 ft on each edge. What is the volume of a cubic foot in cubic inches?

★ 14. What is the volume of this building?

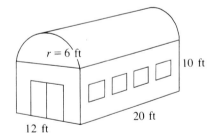

THINK MATH

Guess and Check

A cube has a volume of 29,791 in.³. How long is each edge?

Guess the length, then use a calculator to check your guess.

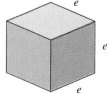

More Practice, page 441, Set C

Liquid Measure

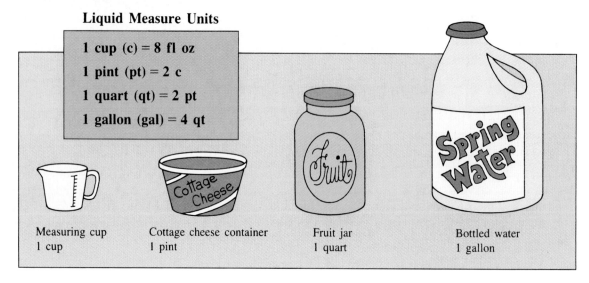

Liquid Measure Units

1 cup (c) = 8 fl oz
1 pint (pt) = 2 c
1 quart (qt) = 2 pt
1 gallon (gal) = 4 qt

Measuring cup
1 cup

Cottage cheese container
1 pint

Fruit jar
1 quart

Bottled water
1 gallon

George is making orange juice. He will make 24 fluid ounces (fl oz) of juice. How many cups of juice will this be?

There are 8 fl oz in 1 c.
24 ÷ 8 = 3

George will make 3 c of orange juice.

Other Examples

Lydia's car needs $1\frac{1}{2}$ qt of oil. How many pints of oil is this? There are 2 pt in 1 qt.
$2 \times 1\frac{1}{2} = 3$
The car needs 3 pt of oil.

The Greenbergs use 3 gal of milk a week. How many quarts of milk do they use? There are 4 qt in 1 gal.
4 × 3 = 12

The Greenbergs use 12 qt of milk a week.

Give the missing numbers.

1. 16 fl oz = ▓ c
2. 2 pt = ▓ c
3. 2 gal = ▓ qt
4. 1 gal = ▓ pt
5. 1 qt = ▓ fl oz
6. 6 pt = ▓ qt
7. 10 gal = ▓ qt
8. 4 fl oz = ▓ c
9. 1 pt = ▓ qt
10. 32 fl oz = ▓ pt
11. 24 pt = ▓ gal
12. 6 gal = ▓ qt

13. How many fluid ounces are there in $2\frac{3}{4}$ c?
14. How many cups are there in $2\frac{1}{4}$ qt?

PROBLEM SOLVING: Practice

QUESTION
DATA
PLAN
ANSWER
CHECK

Solve.

1. Rachel made 2 gal of orange juice. How many quarts is this?

2. Joyce bought 3 qt of milk. How many fluid ounces of milk is this?

3. Todd made 3 qt of lemonade. How many 1-c servings is this?

4. The gasoline tank in Nina's car has a capacity of 16 gal. Nina filled her tank with 11.5 gal. How much gasoline was already in the tank?

5. One bottle of shampoo held 10 fl oz. Another bottle held $\frac{3}{4}$ pt. Which amount is greater? How much greater?

6. Raul drank $\frac{1}{2}$ pt of milk with his lunch. How many fluid ounces of milk did Raul drink?

7. A recipe for 10 pancakes calls for 2 c of pancake mix and $1\frac{1}{3}$ c of milk. How much milk will be needed for 6 c of pancake mix?

8. How much more expensive is salad dressing at 99¢ a quart than at 93¢ for 32 fl oz?

9. Lemon juice is on sale at 89¢ for 16 fl oz. What would 1 qt of lemon juice cost at this rate?

10. A 48-fl oz can of tomato juice was on sale for 75¢. How many quarts is this?

11. The answer to a certain question is, "The total cost is $4." Using liquid measurements, make up a problem that will have this answer.

12. One day Celeste drank 6 fl oz of orange juice at breakfast, 8 fl oz of milk at lunch, and 10 fl oz of water at dinner. How many pints of liquid did she drink that day?

13. Gasoline costs $1.549 per gal. What is the cost (to the nearest cent) of 16.7 gal of gasoline?

14. **Try This** Helen filled the gasoline tank of her car and bought 1 qt of oil. The gasoline cost 10 times as much as the oil. The total cost was $19.25. What was the cost of the oil? What was the cost of the gasoline?

401

Units of Weight

Units of Weight
16 ounces (oz) = 1 pound (lb)
2,000 lb = 1 ton (T)

Clydesdale horse
1 ton

Kitten
1 pound

Tropical fish
1 ounce

Choose the best estimate of weight.

1. A compact car
   A 2 T **B** 20 T **C** 20 lb

2. A textbook
   A 2 oz **B** 2 lb **C** 2 T

3. An orange
   A 6 oz **B** 6 lb **C** 60 lb

4. A bowling ball
   A 16 T **B** 16 oz **C** 16 lb

5. A hippopotamus
   A 2 T **B** 20 lb **C** 200 T

6. A hamburger patty
   A 400 oz **B** 4 oz **C** 40 oz

Give the missing numbers.

7. 2 lb = ▓ oz
8. 48 oz = ▓ lb
9. 2 T = ▓ lb
10. 5 lb = ▓ oz
11. 1 lb 4 oz = ▓ oz
12. 2 lb 8 oz = ▓ oz
13. 128 oz = ▓ lb
14. 9 T = ▓ lb
15. $2\frac{1}{2}$ lb = ▓ oz
16. 10,000 lb = ▓ T
17. $1\frac{1}{4}$ T = ▓ lb
18. $\frac{3}{4}$ lb = ▓ oz

19. One package of frying chicken weighs 3 lb 4 oz. A second package weighs 2 lb 11 oz. How many more ounces are in the first package?

20. Mr. Liu is driving a truck with a total weight of 17,250 lb. A bridge has a load limit of 8 T. How much over the load limit is the total weight of the truck?

Fahrenheit Temperature

Weather temperatures and cooking temperatures are often given in **degrees Fahrenheit** (**°F**). Water freezes at 32°F and boils at 212°F.

Choose the best estimate of temperature.

1. A warm summer day
   A 8°F   **B** 28°F   **C** 80°F
2. A snowy winter day
   A 50°F   **B** 40°F   **C** 30°F
3. Moderate oven temperature
   A 150°F   **B** 350°F   **C** 500°F
4. Water in a mountain stream
   A 40°F   **B** 70°F   **C** 100°F
5. A cup of hot chocolate
   A 150°F   **B** 250°F   **C** 50°F

Estimate each temperature shown.

6. Oven gauge

7. House thermostat

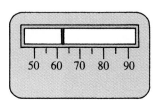

8. Meat thermometer

Solve.

9. Cooking instructions: Cook at 400°F for 10 min, then reduce temperature to 325°F. By how much should the temperature be reduced?

10. The high temperature on a summer day was 92°F. The low temperature was 64°F. What was the mean or average temperature for the day?

SKILLKEEPER

1. 2 ft 5 in.
   + 6 ft 7 in.

2. 3 ft 3 in.
   + 7 ft 13 in.

3. 7 ft 8 in.
   + 6 ft 5 in.

4. 1 ft 3 in.
   + 5 ft 14 in.

5. 3 yd 2 ft
   − 1 yd

6. 7 ft
   −   4 in.

7. 6 yd 3 ft
   − 4 yd 4 ft

8. 12 ft 1 in.
   − 7 ft 3 in.

PROBLEM SOLVING: Using a Calculator

QUESTION
DATA
PLAN
ANSWER
CHECK

Mrs. Ramberg shops at a store that shows unit prices on all grocery items. What is the unit price on a 7-lb bag of pancake mix that sells for $3.92?

To find the unit price, divide the total cost by the number of units. Use a calculator.

Unit price = $\dfrac{\text{total cost}}{\text{number of units}}$ ← pounds

Unit price = $\dfrac{\$3.92}{7}$ = $0.56

The unit price is 56¢ per pound.

A Unit Price Card

Unit price × number of units = total cost
4.9¢ per fl oz × 32 fl oz = 156.8¢
= $1.57

Rounded to the nearest cent.

Solve.

1. Val bought a 2-lb box of crackers for $1.35. What is the unit price to the nearest tenth of a cent?

2. The unit price on a 12-oz can of peanuts is 18.6¢ per oz. What is the cost of this can of peanuts?

3. Jody bought 4 lb of spaghetti for $1.49. What is the unit price of the spaghetti?

4. A 42-oz box of oatmeal is priced at $1.24. What is the unit price (to the nearest tenth of a cent) of the oatmeal?

5. The unit price on a 32-oz bottle of catsup is 3.71¢ per oz. What is the cost of the catsup?

6. A 14-oz bottle of catsup has a unit price of 4.86¢ per oz. What is the cost of the catsup?

9. Kathleen chose a 22-oz bottle of liquid detergent. The price was not on the bottle but the unit price shown was 6.5¢ per oz. Kathleen used her calculator to find the price of the detergent. What is the price of the detergent?

10. Pam Fleetdeer checked the sizes and prices of baby shampoo: 7 oz for $1.99, 11 oz for $2.79, 16 oz for $3.29. Which size has the lowest unit price? Find the unit price (to the nearest hundredth of a cent) for each size.

11. Cole paid 54¢ for a quart of non-fat milk. He could have bought $\frac{1}{2}$ gal for $1.08. Find the cost per gallon for each. Is the unit price of one less than the unit price of the other?

12. Katie bought a 48-oz can of vegetable juice that had a unit price of 2.09¢ per oz. She also had a coupon for 25¢ off the price of the juice. To the nearest cent, what did she have to pay for the can of juice?

13. A 6-oz can of frozen orange juice concentrate costs 54¢. What is the unit price of the concentrate? When the concentrate is mixed with water, it makes 24 oz of orange juice. Not including the price of water, what is the unit price of the orange juice?

14. Try This A package of dried spaghetti and a can of spaghetti sauce cost $2.58. The spaghetti sauce cost $1.00 more than the dried spaghetti. How much did each item cost?

7. Blair saw a box of 8 waffles for 97¢ and a box of 10 waffles for $1.09. Which size box has the lower unit price? Find each unit price to the nearest tenth of a cent.

8. A 2-lb box of flour has a price of $1.39. Write the unit price of the flour in cents per ounce.

APPLIED PROBLEM SOLVING

QUESTION
DATA
PLAN
ANSWER
CHECK

You are in a supermarket. You need to decide which one of several different sizes of tubes of toothpaste would be most economical to buy.

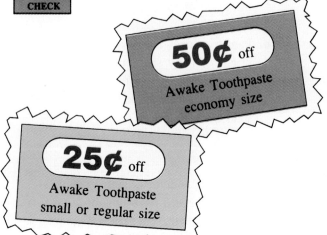

Awake Toothpaste		
small	1.4 oz	$0.65
regular	4.6 oz	$1.39
family	6 oz	$1.75
economy	8.2 oz	$1.95
Generic	6.4 oz	$0.84

Some Things to Consider

- You have enough money to buy any size tube of toothpaste you wish.

- The supermarket does not have unit prices displayed.

- You want to buy only one tube of toothpaste.

- You have coupons for 25¢ off the small or regular size, and 50¢ off the economy size.

Some Questions to Answer

1. What is the price of the small size if you use the coupon?

2. What is the unit price for the small size if you use the coupon?

3. What is the price of the regular size if you use the coupon?

4. What is the unit price of the regular size if you use the coupon?

5. Which size is more economical to buy, the small or the regular size?

6. What are the unit prices for the family and generic sizes?

What Is Your Decision?

Which of the five sizes of tubes of toothpaste will you buy?

CHAPTER REVIEW/TEST

Write the missing numbers.

1. 2 yd = ▒ ft
2. 5,280 ft = ▒ mi
3. 48 in. = ▒ ft
4. 30 ft = ▒ yd
5. 5 ft = ▒ in.
6. 2 ft 9 in. = ▒ in.

Add or subtract.

7. 3 ft 7 in.
   + 1 ft 9 in.

8. 4 ft 10 in.
   − 1 ft 6 in.

9. 6 ft 6 in.
   − 2 ft 11 in.

10. 2 ft 8 in.
    + 5 ft 9 in.

11. 8 ft
    − 3 ft 9 in.

12. 2 yd 2 ft 9 in.
    +   1 ft 6 in.

Find the area. Use 3.14 for π when necessary.

13.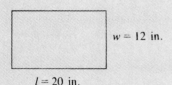
$w = 12$ in.
$l = 20$ in.

14.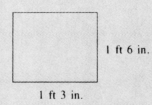
1 ft 6 in.
1 ft 3 in.

15.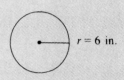
$r = 6$ in.

Find the volume. Use 3.14 for π when necessary.

16.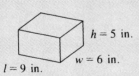
$h = 5$ in.
$w = 6$ in.
$l = 9$ in.

17.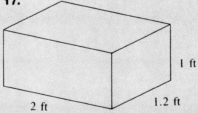
1 ft
2 ft
1.2 ft

18.
$h = 8$ in.
$r = 3$ in.

Write the missing numbers.

19. 2 pt = ▒ c
20. 1 gal = ▒ qt
21. 16 pt = ▒ gal
22. 32 oz = ▒ lb
23. 2 T = ▒ lb
24. 5 lb = ▒ oz

25. A 32-oz package of popcorn costs $1.92. What is the unit cost in cents per ounce?

26. Normal body temperature is 98.6°F. How many degrees Fahrenheit is this above the freezing temperature of water?

ANOTHER LOOK

12 in. = 1 ft
3 ft = 1 yd
5,280 ft = 1 mi

Write the missing numbers.

1. 2 ft = ▥ in.
2. 3 yd = ▥ ft
3. 3 mi = ▥ ft
4. 10 ft = ▥ yd ▥ ft
5. 16 in. = ▥ ft ▥ in.
6. 15 ft = ▥ yd

```
    1
  4 ft 7 in.
+ 3 ft 9 in.
  ──────────
  8 ft 4 in.

9 in. + 7 in. = 16 in.
          16 in. = 1 ft 4 in.
```

Add or subtract.

7.   2 ft 10 in.
   + 3 ft 8 in.

8.   4 ft 4 in.
   − 2 ft 9 in.

9.   2 yd 1 ft 9 in.
   + 3 yd 2 ft 3 in.

10. 5 yd 1 ft 1 in.
   −   3 ft 4 in.

11. 3 yd 1 ft 9 in.
   − 1 yd 3 ft 5 in.

12. 16 yd 12 ft
   + 12 yd 5 ft

1 c   = 8 fl oz
1 pt = 2 c
1 qt = 2 pt
1 gal = 4 qt
16 oz = 1 lb
2,000 lb = 1 T

Write the missing numbers.

13. 4 c = ▥ pt
14. 2 gal = ▥ qt
15. 10 pt = ▥ qt
16. 12 qt = ▥ gal
17. 8 fl oz = ▥ c
18. 5 gal = ▥ qt
19. 2 lb = ▥ oz
20. 6,000 lb = ▥ T
21. 64 oz = ▥ lb

Water boils at 212°F.
Water freezes at 32°F.

Write the temperatures.

22. 3° below the boiling point of water
23. 10° below the freezing point of water

24. Room temperature is 68°F. How much above the freezing temperature of water is this?

25. Moderate oven temperature is 350°F. How much above the boiling temperature of water is this?

ENRICHMENT

Other Number Systems

The problems on this page use a new kind of arithmetic called an eight-clock arithmetic. There are only eight numbers in this arithmetic. You can think of these eight numbers spaced equally around a clock face.

Study the examples below to see how to add and subtract these eight-clock numbers.

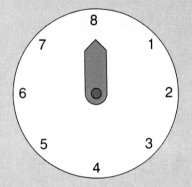
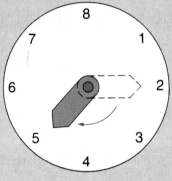
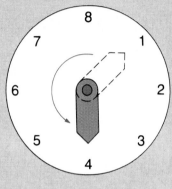

$2 + 3 = 5$

To add, start at 2, then count 3 spaces **clockwise** to get to 5.

$1 - 5 = 4$

To subtract, start at 1, then count 5 spaces **counterclockwise** to get to 4.

Find the sums and differences using eight-clock arithmetic.

1. $4 + 2 = g$
2. $4 + 3 = x$
3. $4 + 4 = b$
4. $4 + 5 = c$
5. $4 + 6 = l$
6. $4 + 7 = s$
7. $4 + 8 = m$
8. $5 + 8 = a$
9. $6 + 8 = i$
10. $5 - 2 = q$
11. $5 - 3 = p$
12. $5 - 4 = k$
13. $5 - 5 = w$
14. $5 - 6 = e$
15. $5 - 7 = t$
16. $5 - 8 = f$
17. $6 - 8 = r$
18. $7 - 8 = h$
19. $8 - 8 = n$
20. $1 - 8 = j$
21. $3 - 8 = y$

Decide how you would multiply in eight-clock arithmetic. Make up and solve some multiplication problems.

TECHNOLOGY

Computer Graphics

Logo is a special programming language that can be used to draw geometric shapes on a computer screen.

A small triangle, called a **turtle** (△), acts like a pencil point in drawing figures. The point of the turtle shows the direction in which lines will be drawn.

The table below shows four simple Logo commands. Using just these four commands, a variety of geometric figures can be drawn.

Command	Short Form	Meaning
Forward 20	FD 20	Draws a line forward 20 units.
Back 20	BK 20	Draws a line backward 20 units.
Right 60	RT 60	Turns the turtle 60° clockwise.
Left 60	LT 60	Turns the turtle 60° counterclockwise.

This is a program for drawing a parallelogram.

```
TO PARALLELOGRAM
FD 30 RT 60
FD 50 RT 120
FD 30 RT 60
FD 50 RT 120
END
```

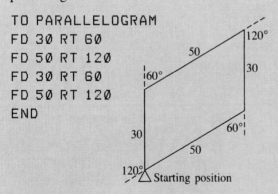

Now when you type PARALLELOGRAM the computer draws the parallelogram.

With the PARALLELOGRAM program stored in the computer's memory, we can use the command REPEAT for the PARALLELOGRAM to draw other shapes.

REPEAT 6 [RT 60 PARALLELOGRAM]

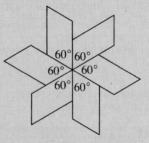

This tells the computer to show six parallelograms each rotated 60° from the starting point of the turtle.

1. Fill in the missing numbers in the program for the right triangle shown.

   ```
   TO TRIANGLE
   FD 20 RT 120
   FD ▨ RT ▨
   FD ▨ RT ▨
   END
   ```

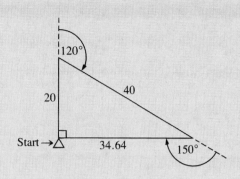

2. Complete the program for the rhombus.

   ```
   TO RHOMBUS
   FD 40 RT 72
   FD ▨ RT ▨
   FD ▨ RT ▨
   FD ▨ RT ▨
   END
   ```

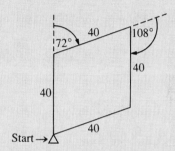

3. Using the rhombus program from exercise 2 and the REPEAT command, the figure at the right can be drawn. What are the two missing numbers in the program?

   REPEAT ▨ [RT ▨ RHOMBUS]

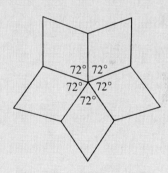

4. Write a program for drawing the SQUARESPIRAL at the right.

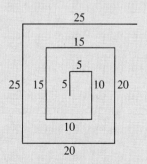

5. Write a Logo program of your own for a geometric figure. Make a sketch of the figure for your program.

CUMULATIVE REVIEW

1. What is the opposite of ⁻7?

   A ⁻14   B 0
   C 7   D not given

2. Which is correct?

   A ⁻12 > ⁻5   B ⁻2 > ⁻5
   C 7 < ⁻4   D not given

3. Find the sum.

   ⁻8 + ⁻25

   A ⁻17   B ⁻33
   C 17   D not given

4. Find the difference.

   ⁻20 − ⁻7

   A ⁻27   B 27
   C ⁻13   D not given

5. Multiply.

   ⁻12 × ⁻4

   A 48   B ⁻16
   C ⁻48   D not given

6. Divide.

   25 ÷ ⁻5

   A 20   B ⁻5
   C 5   D not given

Find the missing numbers.

7. 5 ft = ▓ in.

   A 62   B 15
   C 45   D not given

8. 4 pt = ▓ c

   A 2   B 8
   C 4   D not given

9. 48 oz = ▓ lb

   A 4   B 12
   C 3   D not given

10. 32 gal = ▓ qt

    A 16   B 8
    C 12.8   D not given

11. Add.

    3 ft 7 in.   A 8 ft 4 in.
    + 4 ft 9 in.   B 8 ft 16 in.
                      C 7 ft 4 in.
                      D not given

12. Subtract.

    12 yd 3 ft   A 7 yd 9 ft
    − 4 yd 4 ft   B 7 yd 2 ft
                      C 8 yd 2 ft
                      D not given

13. The low temperature for the month of January was 12°F. The high temperature that month was 22°F. What was the mean or average temperature that month?

    A 10°F   B 34°F
    C 17°F   D not given

14. If positive 12 is subtracted from negative 10, what is the difference?

    A 2   B ⁻22
    C ⁻2   D not given

Appendix

DATA BANK

No. 884. Commercial Broadcast Stations—States and Other Areas

STATE	AM	FM	TV	STATE	AM	FM	TV
Ala	140	74	17	Mont	45	24	12
Alaska	22	11	7	Nbr	49	37	14
Ariz	61	31	13	Nev	22	12	8
Ark	90	54	9	N.H.	28	15	2
Calif	232	204	54	N.J.	39	34	5
Colo	72	51	12	N. Mex	57	29	9
Conn	39	25	5	N.Y.	161	119	30
Del	10	7	—	N.C.	211	88	19
D.C.	7	9	5	N. Dak	26	10	11
Fla	199	115	31	Ohio	123	128	24
Ga	186	90	19	Okla	67	53	12
Hawaii	26	7	10	Oreg	80	35	12
Idaho	44	23	8	Pa	177	124	27
Ill	127	132	23	R.I.	15	7	2
Ind	86	95	20	S.C.	108	56	11
Iowa	76	75	13	S. Dak	33	18	11
Kans	59	45	12	Tenn	158	80	19
Ky	121	87	11	Tex	289	197	58
La	95	64	15	Utah	35	20	4
Maine	37	33	7	Vt	19	12	2
Md	50	37	6	Va	137	72	16
Mass	66	39	12	Wash	92	47	14
Mich	128	117	24	W. Va	63	34	9
Minn	93	69	12	Wis	102	92	17
Miss	106	72	10	Wyo	30	10	5
Mo	111	78	24	P. Rico	54	34	8
				Guam, V. Is.	7	5	3

Nail Chart

Size (Penny)	Length in inches
2	1
3	$1\frac{1}{4}$
4	$1\frac{1}{2}$
5	$1\frac{3}{4}$
6	2
7	$2\frac{1}{4}$
8	$2\frac{1}{2}$
9	$2\frac{3}{4}$
10	3
12	$3\frac{1}{4}$
16	$3\frac{1}{2}$
20	4
30	$4\frac{1}{2}$
40	5
50	$5\frac{1}{2}$
60	6

Adult Trout Length

Arctic trout
   Salvelinus alpinus.................. 30 cm
Brook trout
   Salvelinus fontinalis .... 45 cm
Brown trout
   Salmo trutta .... 25 cm
Cutthroat trout
   Salmo clarki...................... 76 cm

Golden trout
   Salmo aquabonita.................. 25 cm
Lake trout
   Salvelinus namaycush............... 90 cm
Rainbow trout
   Salmo gairdneri................... 110 cm
Sunapee trout
   Salvelinus aureolus................ 50 cm

DATA BANK

Unusual or Old Units of Length

Cubit Length of the forearm from the elbow to the tip of the middle finger; about 18 in.

Hand Unit used in measuring the height of horses; about 4 in.

Fathom Unit used for measuring the depth of water; length of outstretched arms; about 6 ft

League An old unit of distance that varies from country to country; 3 mi

Rod 16.5 ft

Nautical mile Unit of distance used in sea or air navigation; 6,076.115 ft

Barleycorn An old unit of length; about $\frac{1}{3}$ in. Three grains of barley were about 1 inch in length.

Pitchers of the Past

PITCHER	CAREER	YEARS	GAMES
Grover Cleveland Alexander	1911–1930	20	696
Mordecai Brown	1903–1916	12	411
Stan Coveleski	1912–1925	14	450
Bob Feller	1936–1956	18	570
Vernon Gomez	1930–1942	14	368
Sandy Koufax	1955–1966	12	397
Eppa Rixey	1912–1933	21	692
Charles Ruffing	1924–1947	22	624
Warren Spahn	1940–1967	21	750
Early Wynn	1939–1963	23	691
Denton Young	1890–1911	22	906

No. 289. Average Number of Days with Precipitation of 0.01 Inch or More—Selected Cities

State	Station	Length of record (yr.)	Jan	Feb	Mar	Apr	May	Jun	Jul	Aug	Sep	Oct	Nov	Dec	Annual
AL	Mobile	37	11	10	11	7	8	11	17	14	11	6	8	11	125
AK	Juneau	35	18	17	18	17	17	16	17	18	20	24	19	21	222
AZ	Phoenix	39	4	4	3	2	1	1	4	5	3	3	2	4	36
AR	Little Rock	36	10	9	10	10	10	8	9	7	7	6	8	9	103
CA	Los Angeles	43	6	6	5	3	1	1	1	(Z)	1	2	3	5	34
	Sacramento	39	10	9	8	6	3	1	(Z)	(Z)	1	3	7	9	57
	San Francisco	51	11	10	9	6	3	1	(Z)	1	1	4	7	10	63
CO	Denver	44	6	6	8	9	10	9	9	8	6	5	5	5	86
CT	Hartford	24	11	11	11	11	12	12	10	10	10	8	11	12	129
DE	Wilmington	31	11	9	11	11	12	9	9	9	8	7	10	10	116
DC	Washington	37	11	8	11	10	11	9	10	9	8	7	8	9	111
FL	Jacksonville	37	8	8	8	6	8	12	15	15	13	9	6	8	116
	Miami	36	7	6	6	6	11	15	16	17	17	15	8	7	131
GA	Atlanta	44	11	10	12	9	9	10	12	9	7	6	8	10	113
HI	Honolulu	29	10	9	9	9	7	6	8	7	7	9	10	10	101
ID	Boise	39	12	10	9	8	8	6	2	3	4	6	10	11	89
IL	Chicago	20	11	10	13	12	11	10	10	8	10	9	10	11	125
	Peoria	39	9	8	11	12	12	10	9	3	9	7	9	10	114
IN	Indianapolis	39	12	10	13	12	12	10	9	8	8	8	10	12	124
IA	Des Moines	39	7	7	10	10	11	11	9	9	9	7	7	8	105
KS	Wichita	25	5	5	7	8	11	9	8	7	8	6	5	6	85
KY	Louisville	31	12	11	13	12	11	10	11	8	8	7	10	11	124
LA	New Orleans	30	10	9	9	7	7	10	15	13	10	5	7	10	112

Z: Less than $\frac{1}{2}$ day

Winter Olympics (Women) Gold Medalists

500-meter Speed Skating			Time
1960	Helga Haase	Germany	45.9 s
1964	Lydia Skoblikova	U.S.S.R.	45.0 s
1968	Ludmilla Titova	U.S.S.R.	46.1 s
1972	Anne Henning	U.S.A.	43.4 s
1976	Sheila Young	U.S.A.	42.8 s

Alpine Skiing			
Downhill	1948	Hedy Schlunegger	Switzerland
Giant Slalom	1952	Andrea Mead Lawrence	U.S.A.
Slalom	1948	Gretchen Frazer	U.S.A.

Nordic Skiing			
5 km	1964	Klaudia Boyerskikh	U.S.S.R.
10 km	1952	Lydia Wideman	Finland

Free Throw Percentage

PLAYER	ATTEMPT	FT	PERCENT
Rick Barry	4,090	3,675	0.899
Calvin Murphy	3,087	2,730	0.884
Bill Sharman	3,557	3,143	0.884
Mike Newlin	2,428	2,098	0.864
Fred Brown	1,581	1,364	0.863
Larry Siegfried	1,945	1,662	0.854
Flynn Robinson	1,881	1,597	0.849
Dolph Schayes	8,273	6,979	0.844
Jack Marin	2,852	2,405	0.843
Larry Costello	2,891	2,432	0.841

Distances of Planets from the Sun

Planet	Distance in million km	Distance in astronomical units
Mercury	57.9	0.387
Venus	108	0.723
Earth	149.6	1
Mars	227.9	1.524
Jupiter	778.3	5.203
Saturn	1,427	9.539
Uranus	2,869	19.18
Neptune	4,494	30.04
Pluto	5,900	39.44

The Cost of a Kilowatt-Hour of Electricity

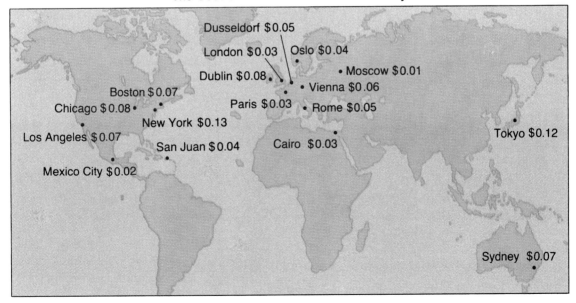

Dusseldorf $0.05
London $0.03
Oslo $0.04
Moscow $0.01
Dublin $0.08
Vienna $0.06
Boston $0.07
Paris $0.03
Rome $0.05
Chicago $0.08
New York $0.13
Tokyo $0.12
Los Angeles $0.07
San Juan $0.04
Cairo $0.03
Mexico City $0.02
Sydney $0.07

DATA BANK

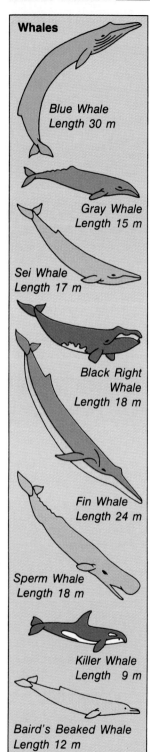

Whales

Blue Whale Length 30 m
Gray Whale Length 15 m
Sei Whale Length 17 m
Black Right Whale Length 18 m
Fin Whale Length 24 m
Sperm Whale Length 18 m
Killer Whale Length 9 m
Baird's Beaked Whale Length 12 m

No. 737 Lowest Annual Temperature of Record—Selected Cities

State	Station	Temp. in degrees C	State	Station	Temp. in degrees C
AL	Mobile	−14	MT	Great Falls	−42
AK	Juneau	−30	NE	Omaha	−30
AZ	Phoenix	−8	NV	Reno	−27
AR	Little Rock	−21	NH	Concord	−38
CA	Sacramento	−7	NJ	Atlantic City	−24
CO	Denver	−34	NM	Albuquerque	−27
CT	Hartford	−32	NY	Albany	−33
DE	Wilmington	−21	NC	Charlotte	−19
DC	Washington	−17	ND	Bismarck	−42
FL	Jacksonville	−11	OH	Cincinnati	−32
GA	Atlanta	−19	OK	Oklahoma City	−20
HI	Honolulu	12	OR	Portland	−19
ID	Boise	−31	PA	Pittsburgh	−28
IL	Peoria	−32	RI	Providence	−25
IN	Indianapolis	−29	SC	Columbia	−16
IA	Des Moines	−31	SD	Sioux Falls	−38
KS	Wichita	−24	TN	Nashville	−26
KY	Louisville	−29	TX	El Paso	−22
LA	New Orleans	−10	UT	Salt Lake City	−34
ME	Portland	−39	VT	Burlington	−34
MD	Baltimore	−22	VA	Richmond	−24
MA	Boston	−24	WA	Spokane	−32
MI	Sault Ste. Marie	−37	WV	Charleston	−24
MN	Duluth	−39	WI	Milwaukee	−31
MS	Jackson	−14	WY	Cheyenne	−37
MO	St. Louis	−26	PR	San Juan	16

Daily Diet of Some Zoo Animals

	Elephant	Giraffe	Gorilla	Hippopotamus
Hay	25 kg	7 kg	—	18 kg
Grain	23 kg	0.5 kg	—	—
Fruit and vegetables	23 kg	—	8 kg	2 kg
Pellets	—	0.5 kg	—	7 kg
Leaves	—	0.5 kg	2 kg	—
Bread	12 kg	—	—	—

MORE PRACTICE

Set A   For use after page 7

Write > or < for each ▩.

1. 2,046 ▩ 2,048
2. 3,642 ▩ 3,462
3. 24,500 ▩ 24,400
4. 620,000 ▩ 62,000
5. 785,610 ▩ 785,611
6. 409,111 ▩ 409,011
7. 3,460,014 ▩ 3,460,104
8. 47,000,000 ▩ 49,000
9. 51,866 ▩ 8,651
10. 92,567,402 ▩ 92,436,402

Set B   For use after page 9

Round to the nearest ten thousand.

1. 34,500
2. 29,600
3. 89,499
4. 45,500
5. 109,000
6. 325,444
7. 186,785
8. 404,651
9. 209,087
10. 335,460

Round to the nearest hundred thousand.

11. 484,085
12. 249,000
13. 857,066
14. 308,500
15. 967,068
16. 4,087,611
17. 6,554,300
18. 2,904,855
19. 197,366
20. 1,856,592

Set C   For use after page 11

Estimate each sum by rounding to the nearest thousand.

1. 24,411 + 36,270
2. 94,115 + 20,641
3. 180,270 + 365,590
4. 4,182 + 3,570
5. 4,061 + 6,506
6. 7,214 + 3,017 + 5,512
7. 23,805 + 16,457
8. 64,805 + 17,667
9. 627,512 + 139,815

Set A   For use after page 13

Add.

1.   29
    + 98
	2. 738
	    + 195
		3. 603
		    + 397
			4. 1,284
			    + 7,671
				5. 3,046
				    + 6,359

6.   21,186
    + 84,398
	7. 66,434
	    + 29,607
		8. 35,419
		    + 66,334
			9. 14,016
			    + 24,495
				10. 64,324
				    + 60,089

11.   61
    3,447
    + 209
	12. 8,604
	    5,250
	    + 1,153
		13. 76,408
		    3,215
		    + 4,479
			14. 61,420
			    33,271
			    + 98,493
				15. 78,447
				    36,110
				    43,408
				    + 92,173

Set B   For use after page 15

Subtract.

1.   408
    − 59
	2. 336
	    − 147
		3. 98
		    − 49
			4. 445
			    − 188
				5. 2,615
				    − 1,408

6.   4,671
    − 3,260
	7. 9,700
	    − 6,833
		8. 2,704
		    − 897
			9. 3,145
			    − 1,846
				10. 2,928
				    − 967

11.   8,400
    − 5,291
	12. 5,462
	    − 1,924
		13. 3,000
		    − 1,280
			14. 9,430
			    − 6,986
				15. 4,000
				    − 3,294

Set C   For use after page 19

Add or subtract the hours and minutes.

1.   3 h 45 min
    + 7 h 20 min

2.   12 h 15 min
    − 10 h 40 min

3.   9 h
    − 3 h 15 min

4.   14 h 17 min
    − 7 h 19 min

5.   9 h 29 min
    + 5 h 48 min

6.   24 h 48 min
    +   50 min

Set A   For use after page 37

Write <, =, or > for each ▦.

1. 0.03 ▦ 0.30
2. 0.6 ▦ 6.0
3. 0.607 ▦ 0.067
4. 64.48 ▦ 64.488
5. 3.00 ▦ 3
6. 0.095 ▦ 0.099
7. 0.001 ▦ 0.0009
8. 5.020 ▦ 5.02
9. 3.664 ▦ 3.646

Set B   For use after page 39

Round to the nearest thousandth.

1. 0.0418
2. 10.9945
3. 31.3546720
4. 1.14121356
5. 0.0008
6. 12.50109
7. 0.883914
8. 0.60982

Round to the nearest whole number.

9. 0.85
10. 25.09
11. 369.943
12. 26.554
13. 3,252.8376
14. 99.9
15. 1.14121356
16. 20.08

Round to 1-digit accuracy.

17. 4.9860
18. 0.00734
19. 0.108
20. 265.412
21. 0.0000564
22. 0.986004
23. 19.9
24. 0.002912

Set C   For use after page 41

Add.

1. 3.04 + 1.99
2. 18.057 + 3.96
3. 0.1040 + 0.7794
4. 12.632 + 0.889
5. 0.7785 + 9.0323

6. 3.216 + 4.805 + 6.664
7. 0.42 + 0.08 + 0.91
8. 35.412 + 7.68 + 16.004
9. 62.487 + 26.113 + 51.857
10. 35.016 + 9.931 + 74.073

11. $33.04 + 17.56 + 2.38
12. $0.62 + 0.38 + 0.90
13. $368.44 + 270.81 + 450.00
14. $32.54 + 0.71 + 0.68 + 17.44
15. $11.11 + 81.00 + 16.43 + 59.98

419

Set A   For use after page 43

Subtract.

1. 68.1
    − 37.5

2. 108.04
    − 67.16

3. 0.406
    − 0.198

4. 2.44
    − 1.87

5. 53.446
    − 18.971

6. 90.00
    − 26.48

7. 3.001
    − 1.942

8. 206.4
    −   9.9

9. 1.048
    − 0.019

10. 3.401
    − 1.267

11. $26.00
    − 5.49

12. $13.46
    − 8.51

13. $100.00
    −   0.35

14. $345.40
    − 208.91

15. $1.46
    − 0.88

Set B   For use after page 45

Estimate each sum or difference by rounding to the nearest tenth.

1. 0.44
    0.65
    0.91
    + 0.04

2. 12.09
    8.40
    19.67
    + 11.54

3. 0.12
    0.69
    0.48
    + 0.32

4. 6.14
    3.27
    8.49
    + 4.50

5. 22.62
    14.08
    12.36
    + 56.79

6. 64.27
    − 35.75

7. 0.64
    − 0.23

8. 2,306.50
    − 1,174.88

9. 407.54
    − 322.89

10. 66.22
    − 35.57

Estimate each sum by rounding to the nearest dollar.

11. $1.08
    4.35
    6.67
    + 9.49

12. $4.18
    5.49
    8.12
    + 6.50

13. $0.86
    1.19
    4.85
    + 6.31

14. $2.45
    3.12
    8.29
    + 9.50

15. $6.06
    1.57
    2.39
    + 4.50

Set C   For use after page 59

Estimate each product by rounding to 1-digit accuracy.

1. 14
    × 31

2. 2,704
    ×   95

3. 1,078
    × 424

4. 8,234
    × 375

5. 2,319
    × 548

6. 2,927
    × 3,389

7. 1,989
    ×   844

8. 15,018
    × 4,299

9. 981
    × 422

10. 692
    × 731

Set A For use after page 61

Multiply.

1. 39 × 4
2. 63 × 5
3. 916 × 5
4. 322 × 7
5. 489 × 4
6. 2,904 × 6
7. 9,118 × 4
8. 3,470 × 5
9. 4,429 × 7
10. 3,008 × 4
11. 36 × 7
12. 408 × 3
13. 9,114 × 5
14. 7,541 × 4
15. 3,143 × 6

Set B For use after page 63

Multiply.

1. 42 × 18
2. 65 × 48
3. 17 × 29
4. 362 × 41
5. 804 × 16
6. 774 × 29
7. 220 × 53
8. 4,047 × 36
9. 2,951 × 48
10. 3,117 × 45
11. 6,497 × 22
12. 5,432 × 84
13. 6,224 × 53
14. 7,914 × 68
15. 5,901 × 92

Set C For use after page 65

Multiply.

1. 436 × 873
2. 240 × 577
3. 179 × 432
4. 344 × 563
5. 243 × 774
6. 3,432 × 262
7. 7,442 × 450
8. 6,322 × 418
9. 4,441 × 226
10. 5,626 × 300
11. 743 × 288
12. 9,244 × 301
13. 7,432 × 227
14. 1,040 × 519
15. 294 × 176

Set A   For use after page 67

Write the factors for each number and find the products.

1. 4^3
2. 5^2
3. 7^3
4. 8^2
5. 6^3

6. 11^2
7. 3^3
8. 8^3
9. 4^4
10. 2^6

Write in exponential notation.

11. $6 \cdot 6 \cdot 6$
12. $15 \cdot 15$
13. $9 \cdot 9$
14. $4 \cdot 4 \cdot 4 \cdot 4$
15. $5 \cdot 5$

16. $3 \cdot 3 \cdot 3 \cdot 3 \cdot 3$
17. $2 \cdot 2 \cdot 2 \cdot 2$
18. $25 \cdot 25$
19. $10 \cdot 10 \cdot 10$
20. $7 \cdot 7 \cdot 7 \cdot 7$

Set B   For use after page 71

Estimate each quotient by rounding to 1-digit accuracy.

1. $529 \div 18$
2. $431 \div 21$
3. $948 \div 291$
4. $598 \div 192$
5. $1{,}900 \div 514$

6. $\dfrac{4{,}388}{1{,}209}$
7. $\dfrac{19{,}301}{5{,}299}$
8. $\dfrac{21{,}494}{988}$
9. $\dfrac{11{,}201}{488}$
10. $\dfrac{3{,}901}{429}$

11. $221 \overline{)788}$
12. $79 \overline{)1{,}648}$
13. $18 \overline{)556}$
14. $581 \overline{)3{,}397}$
15. $66 \overline{)6{,}821}$

Set C   For use after page 72

Find the quotients and remainders.

1. $4 \overline{)657}$
2. $9 \overline{)408}$
3. $5 \overline{)380}$
4. $7 \overline{)641}$
5. $3 \overline{)904}$

6. $4 \overline{)881}$
7. $8 \overline{)5{,}427}$
8. $9 \overline{)6{,}402}$
9. $7 \overline{)4{,}907}$
10. $5 \overline{)3{,}229}$

11. $6 \overline{)5{,}397}$
12. $2 \overline{)1{,}895}$
13. $4 \overline{)8{,}498}$
14. $6 \overline{)3{,}320}$
15. $8 \overline{)4{,}420}$

Set A   For use after page 73

Find the quotients and remainders. Use the short division method.

1. $3\overline{)427}$
2. $8\overline{)109}$
3. $6\overline{)973}$
4. $2\overline{)657}$
5. $3\overline{)444}$

6. $8\overline{)459}$
7. $7\overline{)8,614}$
8. $9\overline{)2,010}$
9. $6\overline{)8,591}$
10. $5\overline{)1,632}$

11. $6\overline{)3,365}$
12. $7\overline{)6,126}$
13. $4\overline{)2,573}$
14. $8\overline{)4,021}$
15. $3\overline{)2,168}$

16. $175 \div 3$
17. $3,044 \div 6$
18. $4,258 \div 7$
19. $2,107 \div 5$

Set B   For use after page 75

Find the quotients and remainders.

1. $12\overline{)679}$
2. $45\overline{)299}$
3. $38\overline{)741}$
4. $92\overline{)460}$
5. $43\overline{)781}$

6. $67\overline{)840}$
7. $92\overline{)543}$
8. $86\overline{)347}$
9. $32\overline{)9,140}$
10. $88\overline{)6,420}$

11. $91\overline{)43,008}$
12. $77\overline{)53,620}$
13. $51\overline{)26,760}$
14. $14\overline{)12,984}$
15. $81\overline{)36,015}$

16. $2,209 \div 81$
17. $19,148 \div 88$
18. $23,762 \div 32$
19. $26,805 \div 31$

Set C   For use after page 77

Find the quotients and remainders.

1. $147\overline{)3,204}$
2. $914\overline{)2,661}$
3. $436\overline{)8,129}$
4. $369\overline{)14,320}$
5. $269\overline{)50,128}$

6. $770\overline{)43,265}$
7. $219\overline{)83,160}$
8. $644\overline{)35,607}$
9. $104\overline{)36,327}$
10. $912\overline{)465,377}$

11. $244\overline{)123,459}$
12. $186\overline{)41,040}$
13. $982\overline{)443,258}$
14. $725\overline{)36,203}$
15. $446\overline{)50,917}$

16. $186,577 \div 324$
17. $113,245 \div 673$
18. $23,406 \div 189$
19. $436,970 \div 847$

Set A   For use after page 93

Find the products.

1. 4.3×1.4
2. 6.9×3.5
3. 0.6×0.4
4. 1.42×1.5
5. 3.17×43

6. 9.26
   $\times 0.08$
7. 72.4
   $\times 6.8$
8. 1.7
   $\times 3.9$
9. 12.44
   $\times 0.08$
10. 2.227
    $\times 1.59$

Set B   For use after page 95

Find the products.

1. 0.0064
   $\times 0.82$
2. 0.43
   $\times 0.08$
3. 24
   $\times 0.09$
4. 3.3
   $\times 0.0008$
5. 0.0038
   $\times 0.26$

6. 4.3
   $\times 0.02$
7. 0.003
   $\times 0.21$
8. 0.4
   $\times 0.04$
9. 5.08
   $\times 0.61$
10. 0.077
    $\times 0.042$

Set C   For use after page 100

Estimate each product.

1. 6.2×2.4
2. 1.43×9.1
3. 91.34×32.06
4. 0.45×1.6
5. 0.143×0.018

6. $\$6.08$
   $\times 0.37$
7. $\$12.00$
   $\times 1.8$
8. $\$0.45$
   $\times 1.27$
9. $\$3.66$
   $\times 0.27$
10. $\$19.38$
    $\times 18$

Set D   For use after page 103

Write each number in scientific notation.

1. 60,000,000
2. 690,000
3. 365,000,000
4. 64,000
5. 3,500,000,000
6. 923,000,000,000
7. 370
8. 6,750,000

Write the standard numeral.

9. 4.8×10^6
10. 2.2×10^2
11. 3.9×10^{12}
12. 8×10^5
13. 1.6×10^4
14. 7.53×10^9
15. 9.3×10^1
16. 9.08×10^3
17. 4.95×10^{10}

Set A   For use after page 117

Find the quotients.

1. $3\overline{)20.1}$
2. $6\overline{)43.8}$
3. $7\overline{)1.68}$
4. $3\overline{)15.9}$
5. $9\overline{)72.9}$
6. $62\overline{)241.8}$
7. $85\overline{)81.6}$
8. $33\overline{)178.2}$
9. $12\overline{)0.624}$
10. $99\overline{)366.3}$
11. $507.78 \div 63$
12. $2.436 \div 29$
13. $249.48 \div 77$
14. $58.5 \div 9$
15. $37.95 \div 55$

Set B   For use after page 119

Find the quotients. Round to the nearest hundredth.

1. $7\overline{)12}$
2. $4\overline{)15}$
3. $3\overline{)12.4}$
4. $7\overline{)29}$
5. $16\overline{)3.7}$
6. $3\overline{)\$26.45}$
7. $14\overline{)\$25.00}$
8. $12\overline{)\$6.98}$
9. $35\overline{)\$438.20}$
10. $5\overline{)\$11.03}$

Set C   For use after page 125

Find the quotients to the nearest hundredth.

1. $9.1\overline{)46.21}$
2. $0.42\overline{)1.08}$
3. $6.2\overline{)3.74}$
4. $0.32\overline{)0.851}$
5. $1.8\overline{)0.643}$
6. $11\overline{)9}$
7. $0.5\overline{)1.374}$
8. $0.41\overline{)0.904}$
9. $1.3\overline{)0.04}$
10. $2.7\overline{)0.8923}$

Set D   For use after page 129

Estimate each quotient.

1. $4.26\overline{)35.03}$
2. $0.32\overline{)0.2977}$
3. $0.37\overline{)26.41}$
4. $9.3\overline{)0.4374}$
5. $2.3\overline{)0.914}$
6. $4.03\overline{)22.94}$
7. $0.429\overline{)1.48}$
8. $2.7\overline{)0.3364}$
9. $17.201\overline{)256.73}$
10. $0.045\overline{)8.3}$

Set A For use after page 169

Write P (prime), C (composite), or N (neither) for each number.

1. 29
2. 48
3. 37
4. 43
5. 49
6. 77
7. 61
8. 1
9. 59
10. 87

Find the composite number in each list. Find the factors of that number.

11. 19, 29, 39, 59
12. 3, 7, 9, 11
13. 17, 27, 37, 47
14. 19, 91, 17, 71
15. 57, 97, 41, 13
16. 33, 42, 90, 102

Set B For use after page 171

Give the prime factorization of each number.

1. 98
2. 56
3. 68
4. 150
5. 220
6. 90
7. 160
8. 180
9. 210
10. 60
11. 360
12. 7,350
13. 405
14. 280
15. 750

Set C For use after page 173

Find the GCF of each pair of numbers.

1. 30, 135
2. 28, 20
3. 35, 42
4. 25, 36
5. 200, 75
6. 63, 42
7. 12, 30
8. 17, 34
9. 18, 35
10. 64, 80
11. 42, 63
12. 45, 42
13. 100, 98
14. 108, 84
15. 200, 64

Set D For use after page 175

Find the LCM of each pair of numbers.

1. 4, 10
2. 6, 9
3. 15, 10
4. 14, 4
5. 7, 6
6. 25, 10
7. 18, 45
8. 11, 6
9. 36, 45
10. 42, 12
11. 10, 14
12. 48, 96
13. 6, 15
14. 7, 15
15. 22, 33

Set A   For use after page 177

Copy and complete each table by evaluating the expressions.

	x	$x + 3$
1.	0	
2.	1	
3.	8	
4.	17	

	m	$8m$
5.	1	
6.	5	
7.	10	
8.	11	

	t	$\frac{t}{2}$
9.	4	
10.	10	
11.	8	
12.	12	

Write an expression.

13. 9 less than a number p

14. 5 more than a number w

15. A number k multiplied by 10

16. 15 divided by a number n

17. A number r divided by 3

18. 17 decreased by a number h

Set B   For use after page 179

Solve.

1. $x + 9 = 25$
2. $s + 12 = 40$
3. $t + 14 = 24$
4. $m + 3 = 15$
5. $x - 3 = 8$
6. $z - 20 = 40$
7. $x - 9 = 18$
8. $k - 28 = 19$
9. $y + 4 = 10$
10. $l - 8 = 9$
11. $t - 11 = 22$
12. $x + 5 = 28$

Set C   For use after page 181

Solve.

1. $6x = 54$
2. $10y = 80$
3. $7p = 49$
4. $2n = 32$
5. $\frac{t}{8} = 9$
6. $\frac{x}{13} = 2$
7. $\frac{b}{6} = 6$
8. $\frac{m}{5} = 100$
9. $\frac{n}{18} = 1$
10. $18z = 180$
11. $8r = 88$
12. $\frac{x}{4} = 13$

Set A For use after page 197

Write the missing numerator or denominator.

1. $\frac{1}{4} = \frac{\square}{8}$
2. $\frac{2}{3} = \frac{\square}{6}$
3. $\frac{4}{9} = \frac{12}{\square}$
4. $\frac{3}{5} = \frac{15}{\square}$
5. $\frac{3}{11} = \frac{\square}{33}$
6. $\frac{3}{4} = \frac{15}{\square}$
7. $\frac{3}{8} = \frac{\square}{24}$
8. $\frac{1}{7} = \frac{\square}{49}$
9. $\frac{7}{10} = \frac{28}{\square}$
10. $\frac{5}{12} = \frac{\square}{36}$
11. $\frac{6}{7} = \frac{18}{\square}$
12. $\frac{5}{8} = \frac{25}{\square}$
13. $\frac{2}{9} = \frac{10}{\square}$
14. $\frac{2}{13} = \frac{\square}{26}$
15. $\frac{7}{15} = \frac{21}{\square}$

Set B For use after page 199

Write each fraction in lowest terms.

1. $\frac{6}{10}$
2. $\frac{9}{27}$
3. $\frac{80}{100}$
4. $\frac{22}{44}$
5. $\frac{14}{30}$
6. $\frac{18}{24}$
7. $\frac{15}{36}$
8. $\frac{10}{50}$
9. $\frac{9}{12}$
10. $\frac{12}{16}$
11. $\frac{5}{9}$
12. $\frac{16}{40}$
13. $\frac{14}{21}$
14. $\frac{14}{70}$
15. $\frac{480}{1,000}$
16. $\frac{10}{36}$
17. $\frac{64}{100}$
18. $\frac{18}{42}$
19. $\frac{25}{60}$
20. $\frac{48}{80}$
21. $\frac{27}{36}$
22. $\frac{40}{100}$
23. $\frac{14}{42}$
24. $\frac{360}{1,000}$

Set C For use after page 201

Write each improper fraction as a mixed number or a whole number.

1. $\frac{9}{7}$
2. $\frac{36}{12}$
3. $\frac{10}{4}$
4. $\frac{14}{5}$
5. $\frac{30}{5}$
6. $\frac{15}{4}$
7. $\frac{15}{10}$
8. $\frac{10}{3}$
9. $\frac{14}{7}$
10. $\frac{11}{5}$
11. $\frac{26}{8}$
12. $\frac{17}{9}$
13. $\frac{37}{7}$
14. $\frac{14}{6}$
15. $\frac{27}{9}$
16. $\frac{45}{45}$
17. $\frac{100}{25}$
18. $\frac{19}{3}$

Write each mixed number as an improper fraction.

19. $1\frac{5}{8}$
20. $3\frac{3}{4}$
21. $1\frac{1}{9}$
22. $3\frac{3}{7}$
23. $8\frac{2}{3}$
24. $6\frac{2}{5}$
25. $8\frac{1}{4}$
26. $1\frac{7}{15}$
27. $12\frac{1}{5}$
28. $6\frac{3}{8}$
29. $12\frac{1}{5}$
30. $8\frac{1}{2}$
31. $33\frac{1}{3}$
32. $9\frac{1}{2}$
33. $49\frac{1}{2}$
34. $4\frac{3}{7}$
35. $6\frac{1}{4}$
36. $2\frac{3}{100}$

Set A For use after page 203

Compare the fractions. Write > or < for each ●.

1. $\frac{7}{12}$ ● $\frac{5}{5}$
2. $\frac{2}{3}$ ● $\frac{5}{9}$
3. $\frac{3}{8}$ ● $\frac{1}{3}$
4. $\frac{1}{6}$ ● $\frac{2}{9}$
5. $\frac{2}{3}$ ● $\frac{3}{4}$
6. $\frac{1}{2}$ ● $\frac{3}{8}$
7. $\frac{6}{7}$ ● $\frac{2}{3}$
8. $\frac{5}{7}$ ● $\frac{9}{14}$
9. $\frac{1}{2}$ ● $\frac{5}{16}$
10. $\frac{1}{3}$ ● $\frac{2}{9}$
11. $\frac{3}{5}$ ● $\frac{8}{15}$
12. $\frac{11}{24}$ ● $\frac{5}{12}$

Set B For use after page 207

Add.

1. $\frac{5}{6} + \frac{1}{6}$
2. $\frac{2}{9} + \frac{1}{9}$
3. $\frac{5}{12} + \frac{5}{12}$
4. $\frac{7}{10} + \frac{7}{10}$
5. $\frac{3}{16} + \frac{5}{16}$

6. $\frac{3}{8} + \frac{3}{4}$
7. $\frac{3}{7} + \frac{5}{14}$
8. $\frac{5}{9} + \frac{5}{6}$
9. $\frac{5}{6} + \frac{3}{4}$
10. $\frac{1}{5} + \frac{7}{10}$

11. $\frac{3}{8} + \frac{3}{16}$
12. $\frac{1}{5} + \frac{7}{15}$
13. $\frac{5}{12} + \frac{5}{6}$
14. $\frac{1}{10} + \frac{3}{20}$
15. $\frac{5}{9} + \frac{2}{3}$

Set C For use after page 209

Subtract.

1. $\frac{7}{8} - \frac{5}{8}$
2. $\frac{7}{9} - \frac{4}{9}$
3. $\frac{11}{12} - \frac{5}{12}$
4. $\frac{9}{10} - \frac{3}{10}$
5. $\frac{5}{8} - \frac{1}{8}$

6. $\frac{5}{6} - \frac{1}{3}$
7. $\frac{7}{8} - \frac{1}{4}$
8. $\frac{3}{4} - \frac{1}{3}$
9. $\frac{9}{10} - \frac{3}{5}$
10. $\frac{3}{5} - \frac{1}{2}$

11. $\frac{5}{12} - \frac{3}{8}$
12. $\frac{5}{6} - \frac{3}{5}$
13. $\frac{7}{10} - \frac{1}{4}$
14. $\frac{2}{3} - \frac{2}{5}$
15. $\frac{7}{9} - \frac{2}{3}$

Set A For use after page 211

Add.

1. $3 + 7\frac{1}{4}$
2. $6 + 6\frac{1}{2}$
3. $8\frac{5}{9} + 9$
4. $13 + 7\frac{3}{8}$
5. $3\frac{1}{2} + 3\frac{1}{2}$

6. $3\frac{5}{9} + 4\frac{1}{6}$
7. $4\frac{3}{5} + 5\frac{11}{15}$
8. $6\frac{1}{7} + 6\frac{2}{3}$
9. $2\frac{2}{3} + 6\frac{5}{6}$
10. $14\frac{7}{8} + 3\frac{1}{8}$

Set B For use after page 213

Subtract.

1. $7\frac{1}{4} - 3\frac{1}{8}$
2. $9\frac{10}{12} - 3\frac{2}{4}$
3. $2\frac{2}{4} - 1\frac{1}{9}$
4. $3\frac{2}{3} - 1\frac{1}{7}$
5. $8\frac{2}{3} - \frac{2}{5}$

6. $8\frac{1}{2} - 4\frac{2}{3}$
7. $4\frac{2}{8} - 3\frac{1}{2}$
8. $7\frac{1}{10} - 4\frac{3}{5}$
9. $5\frac{3}{9} - 1\frac{10}{12}$
10. $6\frac{3}{4} - 2\frac{12}{16}$

11. $15\frac{9}{14} - 9\frac{3}{7}$
12. $21\frac{2}{4} - 7\frac{1}{3}$
13. $40\frac{4}{9} - 8\frac{2}{6}$
14. $82\frac{1}{3} - 19\frac{1}{5}$
15. $17\frac{4}{12} - 4\frac{2}{12}$

Set C For use after page 215

Subtract.

1. $1 - \frac{4}{9}$
2. $6\frac{1}{8} - \frac{7}{10}$
3. $4\frac{1}{4} - 1\frac{1}{2}$
4. $2\frac{2}{9} - \frac{10}{18}$
5. $7 - 2\frac{1}{2}$

6. $5\frac{1}{3} - 4\frac{4}{8}$
7. $2\frac{1}{9} - \frac{5}{9}$
8. $10\frac{1}{7} - 8\frac{6}{7}$
9. $12\frac{1}{3} - 10\frac{1}{2}$
10. $8\frac{7}{8} - 7\frac{1}{2}$

11. $1\frac{1}{8} - \frac{4}{6}$
12. $73\frac{1}{2} - 70\frac{3}{4}$
13. $51\frac{3}{9} - 48$
14. $11\frac{1}{3} - 10\frac{3}{9}$
15. $24 - \frac{9}{10}$

Set A For use after page 227

Find the products.

1. $\frac{3}{5} \times \frac{1}{9}$
2. $\frac{1}{2} \times \frac{4}{5}$
3. $\frac{3}{4} \times \frac{2}{3}$
4. $\frac{5}{12} \times \frac{3}{10}$
5. $\frac{3}{7} \times \frac{14}{27}$
6. $\frac{2}{3} \times \frac{5}{6}$
7. $\frac{5}{12} \times \frac{8}{15}$
8. $\frac{1}{3} \times \frac{9}{10}$
9. $\frac{3}{4} \times \frac{1}{6}$
10. $\frac{5}{9} \times \frac{9}{10}$
11. $\frac{3}{7} \times \frac{2}{3}$
12. $\frac{1}{12} \times \frac{8}{9}$
13. $\frac{3}{5} \times \frac{5}{6}$
14. $\frac{2}{5} \times \frac{1}{7}$
15. $\frac{1}{6} \times \frac{9}{10}$

Set B For use after page 229

Find the products.

1. $3\frac{1}{3} \times \frac{1}{2}$
2. $2\frac{1}{2} \times 1\frac{1}{3}$
3. $3\frac{1}{5} \times 1\frac{1}{4}$
4. $1\frac{7}{8} \times 2\frac{2}{5}$
5. $2\frac{1}{4} \times 5\frac{1}{3}$
6. $1\frac{5}{6} \times 2\frac{2}{5}$
7. $\frac{1}{7} \times 2\frac{4}{5}$
8. $2\frac{2}{3} \times \frac{5}{8}$
9. $5\frac{1}{4} \times 1\frac{1}{7}$
10. $4\frac{1}{5} \times 1\frac{1}{9}$
11. $6 \times 2\frac{3}{4}$
12. $3\frac{1}{2} \times 4\frac{2}{3}$
13. $2\frac{7}{10} \times \frac{5}{6}$
14. $4\frac{1}{5} \times 2\frac{3}{7}$
15. $3\frac{1}{9} \times 2\frac{5}{8}$

Set C For use after page 231

Give the reciprocal of each number.

1. $\frac{1}{4}$
2. $\frac{6}{8}$
3. 7
4. $\frac{5}{3}$
5. $\frac{1}{6}$
6. $3\frac{1}{5}$
7. $1\frac{1}{3}$
8. 24
9. $\frac{2}{7}$
10. $2\frac{10}{12}$

Set D For use after page 233

Find the quotients.

1. $\frac{2}{3} \div \frac{1}{8}$
2. $\frac{3}{5} \div \frac{2}{3}$
3. $\frac{7}{8} \div \frac{1}{4}$
4. $\frac{3}{10} \div \frac{2}{5}$
5. $\frac{1}{4} \div \frac{3}{4}$
6. $\frac{3}{5} \div 6$
7. $\frac{2}{7} \div \frac{4}{5}$
8. $\frac{3}{8} \div \frac{3}{4}$
9. $\frac{4}{9} \div \frac{2}{3}$
10. $\frac{1}{5} \div \frac{3}{10}$
11. $\frac{5}{8} \div \frac{3}{16}$
12. $\frac{2}{5} \div 6$
13. $\frac{4}{5} \div \frac{1}{4}$
14. $\frac{5}{16} \div \frac{3}{8}$
15. $\frac{3}{10} \div \frac{2}{5}$

Set A For use after page 235

Find the quotients.

1. $3\frac{1}{3} \div 2\frac{2}{3}$
2. $4\frac{2}{3} \div 2\frac{1}{3}$
3. $3\frac{3}{4} \div 3$
4. $\frac{5}{6} \div 1\frac{2}{3}$
5. $4\frac{1}{2} \div 1\frac{5}{8}$
6. $9 \div 2\frac{2}{3}$
7. $1\frac{4}{5} \div 2\frac{1}{10}$
8. $7\frac{1}{2} \div 3\frac{3}{4}$
9. $\frac{2}{9} \div 1\frac{1}{3}$
10. $7\frac{1}{3} \div \frac{11}{12}$
11. $4\frac{2}{5} \div 5\frac{1}{3}$
12. $2\frac{3}{16} \div 3\frac{3}{4}$
13. $9 \div 2\frac{2}{5}$
14. $4\frac{1}{3} \div 5\frac{1}{3}$
15. $\frac{5}{7} \div 1\frac{2}{3}$

Set B For use after page 239

Write each decimal as a lowest-terms fraction or a mixed number.

1. 0.8
2. 0.3
3. 0.09
4. 0.03
5. 0.12
6. 3.04
7. 17.017
8. 3.95
9. 2.125
10. 14.75

Write each fraction as a decimal.

11. $\frac{3}{5}$
12. $\frac{27}{4}$
13. $\frac{15}{16}$
14. $\frac{7}{25}$
15. $\frac{1}{5}$
16. $\frac{9}{16}$
17. $\frac{15}{30}$
18. $\frac{25}{100}$
19. $\frac{3}{4}$
20. $\frac{80}{100}$

Set C For use after page 241

Write the decimal for each fraction. Use a bar to show repeating decimals.

1. $\frac{3}{5}$
2. $\frac{7}{9}$
3. $\frac{5}{30}$
4. $\frac{4}{5}$
5. $\frac{6}{9}$
6. $\frac{3}{9}$
7. $\frac{3}{8}$
8. $\frac{10}{22}$
9. $\frac{14}{15}$
10. $\frac{68}{5}$

Compare the fractions by comparing their decimals. Use > or <.

11. $\frac{4}{8} \bigcirc \frac{5}{9}$
12. $\frac{4}{25} \bigcirc \frac{7}{28}$
13. $\frac{3}{5} \bigcirc \frac{2}{3}$
14. $\frac{1}{7} \bigcirc \frac{2}{21}$
15. $\frac{4}{18} \bigcirc \frac{3}{15}$
16. $\frac{5}{17} \bigcirc \frac{2}{7}$
17. $\frac{3}{21} \bigcirc \frac{4}{29}$
18. $\frac{3}{28} \bigcirc \frac{4}{32}$
19. $\frac{36}{40} \bigcirc \frac{17}{19}$
20. $\frac{1}{9} \bigcirc \frac{3}{26}$

Set A For use after page 253

Find the perimeter of each rectangle.

1. $l = 6.2$ m
   $w = 2.7$ m
2. $l = 31.6$ cm
   $w = 7.2$ cm
3. $l = 8.1$ km
   $w = 3.53$ km
4. $l = 4.96$ dm
   $w = 2.42$ dm
5. $l = 6.05$ m
   $w = 4.78$ m

6. $l = 11.66$ km
   $w = 8.12$ km
7. $l = 9.5$ dm
   $w = 6.38$ dm
8. $l = 27.9$ m
   $w = 6.4$ m
9. $l = 8.62$ hm
   $w = 2.1$ hm
10. $l = 13.7$ mm
    $w = 4.6$ mm

Set B For use after page 254

Find the area of each rectangle.

1. $l = 26.8$ cm
   $w = 8.3$ cm
2. $l = 84$ km
   $w = 30$ km
3. $l = 51.4$ m
   $w = 12$ m
4. $l = 16.7$ dm
   $w = 1.9$ dm
5. $l = 70$ mm
   $w = 33$ mm

6. $l = 54$ m
   $w = 32$ m
7. $l = 10.6$ mm
   $w = 8.3$ mm
8. $l = 71.5$ cm
   $w = 26.2$ cm
9. $l = 8.9$ m
   $w = 4.6$ m
10. $l = 18$ km
    $w = 9$ km

Set C For use after page 257

Find the area of each triangle.

1. $b = 47$ km
   $h = 5$ km
2. $b = 25$ cm
   $h = 7$ cm
3. $b = 43$ m
   $h = 6$ m
4. $b = 57$ mm
   $h = 41$ mm
5. $b = 68$ dm
   $h = 53$ dm

6. $b = 20$ mm
   $h = 6$ mm
7. $b = 61$ dm
   $h = 33$ dm
8. $b = 27$ cm
   $h = 8$ cm
9. $b = 59$ km
   $h = 13$ km
10. $b = 108$ m
    $h = 42$ m

Set D For use after page 261

Find the volume of each rectangular prism.

1. $l = 12$ cm
   $w = 9$ cm
   $h = 7$ cm
2. $l = 9$ dm
   $w = 6$ dm
   $h = 5$ dm
3. $l = 3$ m
   $w = 2.1$ m
   $h = 0.7$ m
4. $l = 8$ mm
   $w = 7$ mm
   $h = 3$ mm

5. $l = 7.6$ km
   $w = 3.8$ km
   $h = 4.2$ km
6. $l = 30$ m
   $w = 20$ m
   $h = 14$ m
7. $l = 0.6$ cm
   $w = 0.5$ cm
   $h = 0.3$ cm
8. $l = 25$ dm
   $w = 11$ dm
   $h = 2$ dm

Set A For use after page 279

Write = or ≠ for each ◉.

1. $\frac{2}{6}$ ◉ $\frac{5}{15}$
2. $\frac{3}{4}$ ◉ $\frac{9}{10}$
3. $\frac{6}{12}$ ◉ $\frac{5}{10}$
4. $\frac{3}{5}$ ◉ $\frac{4}{7}$
5. $\frac{1}{8}$ ◉ $\frac{4}{32}$
6. $\frac{2}{9}$ ◉ $\frac{3}{10}$
7. $\frac{1}{7}$ ◉ $\frac{4}{25}$
8. $\frac{15}{6}$ ◉ $\frac{5}{2}$
9. $\frac{4}{5}$ ◉ $\frac{7}{9}$
10. $\frac{8}{12}$ ◉ $\frac{6}{9}$

Write each ratio as a fraction in lowest terms.

11. 3 out of 18
12. 8:12
13. 10 to 35
14. 9 out of 24
15. 6 to 14
16. 60:20
17. 25 out of 100
18. 60:80
19. $\frac{45}{60}$
20. $\frac{80}{100}$

Set B For use after page 281

Solve the proportions.

1. $\frac{2}{5} = \frac{4}{x}$
2. $\frac{70}{100} = \frac{x}{10}$
3. $\frac{9}{3} = \frac{12}{x}$
4. $\frac{7}{14} = \frac{2}{x}$
5. $\frac{3}{25} = \frac{9}{x}$
6. $\frac{5}{4} = \frac{40}{x}$
7. $\frac{4}{8} = \frac{5}{x}$
8. $\frac{3}{15} = \frac{x}{10}$
9. $\frac{x}{12} = \frac{4}{6}$
10. $\frac{12}{x} = \frac{30}{5}$

Set C For use after page 297

Write the lowest-terms fraction for each percent.

1. 35%
2. 90%
3. 12%
4. 150%
5. 75%
6. 80%
7. 16%
8. 32%
9. 4%
10. 250%

Write the percent for each fraction.

11. $\frac{16}{50}$
12. $\frac{3}{20}$
13. $\frac{8}{10}$
14. $\frac{47}{100}$
15. $\frac{7}{4}$
16. $\frac{2}{25}$
17. $\frac{3}{5}$
18. $\frac{34}{50}$
19. $\frac{20}{10}$
20. $\frac{17}{25}$

Write each fraction in lowest terms. Then write the percent for the fraction.

21. $\frac{3}{6}$
22. $\frac{6}{15}$
23. $\frac{12}{48}$
24. $\frac{18}{30}$
25. $\frac{21}{7}$
26. $\frac{12}{75}$
27. $\frac{12}{120}$
28. $\frac{16}{800}$
29. $\frac{66}{275}$
30. $\frac{120}{48}$

Set A For use after page 298

Write each percent as a decimal.

1. 67%
2. 150%
3. 6.5%
4. 74%
5. 3%
6. 0.18%
7. 63.4%
8. 500%
9. 18%
10. 0.25%
11. 10.35%
12. 0.6%
13. 6.75%
14. 95%
15. 10%

Write each decimal as a percent.

16. 1.75
17. 0.33
18. 0.012
19. 0.0099
20. 0.0377
21. 0.6
22. 7.5
23. 0.18
24. 0.301
25. 0.003
26. 0.038
27. 0.903
28. 0.2
29. 5.00
30. 0.1227

Set B For use after page 301

Find the percent for each fraction. Round the percent to the nearest whole percent.

1. $\frac{3}{7}$
2. $\frac{4}{9}$
3. $\frac{5}{6}$
4. $\frac{1}{16}$
5. $\frac{3}{11}$
6. $\frac{5}{12}$
7. $\frac{2}{3}$
8. $\frac{7}{9}$
9. $\frac{4}{15}$
10. $\frac{7}{24}$
11. $\frac{13}{16}$
12. $\frac{8}{9}$
13. $\frac{11}{13}$
14. $\frac{6}{7}$
15. $\frac{5}{14}$
16. $\frac{9}{11}$
17. $\frac{14}{17}$
18. $\frac{5}{18}$
19. $\frac{19}{27}$
20. $\frac{23}{45}$

Set C For use after page 303

Find the percent of each number.

1. $33\frac{1}{3}$% of 81
2. 25% of 64
3. 40% of 250
4. 75% of 28
5. 90% of 60
6. 20% of 10
7. 50% of 900
8. 60% of 75
9. 16% of 52
10. 99% of 25
11. 250% of 120
12. 5.5% of 66
13. 7% of 115
14. 85% of 150
15. 10.5% of 76
16. 110% of 3

Set A For use after page 307

Find the percents.

1. 35 is what percent of 70?
2. What percent of 4 is 6?
3. What percent of 20 is 8?
4. What percent of 32 is 24?
5. 18 is what percent of 30?
6. 45 is what percent of 36?
7. What percent is 14 out of 20?
8. What percent is 11 out of 44?
9. What percent of 6 is 15?
10. What percent is 6 out of 150?

Find the percents. Round each percent to the nearest whole percent.

11. 15 is what percent of 95?
12. What percent of 150 is 175?
13. What percent is 6 out of 14?
14. 135 is what percent of 1,080?
15. What percent is 10 out of 12?
16. 98 is what percent of 40?
17. What percent of 75 is 325?
18. 312 is what percent of 400?

Set B For use after page 311

Solve.

1. $75\% \times n = 12$
2. $36\% \times n = 27$
3. $16\% \times n = 8$
4. $30\% \times n = 18$
5. $72\% \times n = 36$
6. $80\% \times n = 12$
7. $38\% \times n = 19$
8. $50\% \times n = 28$
9. $15\% \times n = 21$
10. $6\% \times n = 27$
11. $55\% \times n = 11$
12. $20\% \times n = 120$
13. $18\% \times n = 9$
14. $40\% \times n = 28$
15. $48\% \times n = 12$

Set C For use after page 315

Write and solve a proportion for each question.

1. What is 8% of 300?
2. 9 is 20% of what number?
3. What percent of 60 is 12?
4. What is 82% of 450?
5. 75 is 150% of what number?
6. What percent of 350 is 14?
7. What percent of 250 is 55?
8. 90 is 120% of what number?

Set A   For use after page 323

Find the circumference of each circle. Use 3.14 for π.

1. $r = 5$ cm
2. $d = 13$ m
3. $r = 60$ km
4. $r = 25$ mm
5. $d = 5.6$ dm
6. $d = 32$ mm
7. $r = 0.4$ cm
8. $r = 7$ m
9. $d = 0.3$ km
10. $d = 10$ m
11. $r = .08$ dm
12. $r = 6$ km
13. $d = 11$ cm
14. $r = 30$ mm
15. $d = 33.4$ m

Set B   For use after page 325

Find the area of each circle. Use 3.14 for π.

1. $r = 5$ m
2. $d = 3.2$ mm
3. $r = 0.7$ cm
4. $d = 14$ mm
5. $r = 1.4$ cm
6. $d = 28$ mm
7. $d = 50$ cm
8. $d = 6.6$ m
9. $r = 5.4$ cm
10. $r = 0.8$ mm
11. $d = 200$ mm
12. $r = 30$ m
13. $d = 2.2$ cm
14. $r = 6$ m
15. $r = 19$ m

Set C   For use after page 329

Find the total surface area of each cylinder. Use 3.14 for π.

1. $r = 40$ mm, $h = 71$ mm
2. $r = 2.9$ cm, $h = 5.1$ cm
3. $r = 0.8$ m, $h = 1.3$ m
4. $r = 6$ mm, $h = 1$ mm
5. $r = 4$ cm, $h = 6$ cm
6. $r = 7$ m, $h = 8$ m
7. $r = 2$ mm, $h = 30$ mm
8. $r = 3$ cm, $h = 5$ cm
9. $r = 11$ m, $h = 3$ m
10. $r = 1$ m, $h = 19$ m

Set D   For use after page 331

Find the volume of each cylinder. Use π for 3.14.

11. $r = 9$ cm, $h = 10$ cm
12. $r = 7$ m, $h = 5$ m
13. $r = 30$ dm, $h = 20$ dm
14. $r = 1.2$ mm, $h = 10$ mm
15. $r = 8$ dm, $h = 3$ dm
16. $r = 5$ m, $h = 3.6$ m
17. $r = 7$ mm, $h = 6$ mm
18. $r = 8$ cm, $h = 0.1$ cm
19. $r = 2$ km, $h = 3$ km
20. $r = 20$ m, $h = 9$ m

Set A   For use after page 353

Find the mode and range of each list of numbers.

1. 37, 14, 43, 38, 29, 14, 15, 34, 19, 14
2. 1, 3, 7, 9, 12, 8, 7, 6, 8, 7, 10, 11
3. 21, 42, 55, 23, 17, 21, 40, 35, 21, 18
4. 17, 17, 13, 18, 14, 15, 16, 10

5. 98, 92, 95, 92, 95, 89, 90, 91, 92
6. 73, 73, 74, 78, 77, 89, 70, 71
7. 102, 94, 96, 98, 99, 80, 99, 98
8. 3, 3, 6, 6, 7, 8, 1, 2, 3, 9

Set B   For use after page 354

Find the arithmetic mean and the median of each list of numbers.

1. 36, 42, 57
2. 130, 150, 200
3. 97, 111, 173
4. 6.3, 8.1, 10.8

5. 0.009, 0.052, 0.161
6. 18, 21, 26, 39, 40
7. 42, 47, 51, 52, 60, 72
8. 408, 113, 310, 225

9. 12, 7, 3, 24, 4, 10
10. 1.1, 2.3, 2.1, 1.7
11. 0.12, 1.34, 2.4, 0.09, 1.05, 0.2
12. 43.1, 42.4, 40, 38.9, 36.4, 41, 42.1, 40

Set C   For use after page 369

Write < or > for each ●.

1. $^+4$ ● $^-5$
2. $^-3$ ● $^-2$
3. $^+8$ ● $^-8$
4. $^+4$ ● 0
5. $^+16$ ● $^-18$
6. $^-8$ ● $^+8$
7. $^-16$ ● $^-17$
8. $^-32$ ● $^-15$
9. $^+9$ ● $^+8$
10. $^+1$ ● $^-10$
11. $^-3$ ● $^-4$
12. 0 ● $^-16$
13. $^+24$ ● $^+23$
14. $^+1$ ● $^-1$
15. $^+17$ ● $^-16$
16. $^+18$ ● $^-42$

Set A For use after page 371

Find the missing integers.

1. $^+5 + {^-5} = \blacksquare$
2. $^-3 + {^+3} = \blacksquare$
3. $^+10 + 0 = \blacksquare$
4. $^+4 + \blacksquare = 0$
5. $\blacksquare + {^-6} = {^-6}$
6. $^-7 + {^+7} = \blacksquare$
7. $^+8 + {^-8} = \blacksquare$
8. $^-20 + \blacksquare = 0$
9. $0 + \blacksquare = {^+70}$
10. $^+18 + \blacksquare = 0$
11. $\blacksquare + {^+9} = 0$
12. $\blacksquare + {^+11} = {^+11}$
13. $^-12 + \blacksquare = {^-12}$
14. $^+32 + {^-32} = \blacksquare$
15. $^+63 + {^-63} = \blacksquare$
16. $\blacksquare + 0 = {^+18}$
17. $\blacksquare + {^+48} = 0$
18. $\blacksquare + {^-13} = 0$

Set B For use after page 373

Find the sums.

1. $^-3 + {^-1}$
2. $^-7 + {^-2}$
3. $1 + 6$
4. $^-3 + {^-8}$
5. $9 + 5$
6. $^-7 + {^-5}$
7. $^-8 + {^-8}$
8. $7 + 9$
9. $^-5 + {^-5}$
10. $^-6 + {^-2}$
11. $^-5 + {^-9}$
12. $^-1 + {^-1}$
13. $^-6 + {^-7}$
14. $^-4 + {^-8}$
15. $7 + 7$
16. $^-4 + {^-6}$
17. $^-1 + {^-3}$
18. $8 + 2$
19. $^-5 + {^-1}$
20. $^-8 + {^-7}$
21. $^-4 + {^-5}$
22. $^-7 + {^-7}$
23. $5 + 2$
24. $^-1 + {^-8}$

Set C For use after page 375

Find the sums.

1. $1 + {^-3}$
2. $4 + {^-4}$
3. $^-3 + 1$
4. $^-8 + {^-2}$
5. $6 + {^-7}$
6. $^-2 + 3$
7. $4 + {^-5}$
8. $^-6 + 6$
9. $8 + {^-7}$
10. $^-4 + {^-3}$
11. $^-8 + 5$
12. $^-3 + 8$
13. $10 + {^-12}$
14. $^-8 + 12$
15. $^-6 + 11$
16. $^-8 + 8$
17. $5 + {^-9}$
18. $^-7 + 14$
19. $^-6 + {^-3}$
20. $8 + {^-12}$
21. $6 + {^-14}$
22. $^-10 + 16$
23. $^-4 + {^-8}$
24. $7 + {^-15}$

Set A For use after page 377

Subtract.

1. ⁻5 − ⁻1
2. 6 − ⁻3
3. ⁻8 − 5
4. 6 − ⁻4
5. ⁻3 − ⁻3
6. 8 − 14
7. 3 − ⁻1
8. ⁻10 − ⁻10
9. 6 − ⁻9
10. ⁻5 − 3
11. ⁻6 − 0
12. 4 − 9
13. 6 − ⁻6
14. 4 − 8
15. 7 − ⁻5
16. ⁻8 − 3
17. 2 − 9
18. ⁻9 − ⁻5
19. ⁻8 − 8
20. 17 − 8
21. 4 − ⁻6
22. ⁻7 − 7
23. ⁻12 − ⁻6
24. 7 − ⁻9

Set B For use after page 381

Multiply.

1. 7 · ⁻3
2. ⁻6 · ⁻1
3. 5 · ⁻3
4. ⁻9 · ⁻5
5. 3 · 9
6. ⁻6 · ⁻5
7. 3 · ⁻7
8. ⁻3 · ⁻3
9. ⁻5 · 0
10. ⁻8 · 6
11. 6 · ⁻9
12. ⁻5 · ⁻4
13. 12 · 8
14. 20 · ⁻3
15. 26 · ⁻5
16. ⁻17 · ⁻3
17. ⁻3 · 5
18. 6 · (⁻5 · ⁻2)
19. ⁻3 · (4 · 0)
20. (⁻2 · 2) · 7
21. (⁻6 · 1) · 6
22. 3 · (3 · 3)
23. (⁻4 · 2) · ⁻7
24. (⁻5 · ⁻5) · ⁻5

Set C For use after page 383

Find the quotients.

1. 14 ÷ ⁻2
2. ⁻30 ÷ ⁻6
3. 45 ÷ ⁻9
4. 64 ÷ ⁻8
5. ⁻35 ÷ 7
6. ⁻9 ÷ ⁻3
7. 21 ÷ ⁻7
8. ⁻2 ÷ 2
9. 48 ÷ ⁻6
10. ⁻24 ÷ ⁻3
11. ⁻18 ÷ 9
12. ⁻25 ÷ ⁻5
13. 40 ÷ ⁻8
14. 12 ÷ 2
15. 32 ÷ ⁻8
16. ⁻16 ÷ ⁻2

Set A For use after page 395

Add the lengths.

1. 3 ft 3 in.
   + 5 ft 9 in.

2. 6 ft 11 in.
   + 4 ft 9 in.

3. 2 yd 2 ft
   + 6 yd 1 ft

4. 13 ft 10 in.
   + 9 ft 6 in.

Subtract the lengths.

5. 7 ft 4 in.
   − 5 ft 2 in.

6. 6 ft 2 in.
   − 3 ft 8 in.

7. 4 ft
   − 2 ft 9 in.

8. 9 yd
   − 5 yd 2 ft

Set B For use after page 397

Find the area of each rectangle.

1. $l = 30$ in.
   $w = 9$ in.

2. $l = 8.6$ ft
   $w = 5$ ft

3. $l = 4\frac{1}{2}$ in.
   $w = 2\frac{1}{2}$ in.

4. $l = 7.4$ mi
   $w = 2.1$ mi

Find the area of each circle. Use 3.14 for π.

5. $r = 4$ in.

6. $r = 3.5$ ft

7. $r = 7$ mi

8. $d = 0.6$ ft

Set C For use after page 399

Find the volume of each rectangular prism. Use $V = lwh$.

1. $l = 5$ in.
   $w = 4$ in.
   $h = 10$ in.

2. $l = 3.5$ ft
   $w = 2$ ft
   $h = 4.5$ ft

3. $l = 6\frac{1}{2}$ in.
   $w = 3\frac{1}{2}$ in.
   $h = 6$ in.

4. $l = 2$ ft 3 in.
   $w = 2$ ft 1 in.
   $h = 6$ in.

Find the volume of each cylinder. Use $V = \pi r^2 h$. Use 3.14 for π.

5. $r = 10$ in.
   $h = 5$ in.

6. $r = 4$ ft
   $h = 13$ ft

7. $r = 6$ in.
   $h = 1$ ft 3 in.

8. $r = 7$ ft
   $h = 2$ ft

TABLE OF MEASURES

Metric System   Customary System

Length

1 meter (m)	1,000 millimeters (mm) 100 centimeters (cm) 10 decimeters (dm)
1 kilometer (km)	1,000 meters (m)
1 hectometer (hm)	100 meters (m)
1 dekameter (dam)	10 meters (m)
1 decimeter (dm)	0.1 meter (m)
1 centimeter (cm)	0.01 meter (m)
1 millimeter (mm)	0.001 meter (m)

1 foot (ft)	12 inches (in.)
1 yard (yd)	36 inches (in.) 3 feet (ft)
1 mile (mi)	5,280 feet (ft) 1,760 yards (yd)
1 nautical mile	6,076 feet (ft) 1,852 meters (m)

Area

1 square meter (m²)	100 square decimeters (dm²) 10,000 square centimeters (cm²)
1 hectare (ha)	0.01 square kilometer (km²) 10,000 square meters (m²)
1 square kilometer (km²)	1,000,000 square meters (m²) 100 hectares (ha)

1 square foot (ft²)	144 square inches (in.²)
1 square yard (yd²)	9 square feet (ft²) 1,296 square inches (in.²)
1 acre (a.)	43,560 square feet (ft²) 4,840 square yards (yd²)
1 square mile (mi²)	640 acres (a.)

Volume

1 cubic decimeter (dm³)	0.001 cubic meter (m³) 1,000 cubic centimeters (cm³) 1 liter (L)
1 cubic meter (m³)	1,000,000 cubic centimeters (cm³) 1,000 cubic decimeters (dm³)

1 cubic foot (ft³)	1,728 cubic inches (in.³)
1 cubic yard (yd³)	27 cubic feet (ft³) 46,656 cubic inches (in.³)

Capacity

1 teaspoon	5 milliliters (mL)
1 tablespoon	12.5 milliliters (mL)
1 liter (L)	1,000 milliliters (mL) 1,000 cubic centimeters (cm³) 1 cubic decimeter (dm³) 4 metric cups
1 kiloliter (kL)	1,000 liters (L)

1 cup (c)	8 fluid ounces (fl oz)
1 pint (pt)	16 fluid ounces (fl oz) 2 cups (c)
1 quart (qt)	32 fluid ounces (fl oz) 4 cups (c) 2 pints (pt)
1 gallon (gal)	128 fluid ounces (fl oz) 16 cups (c) 8 pints (pt) 4 quarts (qt)

Weight

1 gram (g)	1,000 milligrams (mg)
1 kilogram (kg)	1,000 grams (g)
1 metric ton (t)	1,000 kilograms (kg)

1 pound (lb)	16 ounces (oz)
1 ton (T)	2,000 pounds (lb)

MATHEMATICAL SYMBOLS

$=$	Is equal to	$\overleftrightarrow{AB}$	Line through points A and B
$\neq$	Is not equal to	$\overrightarrow{AB}$	Ray AB
$>$	Is greater than	$\overline{AB}$	Segment with endpoints A and B
$<$	Is less than	$\angle ABC$	Angle ABC
$\geq$	Is greater than or equal to	$m\angle ABC$	Measure of angle ABC
$\leq$	Is less than or equal to	$\triangle ABC$	Triangle ABC
$\approx$	Is approximately equal to	$\overset{\frown}{RS}$	Arc with endpoints R and S
$\cong$	Is congruent to	$\overleftrightarrow{AB} \perp \overleftrightarrow{CD}$	Line AB perpendicular to line CD
$\sim$	Is similar to	$\overleftrightarrow{AB} \parallel \overleftrightarrow{CD}$	Line AB is parallel to line CD
$\%$	Percent	$35°$	Thirty-five *degrees*
π	Pi		
$0.\overline{6}$	Repeating decimal		

METRIC SYSTEM PREFIXES

tera	T	one trillion	deci	d	one tenth
giga	G	one billion	centi	c	one hundredth
mega	M	one million	milli	m	one thousandth
kilo	k	one thousand	micro	μ	one millionth
hecto	h	one hundred	nano	n	one billionth
deka	da	ten	pico	p	one trillionth

FORMULAS

$P = a + b + c$	Perimeter of triangle	$C = \pi d$	Circumference of circle
$P = 2(l + w)$	Perimeter of rectangle	$A = \pi r^2$	Area of circle
$A = lw$	Area of rectangle	$A = 2\pi rh$	Lateral area of cylinder
$A = bh$	Area of parallelogram	$A = 2\pi r^2 + 2\pi rh$	Surface area of cylinder
$A = \frac{1}{2}(b_1 + b_2)h$	Area of trapezoid	$A = 2(lh + lw + wh)$	Surface area of rectangular prism
$A = \frac{1}{2}bh$	Area of triangle		
$V = lwh$	Volume of rectangular prism	$V = \pi r^2 h$	Volume of cylinder
$V = Bh$	Volume of prism (B = base area)		

GLOSSARY

abundant number Any number for which the sum of its factors (other than the number itself) is greater than the given number.

acute angle An angle that has a measure less than 90°.

acute triangle A triangle in which each angle has a measure less than 90°.

addend A number that is added.

angle Two rays with a single endpoint, called the vertex of the angle.

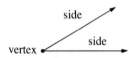

arc A part of a circle.

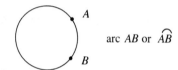

area The measure of a plane region in terms of square units.

arithmetic mean The quotient obtained when the sum of two or more numbers is divided by the number of addends.

associative property The sum (or product) of three or more numbers is the same regardless of grouping:
$(a + b) + c = a + (b + c)$ or
$(a \cdot b) \cdot c = a \cdot (b \cdot c)$

average See *arithmetic mean*.

base (in numeration) The type of grouping involved in a system of numeration. In base eight, 346 means 3 sixty-fours, 4 eights, and 6.

base (of a polygon) Any side of a polygon may be referred to as a base.

Base (of a space figure) See examples below.

Bases of a cylinder   Base of a cone   Base of a pyramid

BASIC A simple programming language.

binary A base-two system of numeration.

bisect To divide into two congruent parts.

bit Binary digit, 0 or 1.

byte String of bits whose length is the smallest accessible unit in computer memory.

capacity The volume of a space figure given in terms of liquid measurement.

centimeter (cm) A metric unit of length. One hundredth of a meter.

central angle An angle that has its vertex at the center of a circle.

chord A segment with both endpoints on a circle.

circle All the points in a plane that are the same distance from one point called the center.

circumference The distance around a circle.

commutative property The sum (or product) of any two numbers is the same regardless of the order in which they are added (or multiplied):
$a + b = b + a$ or $a \cdot b = b \cdot a$

complementary angles Two angles whose measures have a sum of 90°.

composite number A number greater than 0 with more than two different factors.

cone A space figure with a base that is a circular region and one vertex that is not in the same plane as the base.

congruent Two geometric figures are congruent if they have the same size and shape.

coordinate axes Two intersecting perpendicular number lines used for graphing ordered number pairs.

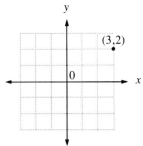

coordinates An ordered number pair matched with a point in the coordinate plane. See figure for *coordinate axes*.

cross products In the equation $\frac{a}{b} = \frac{c}{d}$, the products ad and bc are called cross products. Two ratios $\frac{a}{b} = \frac{c}{d}$ are equal if and only if $ad = bc$.

cross section The intersection of a space figure and a plane.

cube (numeration) A number raised to the third power. 8 is the cube of 2 because $2^3 = 8$. Also, to raise a number to the third power.

cube (space figure) A prism whose faces are all congruent squares.

customary units Units of the system of measurement often used in the United States: pounds, ounces, tons, cups, pints, quarts, gallons, inches, feet, yards, miles, are customary units.

cylinder A space figure with two congruent bases that are circular regions in parallel planes, and a curved face.

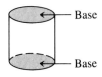

data A collection of unorganized facts that have not yet been processed into information.

decagon A polygon that has ten sides.

decimal (numeral) Any base-ten numeral written using a decimal point.

decimeter (dm) A metric unit of measurement of length equal to 0.1 m.

degree (°) A unit of measure for angles, $\frac{1}{90}$ of a right angle.

degree Celsius (°C) Unit for measuring temperature. On the Celsius scale, water freezes at 0°C and boils at 100°C.

degree Fahrenheit (°F) Unit for measuring temperature. On the Fahrenheit scale, water freezes at 32°F and boils at 212°F.

dekameter (dam) A metric unit of measurement of length equal to 10 m.

denominator For each fraction $\frac{a}{b}$, b is the denominator.

diagonal A segment connecting two non-consecutive vertices of a polygon.

diameter A chord that contains the center of a circle.

difference The number resulting from subtraction.

digits The basic symbols used in a place-value system of numeration. In base ten, the symbols are 0, 1, 2, 3, 4, 5, 6, 7, 8, and 9.

distributive property Connects addition and multiplication when both operations are involved.
$a(b + c) = a \cdot b + a \cdot c$

dividend The number to be divided in a division problem.

```
                       3  ← Quotient
         Divisor →  5)17  ← Dividend
                      15
                       2  ← Remainder
```

divisible A given number is divisible by a second number if the remainder is zero.

divisor See *dividend*.

dodecahedron A polyhedron with twelve faces.

Regular dodecahedron

edge One of the segments making up any of the faces of a space figure.

equally likely outcomes Outcomes that have the same chance of occurring.

equal ratios Ratios that give the same comparison. $\frac{9}{27}$ and $\frac{1}{3}$ are equal ratios.

equation A mathematical sentence using the equality symbol (=). $x - 27 = 102$ is an equation.

equilateral triangle A triangle with all three sides the same length.

equivalent fractions Fractions that represent the same number, such as $\frac{1}{3}$, $\frac{2}{6}$, and $\frac{3}{9}$.

estimate An approximation for a given number. Often used in the sense of a rough calculation.

expanded numeral Representations of numbers as a sum of multiples of 10 such as:

$4{,}325 = 4{,}000 + 300 + 20 + 5$ or

$4 \times 10^3 + 3 \times 10^2 + 2 \times 10 + 5$

exponent A number that tells how many times another number is to be used as a factor.

$$5 \cdot 5 \cdot 5 = 5^3 \quad \begin{array}{l} \leftarrow \text{Exponent} \\ \leftarrow \text{Base} \end{array}$$

exponential notation A system of representing a number using an exponent.

expression (algebraic) Symbols or the combination of symbols, such as numerals, letters, operation symbols, and parentheses used to name a number.

face Any one of the bounding polygonal regions of a space figure.

factors Numbers that are to be multiplied.

factor tree A diagram suggestive of a tree showing the prime factorization of a number.

flowchart A diagram that gives instructions in a logical order.

formula A general fact or rule expressed by using symbols. For example, the area (A) of any parallelogram with base b and height h is given by the formula $A = bh$.

fraction A number in the form $\frac{a}{b}$ when $b \neq 0$.

frequency The number of times a given item occurs in a set of data.

gallon (gal) A customary unit of liquid measure equal to 4 quarts or 16 cups.

gram (g) The basic unit of measurement of weight in the metric system.

greatest common factor (GCF) The greatest number that is a factor of each of two or more numbers.

hectometer (hm) A unit of measurement of length that is equal to 100 m.

heptagon A seven-sided polygon.

hexagon A six-sided polygon.

histogram A bar graph showing frequencies.

icosahedron A space figure with twenty faces.

Regular icosahedron

improper fraction A fraction whose numerator is greater than or equal to its denominator.

integer Any whole number or its opposite, and zero.

intersecting lines Lines that have one common point.

inverse (operation) Two operations that are opposite in effect. Adding 5 is the inverse of subtracting 5.

isosceles triangle A triangle with at least two congruent sides.

kilogram (kg) A metric unit of measurement of weight equal to 1,000 g.

kiloliter (kL) A metric unit of measurement of capacity equal to 1,000 L.

kilometer (km) A metric unit of measurement of length equal to 1,000 m.

least common denominator (LCD) The least common multiple of the denominators of two or more fractions. For example, the least common denominator of $\frac{5}{6}$ and $\frac{3}{4}$ is 12.

least common multiple (LCM) The smallest non-zero number that is a multiple of each of two or more given numbers. The LCM of 4 and 6 is 12.

line A straight path of points that goes on endlessly.

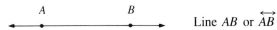

Line AB or $\overleftrightarrow{AB}$

line of symmetry A line on which a figure can be folded so that the two parts fit exactly.

liter (L) The basic unit of capacity in the metric system equal to 1,000 cm³.

loop A command that causes a computer to go back to an earlier step in the program and repeat it.

lowest-terms fraction A fraction for which the greatest common factor (GCF) of the numerator and denominator is 1.

median The middle number of a set of numbers that are arranged in order.

meter (m) The basic unit of measurement of length in the metric system.

metric ton (t) A unit of weight measurement equal to 1,000 kg.

microsecond One millionth of a second.

midpoint A point that divides a segment into two congruent segments.

milligram (mg) A metric unit of measurement of weight equal to 0.001 g.

milliliter (mL) A metric unit of measurement of capacity equal to 0.001 L.

millimiter (mm) A metric unit of measurement of length equal to 0.001 m.

mixed number A number such as $4\frac{2}{3}$ that has a whole number part and a fractional part.

mode In a list of data the mode is the number or item that occurs most often. There may be more than one mode.

multiple A number that is the product of a given number and a whole number.

$3 \cdot 8 = 24$ 24 is a multiple of 3.

negative integer The numbers less than zero that are opposites of natural numbers.

$\{^-1, \ ^-2, \ ^-3, \ldots\}$

nonagon A nine-sided polygon.

numeral A symbol for a number.

numerator For each fraction $\frac{a}{b}$, a is the numerator.

obtuse angle An angle with a measure greater than 90° and less than 180°.

obtuse triangle A triangle with one angle measuring more than 90°.

octagon An eight-sided polygon.

octahedron A space figure with eight faces.

Regular octahedron

opposites property The sum of any number and its opposite is zero. n + ⁻n = 0

ordered pair A number pair arranged in order so there is a first number and a second number. The coordinates of a point in a plane such as (2,7) is an ordered pair of numbers.

ounce (oz) A customary unit of weight measurement. 16 ounces equal 1 pound.

outcome A possible result in a probability experiment.

parallel lines Lines in the same plane that do not intersect.

parallelogram A quadrilateral whose opposite sides are parallel.

pentagon A five-sided polygon.

percent (%) Literally, "per hundred." A way to compare a number with 100.

perimeter The sum of the length of the sides of a polygon.

period Each group of three digits starting with the ones digit is a period.

perpendicular bisector A line that bisects a segment and is perpendicular to it.

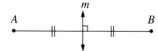

$m \perp$ bis $\overline{AB}$

perpendicular lines Two intersecting lines that form right angles.

pi (π) The ratio of the circumference of a circle to its diameter. The decimal for π is unending and does not repeat.
$\pi = 3.141592\ldots$

pint (pt) A customary unit of liquid measure equal to 2 cups.

place value The value given to the place a digit may occupy in a numeral. In the decimal numeration system, each place of a numeral has ten times the value of the place to its right.

polygon A closed plane figure formed by segments.

polyhedron A closed space figure whose faces are polygonal regions.

positive integer An integer greater than zero {1, 2, 3, . . .}.

pound (lb) A customary unit of weight measurement.

prime factorization The expression of a composite number as the product of prime factors.
$36 = 2^2 \cdot 3^2$

prime number A whole number, greater than 1, whose only factors are itself and 1.

prism A space figure whose bases are congruent polygonal regions in parallel planes and whose faces are parallelograms.

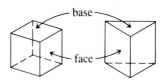

probability The ratio of the number of times a certain outcome can occur to the number of total possible outcomes.

product The number that results when two numbers are multiplied.

program Step-by-step instruction that directs the computer to perform operations.

property of one Any number multiplied by 1 will equal the number:
$a \times 1 = a$ or $a \div 1 = a$

property of zero For every integer a, $a + 0 = a$.

proportion An equation stating that two ratios are equal:
$\frac{3}{18} = \frac{1}{6}$

protractor An instrument for measuring the number of degrees (°) in an angle.

pyramid A space figure with a polygonal base and triangular faces with a common vertex.

448

quadrilateral A four-sided polygon.

quart (qt) A customary unit of liquid measure equal to 2 pints or 4 cups.

quotient See *dividend*.

radius A segment or length of a segment from the center of a circle to a point on the circle.

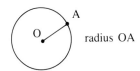
radius OA

range The difference between the greatest number and the least number in a set of data.

rate A ratio that compares different kinds of units.

ratio The ratio of two numbers a and b is their quotient, $\frac{a}{b}$.

ray A part of a line that has one endpoint and extends endlessly in one direction.

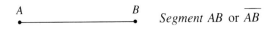
Ray AB or $\overrightarrow{AB}$

reciprocals Two numbers are reciprocals if their product is 1. 7 and $\frac{1}{7}$ are reciprocals.

rectangle A parallelogram with four right angles.

region All the points in the part of a plane bounded by a simple closed curve.

regular polygon A polygon with all sides the same length and all angles the same measure.

relatively prime Two numbers with a greatest common factor (GCF) of 1 are relatively prime to each other. 8 and 27 are relatively prime numbers.

repeating decimal A decimal whose digits from some point on repeat endlessly. 6.2835835 . . . and 0.3333333 . . . are repeating decimals. They may also be written as 6.2$\overline{835}$ and 0.$\overline{3}$ respectively.

rhombus A parallelogram whose sides all have the same length.

right angle An angle that measures 90°.

right triangle A triangle with a right angle.

Roman numerals The numerals I, V, X, L, C, D, M, and combinations of these numerals, used in the Roman numeration system.

scale drawing A drawing made so that distances in the drawing are proportional to actual distances. A scale of 1:10 indicates that distances in the drawing are $\frac{1}{10}$ of the actual distances.

scalene triangle A triangle with all three sides having different measures.

scientific notation A system of writing a number as the product of a power of 10 and a number between 1 and 10.
$2{,}300{,}000 = 2.3 \times 10^6$

segment Two points and all points between them.

$A \bullet \longrightarrow \bullet B$ Segment AB or $\overline{AB}$

similar figures Two or more figures with the same shape but not necessarily the same size.

space figures A (three-dimensional) geometric figure whose points do not all lie in the same plane.

sphere The set of all points in space at a fixed distance from a given point.

square (geometry) A quadrilateral with four right angles and all sides the same length.

square (numeration) A number raised to the second power. 9 is the square of 3. Also, to raise a number to the second power. $3^2 = 9$.

statistics Facts or data of a numerical kind.

straight angle An angle that has a measure of 180°.

string variables Locations in a computer that store data of any kind. A letter and the $ symbol form string variables. A$, B$, etc.

sum The result of the addition operation.

449

supplementary angles Two angles whose measures have a sum of 180°.

surface area The sum of the areas of all the faces of a space figure.

terminating decimal A decimal that represents the quotient of a whole number and a power of 10.

$0.5 = \frac{5}{10}$ $1.28 = \frac{128}{10^2}$ $0.0307 = \frac{307}{10^4}$

tetrahedron A polyhedron with four triangular faces.

ton (T) A customary unit of weight measurement equal to 2,000 pounds.

trapezoid A quadrilateral with one pair of parallel sides.

parallel sides

triangle A three-sided polygon.

unit An amount or quantity used as a standard of measurement.

variable A symbol, usually a letter, used to represent a number in an expression or an equation.

vertex (vertices) A point that two rays of an angle have in common. Also, the common point of any two sides of a polygon or the common point of intersection of three or more faces of a polyhedron.

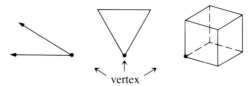

vertex

volume The measure of a space figure in terms of a chosen unit, usually a unit cube.

Venn diagram A special diagram using overlapping circles showing how data are related.

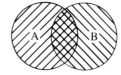

whole number Any number in the set {0, 1, 2, 3, . . .}.

zero property See *property of zero*.

Abundant numbers, 167

Acute triangle, 140–141

Addition
of decimals, 40–41
equations, 178–179
of fractions, 206–207, 233
of integers, 370ff.
basic properties, 370–371
of mixed numbers, 210–211
of units of time, 18–19
of whole numbers, 12ff.

Angle(s), 138–139

Applied problem solving,
316, 336, 362, 388, 406

Arc, 146–147

Area
of circle(s), 324–325, 329, 396–397
of cylinder(s), 328–329
of parallelogram(s), 254, 396
of rectangle(s), 254, 396
surface, 258
of trapezoid(s), 256–257
of triangle(s), 256–257

Associative property
of addition, 370
of multiplication, 56

Average, 126–127, 354

Axis, 384ff.

Base (geometry), 156, 254, 260

Base (numeration)
two, 53
sixteen, 339

Basic properties, 56, 370–371

Binary numbers, 53

Bisect, 154–155

Calculator activities, 28–29, 37, 77, 319, 375, 404

Capacity
customary units, 400
displacement, 271
metric units, 262

Career awareness
agriculture, 358, 359
art and entertainment, 115, 163, 226, 241, 282–283
business services, 23, 81, 105, 206, 245
education, 78, 250, 308
environmental control, 64, 72, 73, 117, 255, 319, 322, 332–333, 343, 358–359, 362, 367
food service, 16–17, 76, 265
health and public service, 18, 55, 166, 184, 193, 352, 362
law and government, 293
marketing, 1, 10, 182, 200, 205, 208, 218, 313, 340
science and technical service, 31, 34, 38, 55, 87, 92, 98, 112–113, 163, 210, 284, 300, 332–333, 334, 350
sports, 74–75, 121, 150
transportation, 21, 29, 48–49, 78–79, 135, 158, 376

Casting out nines, 111

Celsius scale, 266

Centimeter, 248–249

Circle, 146–147
area, 324–325, 329, 396–397
central angle, 146–147
chord, 146–147
circumference, 322–323
diameter, 322–323

Comparing, 6, 36, 202, 368

Compass, 152–156

Complementary angle, 138–139

Composite numbers, 168–169

Computer literacy
BASIC, 67, 75, 112–113
INPUT, 190–191
Decisions, 272–273
Loops, 340–341
Graphics, 410–411

Cone, 156–157

Congruent figures, 148–149

Consumer applications
see also Applied problem solving
area, 255, 324–325
average, 126–127
customary units, 123, 126–127, 401, 402
decimal numbers, 40–41, 44–45, 47, 76–77, 94–95, 101, 118–119, 122–123, 124–125, 326–327
estimation, 58–59, 129
equations, 183, 255, 263, 283, 305
fractions, 205, 236–237, 245
metric units, 76–77, 101, 116–117, 121, 263, 265
percent, 296–297, 304, 305, 313
rate, 122–123, 236–237, 316, 336, 404–405, 406
ratio, 278–279, 282–283
time, 96–97
whole numbers, 8–9, 10–11, 14–15, 16–17, 23, 68–69

Coordinate(s), 384–385

Cross products, 278–279

Cross sections, 334–335

Cubic units, 260ff., 330, 398

Customary units of measure, 394ff.

Cylinder(s), 156–157
area of, 328–329
volume of, 330–331, 398

Data bank, 23, 413–416

Decagon, 144–145

Decimal(s), 32ff.
adding, 40–41
comparing and ordering, 36–37
dividing, 116ff.
estimation with, 44–45, 90–91, 100, 129
fractions and, 238–239, 245

451

multiplication of, 88ff.
numeration system, 4–5, 34
percent(s) and, 294–295, 298
place value, 34–35
repeating, 241
rounding, 38–39, 44–45, 90–91, 100, 118, 129
subtraction of, 42–43
terminating, 241

Decimeter, 248–249

Decision box, 272–273

Degree (angle measure), 138–139

Dekameter, 248–249

Diameter, 146–147, 322

Discount prices and percent, 313

Divisibility, 166–167

Division
of decimals, 116ff.
by a power of ten, 70, 120
equations, 180–181
estimation, 71
of fractions, 232–235
of integers, 382–383
of mixed numbers, 234–235
short division, 73
of whole numbers, 70ff.

Equal ratios, 278–279

Equation(s), 178ff.
addition, 178–179
division, 180–181
graphing of, 386–387
multiplication, 180–181
subtraction, 178–179

Equilateral triangle(s), 140–141, 144–145

Equivalent fractions, 196–197

Estimating
differences, 10–11, 44–45
fractional parts, 204
measurements, 249, 262, 264, 402, 403
percent(s), 299
products, 58–59, 90–91, 100

quotients, 71, 277
sums, 10–11, 44–45
volume, 335

Expanded notation, 4–5

Exponent(s), 66–67

Expression(s), 176–177

Factor(s), 164ff.
common, 172–173
greatest common, 172–173
prime, 170–171
tree, 170–171

Fahrenheit scale, 403

Fibonnaci sequence, 27

Foot, 394–395

Fraction(s), 194ff.
addition, 206–207, 210–211, 233
basic concepts, 194–195
comparing, 202–203
decimals and, 238–239, 245
division, 232–233, 234–235
equivalent, 196–197
estimation, 204
improper, 200–201
lowest-terms, 198–199
mixed numbers, 200–201
multiplication, 224–229
of a whole number, 224–225
percents and, 296–297, 300–301
reciprocals, 231
subtraction, 208–209, 212ff.
unit, 211

Frequency, 352–353

Gallon, 400

Gram, 264

Graph(s)
bar, 356
circle, 359
coordinate, 384–385
of equations, 286–287
line segment, 357
making, 356–359
pictograph, 358

Greatest common factor (GCF), 172–173, 189, 198–199

Hectometer, 248–249

Heptagon, 144–145

Hexagon, 144–145

Icosahedron, 156–157

Inch, 394–395

INPUT, 190–191

Integer(s), 368ff.
addition, 372ff.
basic properties, 370–371.
coordinate graph(s) and, 384–385
division, 382–383
multiplication, 380–381
negative, 368–369, 372–373
opposites, 368–369, 376–377
positive, 368–369, 372–373
subtraction, 376–377

Interest, simple, 304

Kilogram, 264

Kiloliter, 262

Kilometer, 248–249

Least common denominator (LCD), 206

Least common multiple (LCM), 174–175, 206

Length
customary units of, 394–395
metric units of, 248ff.

Line(s)
intersecting, 136–137
parallel, 136–137, 152–153
perpendicular, 152–153
of symmetry, 150–151

Liquid volume, *see* Capacity

Liter, 262

Logical reasoning, 43, 99, 179, 215, 385
problem-solving strategy, 242

Logo, 410–411

Loop, 340–341

Magic square(s), 43, 215, 307

Magic triangle(s), 233, 377

Mathematics history, 2–3, 85, 111, 133, 211, 271, 275

Mean, *see* Average

Measures, *see* Customary units, Metric units

Median, 354

Memory, 28–29

Mental math
  column addition, 13
  decimal products, 98–99
  division, 70, 120
  products, 57, 175
  subtraction, 19

Meter, 248–249

Metric units of measure, 248ff.
  area and, 254, 256–257
  changing units of length, 250–251
  estimation, 249, 262, 264
  prefixes, 251
  surface area, 258
  units of capacity, 262
  units of length, 248ff.
  units of weight, 264
  volume, 260–261

Metric ton, 264

Mile, 394–395

Milligram, 264

Milliliter, 262

Millimeter, 248–249

Mixed numbers, 200ff.
  addition of, 210–211
  comparing, 202–203
  division of, 234–235
  multiplication of, 228–229
  subtraction of, 212ff.

Mode, 352–353

Modular arithmetic, 409

Multiplication
  basic properties, 56, 370
  by 10 and powers of 10, 57
  integers, 380–381
  of decimals, 88–89, 92ff., 98–99
  equations, 180–181
  estimation, 58–59, 90–91, 100
  of fractions, 224–227
  of integers, 380–381
  of mixed numbers, 228–229
  patterns, 175
  of whole numbers, 56ff.

Napier's rods (Bones), 85

Negative integers, 368–369

Network(s), 315

Nonagon, 144–145

Number theory, 111, 164ff.
  abundant numbers, 167
  composite numbers, 168–169
  divisibility, 166–167
  factors, 164ff.
  Fibonnaci, 27
  greatest common factor (GCF), 172–173, 189
  least common multiple (LCM), 174–175
  palindromic, 65
  prime factorization, 170ff.
  prime numbers, 168–169
  square, 11, 117
  triangular, 11

Numerals
  base-sixteen, 339
  base-two (binary), 53
  Chinese, 2
  decimal numeration system, 4–5, 34–35
  Egyptian, 2–3
  Kamba counting, 2
  Mayan, 2–3
  Roman, 2–3

Obtuse angle, 138–139

Obtuse triangle, 140–141

Octagon, 144–145

Octahedron, 156–157

One, property of, 56, 370

Opposite(s), 368–369, 376–377

Opposites property, 370

Ordered pairs, 348–349

Ounce, 402

Outcome(s), 344–345

Palindromic numbers, 65

Parallel lines, 136–137, 152–153

Parallelogram(s), 142–143
  area of, 254, 396–397

Patterns, 7, 11, 65, 117, 169, 239, 375, 381

Pentagon(s), 144–145

Percent(s), 294ff.
  decimals and, 294–295, 298
  discounts and, 313
  estimation of, 299
  finding a number when a percent of it is known, 310–311
  fractions and, 294ff., 300–301
  interest, 304
  of change, 319
  one number is of another, 306–307
  proportion and, 314–315
  ratio and, 294–295, 306–307, 314–315
  of a whole number, 302–303

Perimeter, 252–253

Period(s), 4

Perpendicular lines, 152–153

Pi (π), 322–323, 365, 396

Pint, 400

Place value
  of decimals, 34–35
  of whole numbers, 4–5, 53, 339

Plane, 136–137

Platonic solids, 161

453

Point(s), 136–137
Polygon(s), 144–145
Polyhedrons, regular, 161
Positive integers, 368–369
Pound, 402
Prediction, 205, 336, 334–351, 362
Prime factorization, 170–171
Prime numbers, 168–169
Prism, 156–157
  volume of, 260–261, 390–391
Probability, 344ff.
  chance and, 346–347
  equally likely outcomes, 344–345
  estimation of, 349
  experiment, 365
  graphs and, 356
  ordered pairs, 348–349
  outcome(s), 344–345
Problem solving
  see also Applied problem-solving, Problem-solving applications, Problem-solving data sources, Problem-solving strategies
  answering the question, 21
  choosing the operations, 78–79
  choosing equations, 182–183
  finding averages, 126–127
  finding unit prices, 404–405
  practice, 68–69, 106–107, 121, 217, 255, 259, 263, 267, 326–327, 332–333, 350–351, 379, 401
  understanding questions, 20
  using a calculator, 28–29, 404–405
  using data from an advertisement, 47
  using a data bank, 23
  using a graph, 360–361, 378
  using a map, 48–49, 286

  using a table, 96–97, 105, 122–123, 205, 236–237, 265, 355
  using estimation, 101
  using the 5-Point Checklist, 16–17, 282–283
  using percent(s), 308–309
  using simple interest, 305
  writing questions, 20
  writing and solving equations, 184–185
Problem-solving applications
  area, 255, 259, 326, 333, 362
  averages, 126–127, 135, 362
  circumference, 326–327, 333
  capacity, 263, 401
  customary units, 78–79, 122–123, 127, 217, 282–283, 401, 404–405
  decimal numbers, 46, 47, 69, 79, 96–97, 101, 105, 106–107, 121, 122–123, 126–127, 217, 316, 336, 379
  equations, 182–183, 184–185
  estimation, 101, 129, 287, 362
  fractions, 205, 216, 217, 236–237, 360
  graphs, 360–361, 378
  interest, 305
  integers, 378
  metric units, 20, 121, 255, 259, 263, 265, 278, 332–333, 265, 267, 286, 326–327, 355, 362
  money, 47, 68–69, 96–97, 105, 122–123, 255, 297, 305, 313, 316, 336, 388, 404, 406
  percent, 305, 308–309, 313, 333, 350–351, 361
  probability, 350–351
  ratio, 78, 282–283, 286–287
  statistics, 350–351, 355, 358–359, 360–361
  surface area, 259, 3
  temperature, 267, 361, 378, 379

  time, 20, 48–49, 69, 81, 96–97, 121, 217, 332, 336, 388
  unit prices, 404–405, 406
  volume, 263, 332–333
  weight, 265
  whole numbers, 16–17, 20, 21, 23, 68–69, 78–79, 126–127, 360, 362
Problem-solving data sources
  advertisement, 47
  data bank, 23, 413–416
  drawing, 78, 255, 259
  graph, 283, 360–361, 378
  map, 48–49, 286, 415
  menu, 16–17
  table, 80, 96–97, 104–105, 122–123, 205, 236–237, 265, 355
Problem-solving strategies
  choose the operations, 218
  draw a picture, 108
  find a pattern, 82
  guess and check, 24
  make an organized list, 50
  make a table, 186
  solve a simpler problem, 158
  use logical reasoning, 242
  using the strategies, 268, 288
  work backward, 130
Programs, computer, 112–113, 190–191, 272–273, 340–341, 410–411
Properties
  for integers, 370–371
  for whole numbers, 56
Proportion(s), 278ff.
  percent and, 314–315
Protractor, 138–139
Puzzle, 93, 221
Pyramid(s), 156–157

Quadrilateral(s), 142ff.
Quart, 400

Radius, 146–147, 396ff.

Range (statistics), 352–353

Rate, 96–97, 276ff.

Ratio(s), 276ff.
cross products and, 278–279
equal, 278–279
golden, 291
percent and, 306–307
proportion, 278ff.
rate, 276

Ray, 136–137

Reciprocals, 231

Rectangle(s), 142–143
area of, 254, 396–397

Regular polygon(s), 144–145

Regular polyhedron(s), 161

Repeating decimal(s), 241

Rhombus, 142–143

Right angle, 138–139

Right triangle, 140–141

Roman numerals, 2–3

Rounding
decimals, 38–39, 44–45, 90–91, 118–119
decimal quotients, 118–119
to estimate decimal quotients, 129
to estimate differences, 10–11
to estimate products, 58–59, 90–91
to estimate quotients, 71
to estimate sums, 10–11
whole numbers, 8ff.

RUN, 112–113

REM, 190–191

Scale drawing, 286

Scalene triangle, 140–141

Scientific notation, 102–103

Segment(s), line, 136–137, 154–155

Shape perception, 143, 145, 149, 155, 157, 213

Short division, 73

Similar figures, 284–285

Simple interest, 304

Space figures, 156–157

Space perception, 161, 314
cross sections, 334–335

Sphere, 156–157

Square(s), 142ff.

Square numbers, 11, 117

Standard numerals, 4–5, 66–67, 102–103

Statistics
average, 354
frequency, 352–353
graphs and, 356ff.
median, 354
mode, 352–353
range, 352–353
sample, 362
tally sheet, 349

Subtraction
of decimals, 42–43
equations, 178–179
of fractions, 208–209
of integers, 376–377
of mixed numbers, 212ff.
of units of time, 18–19
of whole numbers, 14–15

Surface area, 258

Tax, 94–95

Technology *see also*
Calculator activities,
Computer literacy
28–29, 67, 75, 77, 85, 112–113, 190–191, 272, 273, 310–311, 340–341

Temperature
Celsius scale, 266
Fahrenheit scale, 403

Terminating decimals, 241

Time, adding and subtracting units of, 18–19, 48–49

Time zones, 48–49

Ton, 402

Trapezoid(s), 142–143
area of, 256–257

Triangle(s), 140–141
area of, 256–257

Triangular numbers, 11

Turtle, 410

Unit fractions, 211

Universal product code, 340–341

Variable(s), 176–177

Venn diagram, 242

Vertex (vertices), 138–139, 157

Volume
and customary units, 398–399
of cylinder(s), 330–331, 335, 398–399
and metric units, 260–261
of prism(s), 260–261, 398–399

Weight
customary units of, 402
metric units of, 264

Whole numbers
addition of, 12–13
comparing and ordering, 6–7
division of, 72ff.
finding a fraction of a, 224–225
finding a percent of a, 302–303
multiplication, 56ff.
rounding, 8ff.
subtraction, 14–15

Yard, 394–395

Zero(s)
annexing, 40ff., 118–119
in products, 94–95
property of, 370–371

455

Acknowledgments cont.:

Lowell Observatory Photograph: 87 (all)
© 1982 **Jeff Lowenthal/Woodfin Camp & Associates:** 238
© 1982 **David Madison/Bruce Coleman Inc.:** 198, 214
Steven Mangold*: 1 top right, 44 right, 70, 71, 81, 88, 90
© 1980 **James Mason/Black Star:** 310
Mike Mazzaschi/Stock, Boston: 164
Coco McCoy/Rainbow: 27
© **Dan McCoy/Rainbow:** 10 right, 34 right, 55 top left, 215, 300, 321 bottom right
© 1980 **Wally McNamee/Woodfin Camp & Associates:** 150
Peter Menzel: 137 left
© 1981 **Peter Menzel:** 182
© **J. Messerschmidt/Bruce Coleman Inc.:** 291
Gary Milburn/Tom Stack & Associates: 136 right, 367 bottom left
© **Kurt Mitchell/Photo Library:** 174, 247 right
© **A. Moldvay/After-Image:** 202
© **Linda Moore/Rainbow:** 223 center, 308–309
Hank Morgan/Rainbow: 223 bottom left
NASA: 6, 102 top
National Semiconductor Corporation: 341
Nelson-Atkins Museum of Art; Kansas City, Missouri (Nelson Fund): 146
Mark Newman/Tom Stack & Associates: 343 center right
The New York Public Library; Astor, Lenox and Tilden Foundations: 1 bottom left, Prints Division; 275 top left; 275 top center, Rare Book Division
D. J. Palke/Atoz Images: 116
Brian Parker/Tom Stack & Associates: 379 top
Maryanne Pendergast, Department of Anatomy, University of North Carolina Medical Center: 34 left
© 1982 **Stacy Pick/Stock, Boston:** 226
Rod Planck/Tom Stack & Associates: 137 right
© 1982 **Marc Pokempner/Black Star:** 282
© **Hans Reinhard/Bruce Coleman Inc.:** 201, 276 bottom right

L. L. T. Rhodes/Atoz Images: 306
© **Ray Richardson/Animals Animals:** 343 bottom right
Michael Rizza/Stock, Boston: 321 top left
© **Dick Rose/Sygma:** 293 center left
© **Leonard Lee Rue III/Atoz Images:** 276 top right, 343 top right
John Running/Photo Library: 288
Bill Ruth/Bruce Coleman Inc.: 267
© 1980 **Sepp Seitz/Woodfin Camp & Associates:** 126
Frank Siteman/Stock, Boston: 193 center right
© **Lester Sloan/Woodfin Camp & Associates:** 293 top center
© **Camilla Smith/Rainbow:** 302
Tom Stack/Tom Stack & Associates: 18, 137 center, 359
© **Gene Stein/West Light:** 193 bottom right
Werner Stoy/Camera Hawaii: 287
© 1982 **Vince Streano/After-Image:** 376
© **William Strode/Woodfin Camp & Associates:** 200
© **Harald Sund:** 8, 108
© **Norman Owen Tomalin/Bruce Coleman Inc.:** 250, 252, 350–351 center
© **Tom Tracy:** 55 bottom left, 76, 123, 158, 321 top right, 374 top
© 1980 **Cheryl A. Traendly/Jeroboam:** 217
Mark Tuschman*: 42, 55 bottom right, 166, 176, 193 top left and top right, 208, 228 right, 231, 248, 258 top, 324, 330, 336, 346 top right, 352, 382, 401, 404–405
B. P. Van Cleve/Atoz Images: 236–237
© **Jerry Wachter/Focus On Sports:** 36
© 1977 **Robert Weinreb/Bruce Coleman Inc.:** 305
© **Cary Wolinsky/After-Image:** 247 bottom right
Cary Wolinsky/Stock, Boston: 247 top left
Baron Wolman*: 163 (all)
© **Adam Woolfitt/Woodfin Camp & Associates:** 100 right
© **John Yates/After-Image:** 106, 223 top right, 393 top left
Carl Zeiss/Bruce Coleman Inc.: 136 left

* Photographs provided expressly for the publisher.

Selected Answers for

ADDISON-WESLEY MATHEMATICS

1 Addition and Subtraction of Whole Numbers

Page 3
1. 34   3. 2,322   14. 12   16. 61
18. 1,159   22. VII   25. LIII
27. MCCC

Page 5
1. 40   3. 60,000   5. 300,000
10. 3,247   12. 775,066
17. forty-two thousand, five hundred ninety-one
23. 50,000 + 400 + 60 + 8   29. 5,461
31. 305,002

Page 6
1. 59,066,927   4. 966,769

Page 7
2. >   4. <   7. >   13. 100
14. 330   21. 3,799   25. 1,100
30. 1,321; 1,329; 1,400
33. Triton to Neptune; 354,000 < 384,500

Page 8
2. 2,060; 2,100   5. 750,000; 1,000,000

Page 9
2. 2,400   6. 5,600   12. 9,000
17. 7,000   22. 60,000   25. 90,000
33. 1,000,000   37. 16,000,000
41. 60,500

Page 10
1. 100   6. 200   9. 5,000

Page 11
Estimates may vary.
2. 200   5. 3,000   11. 600,000
14. 30   17. 500   23. 600,000
25. 260

Page 12
2. 1,446   5. 12,000   8. 3,276

Page 13
3. 1,660   8. 13,923   12. 36,839
17. 18,427   21. 39,546,313
22. 1,875,542   24. 4,114 mi

Page 14
2. 2,557   5. 225

Page 15
2. 223   7. 3,251   12. 10,558
17. 445   21. 2,279   25. 570
29. 1,861   32. 38,544   34. 267

Page 16
1. 45   2. 245

Page 17
2. 730   6. 825   8. 345

Page 18
1. 7 h 55 min   5. 5 h 15 min

Page 19
1. 5 h   3. 14 h 17 min   6. 3 h 50 min
10. 8 h 25 min   12. 7 h 10 min
15. 7 h 35 min   17. 2 h 45 min

Page 21
1. 17,318,000   3. 44,238,000

Page 22
3. 1,491   7. 118,617   13. 4,147,354
17. 619,109   23. 294,949
26. 18,521,509   33. 85,745,220
36. 1,066,448   42. 98,463   45. 11,187

Page 23
2. 68   5. Pennsylvania; 5

Page 24
1. 53

Page 26
1. < **5.** 2,400 **9.** 60,000 **12.** 300
14. 900 **18.** 4,585 **20.** 153
24. 1,794

Page 27
3. 143

2 Addition and Subtraction of Decimals

Page 33
2. 0.29 **5.** 1.9 **7.** 3.5 **10.** 3.16
12. 0.47 **20.** 0.67; 0.33

Page 35
2. 0.3066 **4.** 0.0014 **7.** 18.4627
14. 0.7 **16.** 0.09 **18.** 200
21. 0.0001

Page 36
2. < **6.** 3.15; 3.155; 3.182

Page 37
1. < **3.** < **6.** = **8.** >
20. 5.185 **23.** 0.67; 0.72; 0.79
25. 0.043; 0.24; 0.34 **29.** Jones; 0.723

Page 38
1. 7.9 **6.** 3.000

Page 39
1. 3.7 **3.** 0.6 **9.** 23.03 **11.** 0.55
17. 0.388 **19.** 5.032 **25.** 43
28. 382 **34.** 9 **35.** 0.7 **41.** 1.5

Page 40
2. 0.84 **7.** $29.63

Page 41
3. 34.311 **9.** 145.837 **13.** 1.978
17. $11.21 **21.** 34.22 **27.** 11.355
29. 79.519 **32.** 10.112 **34.** $52.90
36. $2.5852

Page 42
1. 31.77 **8.** 21.572

Page 43
3. 0.486 **7.** 2.3752 **12.** 3.15
16. 2.01 **22.** $12.37 **27.** 12.909
29. 0.325 **33.** 0.186 **35.** 18.555 m/s

Page 44
2. $50 **6.** 1.6

Page 45
1. $8 **4.** $6 **6.** $80 **9.** $60
11. $500 **12.** $700 **16.** 1.1
22. 0.4 **30.** $13

Page 46
1. 112.27 **7.** 134.49 **13.** $759.92
17. 8.7 **21.** 0.0074 **27.** 77.85
31. 83.167 **38.** 111,710.5
41. 775.58 **44.** 146,285.42

Page 47
1. $17.75 **4.** $7.50 **6.** $4.55

Pages 48–49
2. 3:00 a.m. **4.** 1:00 p.m. **8.** 7:30 p.m.
11. 9:07 p.m.

Page 50
1. 9

Page 52
1. 0.8 **4.** 0.018 **6.** < **9.** 0.75
13. 1.4051 **22.** 2.67

Page 53
1. 6 **7.** 64 **10.** 10000 **13.** 11000001

3 Multiplication and Division of Whole Numbers

Page 56
1. Property of One **3.** Distributive **7.** 26
9. 1 **12.** 4

459

Page 57
2. 90   **5.** 360   **7.** 8,600
15. 7,200   **19.** 12,000   **28.** 40,000
32. 30,000   **36.** 60,000   **38.** 8,000
40. 3,600 s

Page 58
1. C   **5.** 540   **9.** 12,000

Page 59
2. 800   **9.** 1,200   **12.** 2,400
17. 2,100   **22.** 36,000   **27.** 2,800
32. 180,000   **36.** $4.80

Page 60
2. 632   **7.** 9,051

Page 61
3. 760   **8.** 4,832   **12.** 104,372
18. 37,650   **25.** 73,494   **28.** 1,962
31. 81

Page 62
1. 3,312   **6.** 6,846

Page 63
3. 2,923   **7.** 14,688   **13.** 90,552
17. 421,142   **23.** 62,910   **28.** 748,524
34. 3,472   **37.** 35,770   **40.** 29,133
41. 896,445   **47.** 1,536

Page 64
1. 36,900   **6.** 1,110,203

Page 65
2. 154,224   **9.** 229,992   **13.** 165,452
18. 1,016,925   **22.** 71,484
25. 1,671,468   **28.** 593,640

Page 66
1. 2 base, 5 exponent   **7.** 3^2 **11.** 36

Page 67
2. 8^4 **7.** 4^2 **12.** 7^7 **17.** 8   **20.** 100
24. 32   **32.** 100   **34.** 98   **37.** 343

Pages 68–69
1. $5   **4.** $3,000.00   **8.** $2.50   **12.** $2.45

Page 70
3. 9   **7.** 40   **10.** 4   **15.** 700
20. 5   **24.** 7   **25.** 80

Page 71
2. 30   **8.** 200   **11.** 50   **14.** 20
19. 10   **22.** 3   **28.** 5   **31.** 40

Page 72
1. 147   **3.** 293 R4
9. 24 kg per person, remainder 2 kg

Page 73
2. 513 R3   **7.** 1,216   **9.** 1,643
12. 548 R2

Page 74
2. 20 R24   **6.** 50 R38

Page 75
3. 19 R27   **6.** 62 R38   **12.** 40
14. 49 R37   **18.** 232 R57   **23.** 407
25. over 37

Page 76
1. 26 R70

Page 77
2. 17 R596   **8.** 37 R100   **10.** 304
16. 84   **18.** 704 R69   **22.** 1,000; 903
27. 70

Pages 78–79
2. 5,250 ft   **5.** 27 in.   **8.** 108.3 mi

Page 80
2. 72,873   **7.** 34,998   **11.** 346,500
17. 782,132   **21.** 1,834 R3
27. 168,041 R25   **29.** 664 R11
33. 77 R725   **38.** 13,082,420
41. 16 R394   **46.** 14,667,668
48. 1,119 R547

Page 81
1. 62 words per minute
4. 37 words per minute

Page 82
2. 119 cm

Page 84
2. 1,600   8. 882   10. 986
14. 397,026   18. 10^5 20. 200
26. 7 R3   32. 217 R57

Page 85
2. 435   7. 2,178

4 Multiplication of Decimals

Page 89
1. 0.9; 0.9   5. 2.0; 2.0   8. $0.8 \times 0.5 = 0.40$
12. $0.2 \times 0.5 = 0.10$ 19. 0.28   22. 0.006

Page 90
2. 50.302   4. $68.15   7. 49.52

Page 91
2. A   6. C   8. C   13. 50.4
16. $79.17   24. 460.6   26. 694.88
30. 208.0 g

Page 92
2. 10.062   5. 0.6364

Page 93
3. 2.7170   8. 4,160.00   14. 2.772
18. $11.25   24. 15.5475   27. 402.00
32. 4.5144   35. 0.4624   38. 242.55 km

Page 94
2. 5; 0.00561

Page 95
3. 0.000096   10. 0.0108   12. 0.0675
19. 0.002108   21. 4.9714
28. $0.9880; $0.99   34. $0.3350; $0.34
36. $0.08 or 8¢

Page 96
2. $0.69   5. $1.24   8. $1.52

Page 97
1. $1.30   4. $3.05   7. $1.63

Page 98
2. 8; 80; 800   7. 19; 1.9; 0.19

Page 99
6. 0.06   9. 7,462   12. 91.3   18. 150
24. 4.8   27. 3,140   33. 72.6
39. 0.0007   42. 0.097   48. 2
51. 7.4683   56. 0.04763   61. 7.6 cm

Page 100
2. 4.2   9. 0.48   12. 6.0 or 6   16. $12
19. $0.80   22. $480

Page 101
2. $12   5. 210 m   7. $650

Page 102
1. 3   4. 9.6   7. 6,300,000

Page 103
1. 2.5×10^4 3. 6.1×10^{11} 8. 5.9×10^5
10. 320,000   13. 3,090,000
16. 36,000,000   19. 1,430,000,000

Page 104
1. 26.52   7. $109.98   11. 0.2366
17. 0.00091   21. 0.62244   27. 55.118
32. 122.534   36. $26.496; $26.50
42. 2.55714   44. 3,692.923

Page 105
1. $23.49   3. $46.48

Pages 106–107
2. 97.50   4. 100.80
8. 63.00; 47.60; 64.80; 67.20; 242.60

Page 108
1. 129 km

461

Page 110
1. 76.8   **7.** 0.0126   **10.** 0.001872
15. 3,141.6   **16.** 2.27   **21.** 81   **24.** $40

5 Division of Decimals

Page 116
1. 0.57

Page 117
1. 3.27   **7.** 57.2   **14.** 0.037   **18.** 3.5
22. 3.07   **27.** 0.94   **31.** 9.96 kg

Page 118
1. 0.391   **4.** $1.97

Page 119
2. 1.6   **7.** 2.1   **10.** 6.7   **14.** 2.367
20. 1.105   **22.** 1.264   **26.** $0.88
31. $0.36   **33.** $0.16

Page 120
1. 0.007   **3.** 23.72   **6.** 31.682
8. 0.397

Page 121
2. 2.187 min   **3.** 27.7 s   **6.** 3.72 s

Page 122
2. $5.25

Page 123
1. $18.07   **4.** $11.04
8. 4,940 ft^3; $18.03

Page 124
1. 14.1   **4.** 4.09

Page 125
1. 2.8   **6.** 81.3   **10.** 5.2   **15.** 5.26
19. 0.78   **22.** 0.67   **25.** 13.889
30. 3.058   **33.** 14

Pages 126–127
2. 81   **5.** 138.2¢   **7.** 105 lb

Page 128
2. 13.3   **5.** 25.3   **9.** 3.5   **17.** 438.27
21. 15.58   **26.** $8.61   **29.** $28.86
34. 1.079   **42.** 0.184

Page 129
1. 5   **4.** 2   **8.** 20   **10.** 0.03
13. $0.20; $0.195

Page 130
1. 61

Page 132
1. 2.4   **5.** 1.7   **10.** 0.72; 0.072; 0.0072
14. 5.24   **20.** 12.972   **25.** 30

Page 133
1. $(15 \times 16) \div 2 = 120$
3. $(10 \times 20) \div 2 = 100$

6 Geometry

Page 137
5. $\overline{AC}, \overline{AB}, \overline{BC}$ **9.** $\overrightarrow{OZ}, \overrightarrow{OY}, \overrightarrow{OX}$
11. $\overleftrightarrow{RS} \parallel \overleftrightarrow{MN}$

Page 139
1. 25°; acute   **5.** 155°; obtuse   **12.** 50°
15. 58°   **20.** 2°   **23.** 60°   **25.** 45°
29. 74°

Page 141
2. right   **5.** scalene   **9.** obtuse; isosceles
11. 121°   **14.** 60°   **16.** obtuse

Page 143
2. rhombus   **5.** 107°   **8.** 49°   **10.** 103°

Page 144
2. pentagon   **4.** triangle   **6.** hexagon

Page 145
2. $\overline{AB}, \overline{BC}, \overline{CD}, \overline{DE}, \overline{EA}$ **4.** 540°
6. quadrilateral   **9.** regular pentagon

Pages 146–147
2. $\overline{YW}$ **5.** ∠ZQW, ∠ZQY   **7.** 3.6 cm
12. 90°   **15.** ∠QPS

Pages 148–149
2. ∠P ≅ ∠R   **5.** ∠Q ≅ ∠O; ∠S ≅ ∠N; ∠R ≅ ∠M; $\overline{QS}$ ≅ $\overline{ON}$; $\overline{SR}$ ≅ $\overline{NM}$; $\overline{RQ}$ ≅ $\overline{MO}$
9. ∠R   **13.** $\overline{RS}$ **15.** 90°

Page 150
2. yes   **6.** no   **7.** yes

Page 151
11. 5   **14.** 10

Page 157
3. V5, F5, E8   **7.** V6, F6, E10
12. triangular pyramid

Page 158
1. 1,057

Page 160
1. acute   **6.** 90°   **8.** parallelogram
11. $\overline{PN}$ **14.** T, U, V, W, X, Y

Page 161
cube: 8, 6, 12, 2

7 Number Theory and Equations

Page 164
1. 1, 2, 5, 10   **5.** yes; 8 × 29

Page 165
1. 1, 2, 4, 7, 14, 28   **7.** 1, 3, 5, 15
15. 1, 3, 5, 9, 15, 45   **22.** no
26. yes; 18 × 37   **34.** yes; 120 × 5
39. 4; 1, 2, 4   **44.** yes; 47,287 × 294

Page 166
1. 2,3   **5.** 3,9

Page 167
3. no   **7.** yes   **13.** yes   **17.** no
25. yes   **28.** yes   **32.** 2, 3, 5, 10
39. 2, 3, 9   **41.** no

Page 168
3. 1, 13; prime   **5.** 1, 3, 5, 15; composite

Page 169
1. C   **4.** P   **11.** P   **14.** P   **18.** 23
21. 89   **24.** 16; 1, 2, 4, 8, 16   **30.** 2

Page 170
5. 2^3 **6.** $2^3 \cdot 3$

Page 171
17. $2^3 \cdot 11^2$ **21.** $5 \cdot 7^3 \cdot 11$
27. 2 × 107 × 127

Page 172
1. 1, 2; 2   **5.** 1; relatively prime

Page 173
2. 1, 2, 7; 7   **6.** 9   **10.** 14
14. 2, 3, 5; 2, 3, 2, 2; 6   **18.** 6
22. 15   **28.** 1   **30.** 13   **33.** 15 in.

Page 174
1. 36   **5.** 12

Page 175
1. 70   **5.** 32   **10.** 60   **15.** 84
17. 24   **22.** 120   **30.** 7,610,526

Page 176
1. n **6.** 35

Page 177
2. 56   **8.** 7   **11.** 29
17. 28, 40, 60, 100   **20.** $x - 8$ **24.** $\frac{k}{7}$
26. $w + 11$

Page 178
1. $x = 15$ **5.** $s = 23$

463

Page 179
1. $x = 7$ 6. $m = 27$ 9. $h = 25$
11. $z = 16$ 14. $r = 21$ 18. $f = 31$
23. $v = 53$ 26. $w = 10$ 31. $c = 100$
35. $n = 200$ 38. $k = 110,365$

Page 180
1. $p = 8$ 5. $p = 150$

Page 181
2. $n = 8$ 6. $t = 3$ 11. $z = 10$
13. $h = 20$ 19. $s = 0$ 23. $b = 161$
26. $h = 30$ 30. $z = 100$ 34. $n = 2$
38. $m = 13,012,032$

Pages 182–183
1. A; $n = 48$ 6. B; $h = 39$ 9. D; $e = 624$

Pages 184–185
2. $h - 339 = 1,286$; $h = 1,625$ m
5. $\frac{h}{4} = 245$; $h = 980$ m
7. $42h = 4,410$; $h = 105$ m

Page 186
1. 1990

Page 188
1. 1, 2, 3, 4, 6, 12   4. C   7. $2 \cdot 3 \cdot 5$
10. 4   13. 30   16. 32   17. 56
18. 80   22. $x = 16$ 25. $z = 78$

Page 189
1. 77   3. 107

8 Addition and Subtraction of Fractions

Page 194
1. $\frac{1}{6}$ 5. $\frac{2}{9}$

Page 195
1. $\frac{2}{3}$ 5. $\frac{3}{8}$ 7. $\frac{4}{5}$ 11. $\frac{5}{6}$ 14. $\frac{8}{25}$
17. $11 \div 12$ 20. $7 \div 10$ 22. $\frac{1}{6}$

Page 196
2. 12   4. $\frac{3}{6}, \frac{4}{8}$

Page 197
1. 8   7. 35   13. 12   18. 144
22. $\frac{4}{40}, \frac{5}{50}, \frac{6}{60}$ 25. $\frac{9}{24}$ 30. $\frac{90}{100}$ 34. $\frac{3}{9}$

Page 198
1. $\frac{3}{4}$ 6. $\frac{8}{9}$

Page 199
3. $\frac{1}{3}$ 8. $\frac{5}{12}$ 14. $\frac{2}{3}$ 17. $\frac{1}{4}$
24. $\frac{14}{25}$ 27. $\frac{7}{8}$ 31. $\frac{1}{2}$

Page 200
1. $3\frac{2}{3}$ 7. $\frac{9}{4}$

Page 201
5. 3   9. $4\frac{4}{5}$ 12. $4\frac{1}{4}$ 20. 2
22. $\frac{22}{3}$ 28. $\frac{67}{12}$ 35. $\frac{79}{12}$ 37. $\frac{153}{10}$
41. $\frac{7}{2}$ 45. $\frac{5,925}{64}$ 49. $1\frac{1}{3}$

Page 202
1. $<$ 3. $>$

Page 203
1. $>$ 7. $<$ 12. $>$ 15. $<$
20. $<$ 22. $>$ 25. $>$ 30. $<$
33. no; $\frac{7}{12} < \frac{2}{3}$

Page 204
1. $\frac{1}{4}$ 4. $\frac{3}{4}$

Page 205
2. $\frac{6}{7}$ 6. 748.75 (749)   10. 45

Page 206
1. $\frac{2}{3}$ 6. $\frac{7}{10}$

Page 207
1. $\frac{1}{2}$ 8. $\frac{1}{2}$ 13. $\frac{9}{10}$ 19. $\frac{39}{40}$ 23. $1\frac{3}{8}$
26. $\frac{2}{3}$

Page 208
1. $\frac{1}{2}$ 6. $\frac{7}{16}$

Page 209
1. $\frac{2}{3}$ 8. $\frac{1}{12}$ 13. $\frac{1}{20}$ 21. $\frac{2}{9}$ 24. $\frac{27}{40}$
26. $\frac{1}{8}$

Page 210
2. $9\frac{4}{5}$ 5. $5\frac{5}{24}$

Page 211
3. 12   6. $3\frac{11}{12}$ 14. $24\frac{11}{16}$ 19. $54\frac{1}{10}$
23. 10   26. $46\frac{5}{8}$ in.

Page 212
1. $4\frac{1}{4}$ 5. 58

Page 213
3. $7\frac{1}{8}$ 9. $\frac{19}{30}$ 13. $7\frac{7}{18}$ 19. $88\frac{1}{8}$
23. $27\frac{2}{3}$ 25. $1\frac{1}{8}$ in.

Page 214
2. $3\frac{1}{6}$ 5. $11\frac{2}{3}$

Page 215
2. $3\frac{1}{2}$ 9. $6\frac{1}{8}$ 13. $1\frac{23}{25}$ 18. $1\frac{5}{8}$
22. $10\frac{7}{8}$ 25. $1\frac{1}{2}$ h

Page 216
1. 1   7. $\frac{11}{12}$ 10. $23\frac{1}{40}$ 16. $\frac{1}{6}$
20. $3\frac{1}{2}$ 28. $1\frac{13}{14}$ 29. $\frac{11}{20}$ 37. $1\frac{6}{7}$
40. $174\frac{7}{18}$

Page 217
1. $1\frac{2}{3}$ ft   4. $1\frac{1}{3}$ innings   6. 6.2 yd

Page 218
1. 13

Page 220
2. 10   4. $\frac{1}{3}$ 7. $3\frac{2}{3}$ 13. $\frac{35}{6}$ 19. >
25. $1\frac{1}{6}$ 29. $1\frac{3}{8}$

Page 221
1. $\frac{2}{3}$ 3. $\frac{5}{12}$ 5. $\frac{25}{5}$

9 Multiplication and Division of Fractions

Page 224
1. 3   5. 2

Page 225
1. 6   5. 3   10. 4   13. 27
19. 2   22. 45   29. 50   33. 4

Page 226
1. $\frac{2}{15}$ 6. $\frac{3}{10}$ 9. $\frac{3}{8}$ 13. 1   18. $\frac{1}{9}$
21. $\frac{1}{2}$

Page 227
2. $\frac{1}{18}$ 5. $\frac{1}{3}$ 9. $1\frac{1}{3}$ 13. $\frac{1}{4}$
18. $\frac{1}{15}$ 21. $\frac{3}{8}$

Page 228
1. $\frac{4}{9}$ 6. $17\frac{1}{3}$

Page 229
2. 26   5. 3   11. $11\frac{9}{10}$ 13. 3
19. $9\frac{1}{2}$ 22. $12\frac{2}{15}$ 25. 18   29. $12\frac{3}{16}$ in.

Page 230
2. $\frac{7}{10}$ 7. $1\frac{1}{6}$ 11. $37\frac{3}{5}$ 14. $\frac{2}{5}$
17. $8\frac{1}{2}$ 22. $4\frac{1}{15}$ 26. $\frac{3}{10}$ 35. 11

Page 231
1. 2   7. $\frac{2}{3}$ 12. $n = \frac{8}{7}$ 20. 6

Page 232
2. $3\frac{3}{4}$ 5. 6

Page 233
1. $3\frac{3}{4}$ 5. $\frac{1}{8}$ 11. 10   17. $\frac{1}{3}$ 23. $\frac{7}{10}$
28. $2\frac{2}{3}$ min

Page 234
2. $\frac{7}{18}$ 5. $9\frac{19}{36}$

Page 235
1. $2\frac{3}{10}$ 7. $\frac{13}{25}$ 10. $\frac{57}{64}$ 14. $\frac{13}{14}$
21. $1\frac{1}{18}$ 29. 16

465

Page 236
1. $3\frac{1}{3}$ kWh   **3.** $4\frac{53}{60}$ kWh   **7.** $18\frac{2}{3}$

Page 238
1. $0.4; \frac{2}{5}$ **6.** $\frac{1}{5}$ **10.** 0.74

Page 239
1. $0.6; \frac{3}{5}$ **8.** $10.8; 10\frac{4}{5}$ **14.** $\frac{1}{10}$
22. $\frac{3}{500}$ **27.** $\frac{1}{8}$ **31.** 0.8   **37.** 0.52
41. 5.80   **45.** $\frac{16}{25}$

Page 240
2. $\frac{1}{11}$ **7.** $\frac{27}{32}$ **9.** 16   **14.** $2\frac{4}{7}$
21. $3\frac{2}{11}$ **25.** $3\frac{3}{10}$ **31.** $14\frac{19}{20}$ **34.** $\frac{4}{5}$
38. $\frac{7}{55}$ **44.** $\frac{3}{20}$ **50.** $1\frac{1}{35}$

Page 241
1. $0.\overline{43}$ **5.** $2.\overline{345}$ **10.** $0.1\overline{6}$
16. 0.75   **21.** $0.91\overline{6}$ **30.** $0.\overline{407} > 0.40625$

Page 242
1. 12

Page 244
1. 2   **4.** 9   **7.** $\frac{1}{10}$ **11.** $\frac{7}{8}$
16. $1\frac{2}{3}$ **20.** $1\frac{1}{3}$ **25.** $0.\overline{6}$

Page 245
2. Computer: $1,350; ITE: $2,600; AMRO: $2,975; New Gas: $4,750

10 Measurement: Metric Units

Page 248
1. 10   **4.** 10

Page 249
1. B   **5.** A   **11.** mm   **16.** cm
20. 100   **24.** 100   **30.** 0.1

Page 250
2. 358.7 cm   **5.** 184 m

Page 251
2. 3.84 m   **7.** 1,842 m   **18.** 6 cm
21. 200 cm   **25.** 2 km   **27.** 0.23 km
37. 30 mm   **43.** 200 mm
49. 3.5 cm; 0.35 dm; 0.035 m

Page 252
1. 47 cm

Page 253
1. 17 m   **5.** 8 m   **9.** 824 km
13. 598 cm

Page 254
2. 250 m^2 **5.** 18.2 cm^2 **9.** 120 cm^2
14. 87.36 m^2

Page 255
1. 14.52 m^2 **4.** $21.81

Page 256
1. 10.8 cm^2

Page 257
2. 37.2 cm^2 **5.** 1 m^2 **8.** 35 cm^2
11. 7 km^2 **13.** 10 m^2

Page 258
2. 127 cm^2 **6.** 23.96 cm^2

Page 259
2. 20 cm   **5.** $240.90

Page 260
1. 864 cm^3

Page 261
3. 2.4 cm^3 **7.** 1,000 km^3 **9.** 180 cm^3
13. 10.6215 m^3

Page 262
1. 1,500   **5.** kL

Page 263
2. 800 L   **5.** Monday morning   **9.** 1,950 mL

Page 264
2. kg   **6.** 500 g   **10.** 3,174   **15.** 2,000   **18.** 700

Page 265
2. 336 g   **5.** 181 g   **8.** 25

Page 266
1. 42°C   **4.** 19°C

Page 267
3. 1,062°C   **5.** 50°C

Page 268
2. 268 cm

Page 270
1. 1.85   **6.** 82 cm   **11.** 20.4 cm^2
13. 300 cm^3 **15.** 1.750 L

Page 271
2. 325 g   **3.** 20 g

11 Ratio and Proportion

Page 276
2. $\frac{1}{7}$ **6.** $8/h

Page 277
1. $\frac{8}{9}$ **8.** $\frac{2}{6}$ **10.** $\frac{1.2}{4.3}$ **13.** $\frac{10}{1}$
18. $0.16/L   **20.** 66 words/min   **27.** $\frac{3}{1}$ or 3:1

Page 278
1. =   **6.** $\frac{4}{3}$

Page 279
2. ≠   **5.** =   **11.** ≠   **17.** $\frac{2}{3}$
23. $\frac{3}{1}$ **26.** $\frac{3}{4}$ **33.** yes

Page 280
2. $x = 18$

Page 281
1. $x = 6$ **6.** $x = 20$ **10.** $b = 15$

Page 283
2. 6   **5.** 6,300 in.   **7.** 120 h

Page 284
1. $\frac{4}{7} = \frac{8}{x}$; $x = 14$ cm

Page 285
1. $x = 24$ cm   **5.** $x = 84$ m   **7.** 15 cm

Page 286
1. 30.1 km   **5.** 91 km

Page 287
2. 170 kg; 168 kg   **4.** 10 m; 9 m

Page 288
2. 20 weeks

Page 290
1. $\frac{6}{4}$ **4.** $\frac{\$15}{3}$ **8.** =   **10.** ≠
13. $x = 20$ **20.** $x = 2$ m

12 Percent

Page 295
1. 35%   **6.** 36%   **10.** 1%   **14.** 25%
19. $\frac{8}{100}$; 0.08; 8%   **26.** 39%

Page 296
1. $\frac{1}{20}$ **7.** $\frac{3}{5}$ **13.** 15%   **17.** 99%

Page 297
2. $\frac{1}{5}$ **8.** $\frac{7}{10}$ **13.** 32%   **19.** 12%
24. $\frac{1}{2}$; 50%   **30.** 2; 200%   **33.** $\frac{1}{20}$

Page 298
1. 0.38   **8.** 0.096   **12.** 1.05   **16.** 48%
23. 2.8%   **26.** 136%

Page 299
1. 18%; 82%   **5.** 25%

Page 300
2. 38%   **6.** 56%   **8.** 42%

Page 301
1. 83%   **7.** 59%   **13.** 71%
21. 46.9%   **25.** 44.9%   **27.** 58%

467

Page 302
1. 8   **5.** 24   **8.** 96.25

Page 303
1. 20   **6.** 27   **10.** 58.50   **13.** 72
18. 107   **21.** 45   **26.** 10   **33.** 30
41. 2

Page 304
1. $720   **4.** $676

Page 305
2. $247.50   **4.** $8,600   **7.** $6,384

Page 306
1. 50%   **3.** 10%

Page 307
1. 20%   **7.** 37.5%   **10.** 2%   **13.** 33%
16. 94%   **21.** 30%

Pages 308–309
1. 95%   **2.** 83   **6.** 34   **12.** 85%

Page 310
1. $n = 50$ **4.** $n = 400$

Page 311
1. $n = 400$ **6.** $n = 20$ **11.** $n = 800$
13. $20\% \times n = 18$; $n = 90$
17. $9\% \times n = 2.7$; $n = 30$ **22.** $n = 30$
27. $n = 360$ **30.** 8,000 ft

Page 312
2. 0.39   **7.** 4   **12.** 18%
21. 60%   **22.** 60%   **32.** 44.4%
36. $\frac{1}{4}$ **42.** $\frac{9}{20}$ **48.** $\frac{19}{25}$
52. 30   **61.** 80

Page 313
2. $37.50; $37.50   **7.** $4.80; $11.20

Page 314
1. $\frac{60}{100} = \frac{n}{35}$; $n = 21$ **3.** $\frac{n}{100} = \frac{24}{60}$; $n = 40$

Page 315
1. $\frac{48}{100} = \frac{n}{75}$; $n = 36$ **3.** $\frac{n}{100} = \frac{20}{25}$; $n = 80$
9. $\frac{12.5}{100} = \frac{18}{n}$; $n = 144$ **17.** 160 cm

Page 316
3. no; $10.15 with chicken, $10.16 with ground beef

Page 318
1. 30%   **3.** 27%   **7.** $\frac{1}{5}$ **11.** 0.28
15. 23%   **19.** 38%   **25.** 60%

13 Circles and Cylinders

Page 322
1. 9.42 cm   **3.** 132 cm

Page 323
1. 25.12 m   **4.** 22.608 m   **7.** 66 mm
10. 4.4 m   **13.** 6 cm   **17.** 8 cm   **19.** 10.676 m

Page 324
1. 314 cm^2 **4.** 1,256 cm^2

Page 325
1. 113.04 cm^2 **5.** 7,850 m^2 **10.** 3.14 m^2
14. 17,662.5 cm^2 **16.** 0.5 m^2

Pages 326–327
1. 207.24 cm   **3.** $4.51   **7.** 4.71 m

Page 328
1. 753.6 cm^2 **5.** 75.36 cm^2

Page 329
2. 251.2 cm^2 **5.** 314 cm^2 **9.** 87.92 m^2
14. 395.64 cm^2 **17.** 565.2 cm^2

Page 330
2. 2,411.52 dm^3

Page 331
1. 942 cm^3 **7.** rectangular prism; 46.44 cm^3

Pages 332–333
1. 37.68 m^3 **3.** 1.6 min   **6.** 62.8

Page 334
1. B

Page 336
1. $364   3. $402.50

Page 338
1. 62.8 m   3. 110 m   5. 78.5 cm^2
9. 75.36 cm^2 11. 18.84 cm^2
13. 254.34 cm^3

Page 339
1. 2A$_{hex}$ 5. 43   10. 224   16. 2,766

14 Probability, Statistics, and Graphs

Page 344
1. equally likely

Page 345
2. Leo, Bess, Pam, Roy; equally likely
7. 3; not equally likely

Page 346
1. 2; $\frac{1}{2}$

Page 347
1. $\frac{1}{8}$ 5. $\frac{5}{8}$ 9. 0   17. $\frac{1}{6}$
20. jazz record

Page 348
3. $\frac{5}{36}$ 6. $\frac{1}{12}$

Page 349
2. $\frac{1}{4}$ 7. $\frac{1}{10}$

Pages 350–351
1. 9   3. about 117   7. 75   10. 5

Page 352
1. 78; 24

Page 353
2. 10   4. 61   9. $35

Page 354
1. 29; 28   6. 29.3; 29.5

Page 355
2. 11 cm; flamefish   7. 118.8 cm   9. about 8.3

Page 356
1. 22; 28   4. 200   7. 2; 2

Page 357
1. 70   4. 100   7. 20

Page 358
2. 500,000   6. 30,000,000   9. 23,000,000

Page 359
1. 2.88 million   4. 61.2°   9. 28.8°

Pages 360–361
1. 340 m   5. Iowa   7. $\frac{1}{16}$
13. 18°   18. 14.5%

Page 362
2. C; F   5. 36 ÷ 8 = 4.5 acorns

Page 364
3. $\frac{1}{4}$ 7. 0   9. 28°; 28°   11. $360

15 Integers

1. $^-30$ 5. negative   11. $^+1$ 13. 0
16. $^+6$

Page 369
1. >   6. <   7. >   12. <   16. $^-15$°
19. $^-2$ km   22. $^-10, ^-7, ^-5, ^-3, 2$

Page 370
1. One   3. Commutative
5. Distributive   8. Opposites

Page 371
2. 0   8. $^-3$ 15. $^+64$
23. $(^-1 \cdot ^-1) + ^+5$ 28. $^+4 \cdot ^+2 + ^+4 \cdot ^+3$

469

Page 372
1. ⁻4   5. 7

Page 373
1. ⁻5   8. 8   9. ⁻12   15. 9
25. ⁻6   29. 9   33. ⁻18   38. positive

Page 374
1. 5   6. ⁻2

Page 375
2. 2   6. ⁻8   10. ⁻19   17. ⁻4
21. ⁻3   24. 0   27. ⁻9   30. 8 + ⁻13 = ⁻5

Page 376
2. 11   7. 3   9. ⁻15

Page 377
1. ⁻8   5. ⁻6   9. 20   13. 5
20. 2   25. ⁻10   31. 4

Page 378
1. ⁻11°C; 4°   5. 8°; more

Page 379
2. 750   5. 156 kWh

Page 380
2. 7   5. ⁻24

Page 381
1. ⁻42   7. 24   9. ⁻48   15. 48
25. 380   28. ⁻162   33. 12
37. ⁻24   43. 12   45. ⁻18

Page 382
2. ⁻5; 5   5. 8   7. ⁻7

Page 383
1. ⁻9   5. 3   13. 1   18. ⁻5
24. 9   29. ⁻15   34. ⁻6   38. 8

Page 384
3. (⁻40,50)   6. (20,⁻10)   10. (⁻30,45)
15. (30,⁻28)

Page 385
3. (1,2)   6. (⁻2,1)   8. (0,⁻2)
11. (⁻5,⁻2)   17. (⁻3,1),(⁻1,3),(⁻3,5),(⁻5,3)
21. square

Page 388
1. $414; $38   3. $216

Page 390
1. >   5. <   7. ⁻7   13. ⁻11
20. ⁻2   22. ⁻6   26. ⁻24   31. 1
35. 3   40. 4

Page 391
1. 6, 0, ⁻6, ⁻9, ⁻15   4. subtract 4

16 Measurement: Customary Units

Page 394
1. 24   6. 7   11. 13,200   13. 440 yd

Page 395
1. 7 ft 11 in.   6. 5 yd 1 ft 1 in.
11. 7 ft 8 in.   14. 2 ft 8 in.   17. $\frac{3}{8}$ in.

Page 396
1. 80 in.^{2} 5. 314 ft^2

Page 397
1. 198 in.^{2} 6. 366 in.^{2}

Page 399
2. 1,400 in.^{3} 6. 2,128 in.^{3} 8. 56.52 ft^3
13. 1,728 in.^{3}

Page 400
1. 2   5. 32   7. 40   11. 3

Page 401
1. 8 qt   5. $\frac{3}{4}$ pt; 2 fl oz   9. $1.78

Page 402
2. B   7. 32   11. 20   14. 18,000
16. 5   19. 9 oz

Page 403
2. C **7.** 325°F **10.** 75°

Pages 404–405
1. $0.675 **5.** $1.18 **9.** $1.43
11. $2.16/gal; no

Page 406
1. 40¢ **5.** regular

Page 408
1. 24 in. **4.** 3 yd 1 ft **6.** 5 yd
8. 1 ft 7 in. **11.** 1 yd 1 ft 4 in. **16.** 3
22. 209°

Page 409
1. $g = 6$ **5.** $\ell = 2$ **8.** $a = 5$
13. $w = 8$ **14.** $e = 7$ **20.** $j = 1$

More Practice

Page 417
Set A: 1. < **3.** > **7.** <
Set B: 1. 30,000 **3.** 90,000
6. 330,000 **12.** 200,000 **14.** 300,000
16. 4,100,000
Set C: 1. 60,000 **3.** 546,000 **5.** 11,000

Page 418
Set A: 1. 127 **3.** 1,000 **5.** 9,405
7. 96,041 **11.** 3,717
Set B: 1. 349 **5.** 1,207 **8.** 1,807 **11.** 3,109
Set C: 1. 11 h 5 min **3.** 5 h 45 min

Page 419
Set A: 1. < **3.** > **5.** =
Set B: 1. 0.042 **4.** 1.141 **9.** 1
11. 370 **17.** 5 **19.** 0.1
Set C: 2. 22.017 **5.** 9.8108 **6.** 14.685
11. $52.98 **14.** $51.37

Page 420
Set A: 1. 30.6 **7.** 1.059 **11.** $20.51
Set B: 1. 2.00 **4.** 22.40
8. 1,131.60 **12.** $24
Set C: 1. 300 **3.** 400,000 **7.** 1,600,000

Page 421
Set A: 1. 156 **3.** 4,580 **6.** 17,424
13. 45,570
Set B: 1. 756 **4.** 14,842 **8.** 145,692
10. 140,265
Set C: 1. 380,628 **6.** 899,184
12. 2,782,444

Page 422
Set A: 1. $4 \cdot 4 \cdot 4 = 64$ **6.** $11 \cdot 11 = 121$
11. 6^3 **14.** 4^4
Set B: 1. 25 **3.** 3 **6.** 4 **12.** 25
Set C: 1. 164 R1 **7.** 678 R3 **9.** 701

Page 423
Set A: 1. 142 R1 **7.** 1,230 R4 **11.** 560 R5
Set B: 1. 56 R7 **6.** 12 R36 **11.** 472 R56
Set C: 1. 21 R117 **6.** 56 R145
11. 505 R239 **17.** 168 R181

Page 424
Set A: 2. 24.15 **7.** 492.32
Set B: 1. 0.005248 **7.** 0.00063
Set C: 1. 12 **5.** 0.002 **7.** $20.00
Set D: 1. 6×10^7 **5.** 3.5×10^9 **13.** 16,000

Page 425
Set A: 2. 7.3 **7.** 0.96 **11.** 8.06
Set B: 1. 1.71 **7.** $1.79
Set C: 1. 5.08 **7.** 2.75
Set D: 2. 1 **5.** 0.45

Page 426
Set A: 1. P **5.** C **11.** 39: 1, 3, 13, 39
Set B: 2. $2 \cdot 2 \cdot 2 \cdot 7$ **6.** $2 \cdot 3 \cdot 3 \cdot 5$
11. $2 \cdot 2 \cdot 2 \cdot 3 \cdot 3 \cdot 5$
Set C: 1. 15 **6.** 21 **13.** 2
Set D: 2. 18 **7.** 90 **11.** 70

Page 427
Set A: 2. 4 **7.** $\frac{80}{n}$ **11.** 4
13. $p - 9$ **16.** $\frac{15}{n}$
Set B: 1. $x = 16$ **5.** $x = 11$ **10.** $l = 17$
Set C: 1. $x = 9$ **6.** $x = 26$ **10.** $z = 10$

Page 428
Set A: 1. 2   **7.** 9   **11.** 21
Set B: 1. $\frac{3}{5}$ **3.** $\frac{4}{5}$ **9.** $\frac{3}{4}$ **15.** $\frac{12}{25}$ **20.** $\frac{3}{5}$
Set C: 1. $1\frac{2}{7}$ **5.** 6   **8.** $3\frac{1}{3}$
12. $1\frac{8}{9}$ **15.** 3   **20.** $\frac{15}{4}$ **26.** $\frac{22}{15}$
31. $\frac{100}{3}$ **34.** $\frac{31}{7}$

Page 429
Set A: 1. <   **3.** >   **7.** >
Set B: 2. $\frac{1}{3}$ **6.** $1\frac{1}{8}$ **12.** $\frac{2}{3}$
Set C: 2. $\frac{1}{3}$ **7.** $\frac{5}{8}$ **12.** $\frac{7}{30}$

Page 430
Set A: 3. $17\frac{5}{9}$ **6.** $7\frac{13}{18}$
Set B: 2. $6\frac{1}{3}$ **7.** $\frac{3}{4}$ **12.** $14\frac{1}{6}$
Set C: 2. $5\frac{17}{40}$ **6.** $\frac{5}{6}$ **11.** $\frac{11}{24}$

Page 431
Set A: 1. $\frac{1}{15}$ **7.** $\frac{2}{9}$ **12.** $\frac{2}{27}$
Set B: 2. $3\frac{1}{3}$ **6.** $4\frac{2}{5}$ **11.** $16\frac{1}{2}$
Set C: 1. $\frac{4}{1}$ **3.** $\frac{1}{7}$ **5.** 6
Set D: 2. $\frac{9}{10}$ **5.** $\frac{1}{3}$ **7.** $\frac{5}{7}$ **13.** $3\frac{1}{5}$

Page 432
Set A: 2. 2   **7.** $\frac{6}{7}$ **11.** $\frac{33}{40}$
Set B: 1. $\frac{4}{5}$ **4.** $\frac{3}{100}$ **8.** $3\frac{19}{20}$
12. 6.75   **15.** 0.2   **16.** 0.5625
Set C: 2. $0.\overline{7}$ **4.** 0.8   **8.** $0.4\overline{5}$
11. <   **14.** >   **17.** >

Page 433
Set A: 2. 77.6 cm   **6.** 39.56 km
Set B: 1. 222.44 cm²   **8.** 1,873.3 cm²
Set C: 1. 117.5 km²   **4.** 1,168.5 mm²
8. 108 cm²
Set D: 2. 270 dm³   **3.** 4.41 m³

Page 434
Set A: 2. ≠   **8.** =   **11.** $\frac{1}{6}$ **16.** 3
Set B: 1. 10   **5.** 75   **8.** 2
Set C: 1. $\frac{7}{20}$ **4.** $\frac{3}{2}$ **6.** $\frac{4}{5}$ **12.** 15%
16. 8%   **22.** 40%   **26.** 16%

Page 435
Set A: 2. 1.50   **7.** 0.634   **14.** 0.95
16. 175%   **18.** 1.2%   **21.** 60%
Set B: 2. 44%   **8.** 78%   **11.** 81%   **18.** 28%
Set C: 1. 27   **3.** 100   **5.** 54   **11.** 300

Page 436
Set A: 1. 50%   **4.** 75%   **7.** 70%
11. 16%   **13.** 43%   **17.** 433%
Set B: 1. 16   **8.** 56   **13.** 50
Set C: 2. 45   **4.** 369   **6.** 4%

Page 437
Set A: 2. 40.82 m   **4.** 157 mm   **8.** 43.96 m
Set B: 1. 78.5 m²   **4.** 153.86 mm²
11. 31,400 mm²
Set C: 2. 145.696 cm²   **6.** 659.4 m²
10. 125.6 m²

Page 438
Set A: 2. 7; 11   **6.** 73; 19
Set B: 2. 160; 150   **5.** 0.074; 0.052
10. 1.8; 1.9
Set C: 2. <   **5.** >   **12.** >   **16.** >

Page 439
Set A: 2. 0   **8.** ⁺20   **10.** ⁻18   **14.** 0
Set B: 2. ⁻9   **5.** 14   **10.** ⁻8
15. 14   **17.** ⁻4   **21.** ⁻9
Set C: 1. ⁻2   **6.** 1   **10.** ⁻7
13. ⁻2   **18.** 7   **21.** ⁻8

Page 440
Set A: 2. 9   **5.** 0   **10.** ⁻8   **15.** 12
18. ⁻4   **22.** ⁻14
Set B: 1. ⁻21   **6.** 30   **9.** 0   **13.** 96
17. ⁻15   **20.** ⁻28   **23.** 56
Set C: 2. 5   **7.** ⁻3   **10.** 8   **14.** 6

Page 441
Set A: 1. 9 ft   **3.** 9 yd   **6.** 2 ft 6 in.
Set B: 2. 43 ft²   **5.** 50.24 in.²
8. 0.2826 ft²
Set C: 2. 31.5 ft³   **5.** 1,570 in.³   **8.** 307.72 ft³